Forest and Plant Bioproducts Division

Presentations at the 2010 AIChE Annual Meeting

Salt Lake City, Utah, USA
7-12 November 2010

ISBN: 978-1-61782-162-2

Printed from e-media with permission by:

Curran Associates, Inc.
57 Morehouse Lane
Red Hook, NY 12571

Some format issues inherent in the e-media version may also appear in this print version.

Printed by Curran Associates, Inc. (2011)

For permission requests, please contact AIChE
at the address below.

AIChE
3 Park Avenue
New York, NY 10016-5991

Phone: (203) 702-7660
Fax: (203) 775-5177

www.aiche.org

Additional copies of this publication are available from:

Curran Associates, Inc.
57 Morehouse Lane
Red Hook, NY 12571 USA
Phone: 845-758-0400
Fax: 845-758-2634
Email: curran@proceedings.com
Web: www.proceedings.com

Forest and Plant Bioproducts Division

Presentations at the 2010 AIChE Annual Meeting

Salt Lake City, Utah, USA
7-12 November 2010

ISBN: 978-1-61782-162-2

AIChE

TABLE OF CONTENTS

(7a) Evaluation of Thermosensitive Microparticle-Hydrogel Composite for Protein Delivery

Carl Beigie[1], Lussier Danielle[1], Sokal Tastiana[1] and **Jennifer Vernengo**[2], (1)Chemical Engineering, Rowan University, Glassboro, NJ, (2)Department of Chemical Engineering, Rowan University, Glassboro, NJ

Over the past several years, in situ-forming polymer systems have attracted interest due to their minimally invasive implantation and ability to a form space-filling solid within tissue. Thermally sensitive poly(N-isopropylacrylamide) (PNIPAAm) is one such system. Below its lower critical solution temperature (LCST), typically around 32°C, the polymer forms a miscible solution with water. Above the LCST, it becomes hydrophobic, so the polymer and water separate, forming a compact gel. We are investigating novel injectable materials composed of PNIPAAm lightly crosslinked with poly(ethylene glycol) (PEG) 4600 and 8000 g/mol as multi-functional scaffolds for repair of the degenerated intervertebral disc. Because of the injectable nature of this PNIPAAm-based system, it can be used to simultaneously deliver molecular and cellular factors necessary for tissue regeneration. In this work, the potential of these copolymers to serve as prolonged release vehicles for therapeutic proteins is investigated.

Poly(lactic-co-glycolic) acid particles loaded with the model protein lysozyme, with an average size of 1.7651 μm and a protein loading efficiency of 51.24% were combined with both PNIPAAm-PEG (4600 and 8000 g/mol) to form a composite system. The composite dampened the typical burst effect of protein release in the first 72 hours of a 28 day release study when compared to freely suspended microparticles. The effect of PEG molecular weight on protein release from the composite was shown to be negligible. Compressive testing was also conducted and it was determined through that the presence of microparticles did not significantly affect the modulus values of the hydrogels. Currently, we are investigating the morphological characteristics of the hydrogels to determine changes in the porosity and pore interconnectivity as the embedded particles degrade. The results of this study demonstrate PNIPAAm-PEG copolymers have the potential to serve as multi-functional scaffolds for intervertebral disc tissue regeneration.

See more of this Session: Bio-Based Composites
See more of this Group/Topical: Materials Engineering and Sciences Division

(7b) High Performance Impact-Tolerant and Abrasion-Resistant Materials: Lessons From Nature

Qianqian Wang[1], Dongsheng Li[1], Michiko Nemoto[1], Brian Weden[1], Shinobu Heier[2], Elaine DiMasi[3] and David Kisailus[1], (1)Department of Chemical and Environmental Engineering, University of California, Riverside, Riverside, CA, (2)Thermo Fisher Scientific, West Palm Beach, FL, (3)National Synchrotron Light Source, Brookhaven National Laboratory, Upton, NY

Chitons are marine mollusks found worldwide in the intertidal or subtidal zones of cold water as well as in tropical waters. These organisms have evolved an amazing feeding structure called a Radula. The Radula is a ribbon-like structure that consists of abrasion resistant teeth anchored to a flexible stylus that the organism uses to abrade rocky substrates to reach endolithic and epilithic algae. In this work, we investigate the structure and mineralization process in Cryptochiton stelleri, the largest of the chitons. Using various microscopy and spectroscopy techniques as well as synchrotron analyses, we have uncovered critical structure-function relationships in the mineralized teeth as well as insights into the mineralization processes in these unique structures. Investigation of the mechanical properties of the fully mineralized teeth have revealed that the combination of ultrahard minerals and templating organics, architected in a unique microstructure, lead to a damage tolerant composite that is of the hardest known biominerals known in nature.

See more of this Session: Bio-Based Composites
See more of this Group/Topical: Materials Engineering and Sciences Division

(7c) Mechanical, Structural and Thermal Properties of Polymer Composites Containing Short Ragweed Pollen Grains

Carson Meredith[1], Jung-Hyun Lee[2], Brandon Suttle[2] and Hyung-Ju Kim[2], (1)School of Chemical and Biomolecular Engineering, Georgia Institute of Technology, Atlanta, GA, (2)Chemical & Biomolecular Engineering, Georgia Institute of Technology, Atlanta, GA

Pollen grains are ubiquitous biological particles possessing a high-strength outer shell with remarkable chemical stability. In addition, the unique 'spiny' ornamentations on many pollens offers intriguing possibilities for bio-based polymer composites with enhanced matrix-filler adhesion. In this work, the effect of the incorporation of short ragweed pollen grains on the mechanical, interfacial and thermal properties of pollen/polymer (PCL: poly(ε-caprolactone) or PS: polystyrene) composites cast by different solvents (THF and chloroform) was investigated under different annealing conditions. The addition of pollen grains resulted in the reduction in both the tensile strength and elongation, while the modulus of PCL composites unchanged. Pollen/PCL composites exhibited detwetted interfaces with interfacial voids due to poor compatibility between the pollen and the PCL matrix. Casting solvents have an influence on the morphological (crystalline density), mechanical and interfacial properties of semicrystalline PCL films. Less ordered structures, with lower crystalline density achieved by THF casting and sub-melting temperature (Tm) annealing, showed reduced interfacial defects. This led to less reduction in the elongation properties by the incorporation of pollens. Thermal characterization revealed that the presence of pollens could narrow the crystallization behavior of the PCL matrix without changing the melting and crystallization temperature. For amorphous PS polymer, the elongation properties of composites were maximized at low pollen loading, whereas the tensile strength decreased continuously with pollen content. Together with an increase in the glass transition temperature of PS composites by incorporation of pollens, the reduced defect formation at pollen-PS interfaces indicates better affinity of the pollen surface with the PS matrix compared to those of PCL composites.

See more of this Session: Bio-Based Composites
See more of this Group/Topical: Materials Engineering and Sciences Division

(7d) Micromechanics of Actin-Microtubule Composites

Moumita Das and F.C. MacKintosh, Department of Physics and Astronomy, Vrije Universiteit, Amsterdam, Netherlands

We present a theory for the micro-mechanics and bulk elastic properties of composites consisting of stiff fibers embedded in an elastic matrix. It has been known, for instance, that novel elastic behavior can result from the embedding of particles or rods in a matrix. In particular, recent experiments published in Physical Review Letters [102, 188303 (2009)] have suggested that the inclusion of stiff microtubules in a softer biopolymer matrix can lead to enhanced compressibility. Our work provides a possible explanation for these observations.

Our results suggest an important role for stiff filaments such as microtubules and stress fibers in cell mechanics---they not only enhance the stiffness of the cell cytoskeleton and its ability to bear large forces, but also can endow it with enhanced compressibility. At the same time, our results have a wider elasticity context, with potential applications to composite elastic media with a wide separation of scales in stiffness of its constituents such as carbon nanotube-polymer composites, which have been shown to have highly tunable mechanics.

See more of this Session: Bio-Based Composites
See more of this Group/Topical: Materials Engineering and Sciences Division

(7e) Polylactic Acid-Clay Nanocomposites Via Solid-State Shear Pulverization

Alexander S. Fielding and Katsuyuki Wakabayashi, Department of Chemical Engineering, Bucknell University, Lewisburg, PA

Bio-based polymers offer a sustainable alternative to traditional polymers from petroleum derivatives, especially in short lifetime products such as packaging materials. Derived from natural organic resources, these types of polymers degrade under natural conditions in a time period ranging from several weeks to several years. Neat forms of many bio-based polymers, however, exhibit many undesirable physical properties, often associated with the low glass transition and melting temperatures. Incorporating small amounts of nanoscale fillers, such as layered silicates (clay), into the biodegradable polymer matrix can improve mechanical, thermal, and other physical properties to make them applicable in commercial applications. Studies involving nanocomposites of polylactic acid (PLA), a common bio-based, compostable polymer, and montmorillonite clay have shown significant property enhancement when complete exfoliation and good dispersion of the clay sheets are achieved [1,2]. However, common processing methods used to prepare PLA-clay nanocomposites, such as *in situ* polymerization and melt processing, have practical limitations [1-4].

We herein propose solid-state shear pulverization (SSSP) as an effective processing method to disperse clay-based nanofillers in bio-based polymer nanocomposites. SSSP has proven to successfully compatibilize immiscible polymer blends [5] and exfoliate silicate sheets in more traditional polymer-clay nanocomposites [6,7]. Unlike conventional nanocomposite fabrication techniques that often require chemical modification, solvents, and/or heat, SSSP processes the polymer nanocomposites simply by maintaining the polymer below its glass transition temperature and subjecting the material to high shear forces to promote filler exfoliation and dispersion. Thus nanocomposites prepared via SSSP are not limited by thermodynamics or transport issues. The continuous operation of SSSP allows for facile scale-up for industrial use in applications such as biodegradable packaging materials.

Commercial grade PLA and pristine and organically-modified clay were processed under different SSSP conditions. Electron microscopy and X-ray diffraction were employed to investigate the exfoliation and dispersion of clay sheets. Differential scanning calorimetry and thermogravimetric analysis were employed for thermal characterization. Mechanical properties were measured using static and dynamic mechanical tests, and the barrier properties were obtained from oxygen permeation and water vapor transmission testing. Results from these tests will be used to compare the performance of PLA/clay nanocomposites processed via SSSP to that of neat PLA and PLA/clay nanocomposites processed via traditional twin-screw extrusion.

[1] S. S. Ray, and M. Bousmina, Progress in Materials Science, 50, 962 (2005) .

[2] S. S. Ray, and M. Okamoto, Macromolecular Rapid Communication, 24, 815 (2003).

[3] J. A. Ratto, D. M. Steeves, E. A. Welsh, and B. E. Powell, Society of Plastics Engineers ANTEC, 57, 1628 (1999).

[4] K. Fukushima, D. Tabuani, and G. Camino, Materials Science and Engineering C, 29, 1433 (2009).

[5] A. H. Lebovitz, K. Khait, and J. M. Torkelson, Macromolecules, 35, 8672 (2002).

[6] K. G. Kasimatis, and J. M. Torkelson, American Chemical Society, Division of Polymeric Materials: Science and Engineering Preprints, 91, 173 (2004).

[7] K. Wakabayashi, C. Pierre, D. A. Dikin, R. S. Ruoff, T. Ramanathan, L. C. Brinson, and J. M. Torkelson, Macromolecules, 41, 1905 (2008).

See more of this Session: Bio-Based Composites
See more of this Group/Topical: Materials Engineering and Sciences Division

(8a) Study of Super-Toughed Poly(lactic acid) Ternary Blends Prepared by Dynamic Vulcanization-Induced Compatibilization

Hongzhi Liu and **Jinwen Zhang**, Composite Materials and Engineering Center, Washington State University, Pullman, WA

In this work, the preparation of poly(lactic acid) (PLA) ternary blends consisting of PLA, an epoxy-containing terpolymer elastomer and an ionomer of ethylene-methyacrylic acid copolymer and investigation of mechanical and physical properties of the resulting blends were investigated. The blending was performed in a twin-screw extruder. The curing of the elastomer by the ionomer took place during compounding and reactive compatibilization between the cured elastomer phase and the PLA matrix was also established. The reaction process under different conditions was assessed by performing the mixing in a torque theometer, the reaction products were isolated and examined by FT-IR. Mechanical, thermal dynamic mechanical properties and morphology of the blends were thoroughly studied. The ionomer/elastomer weight ratio was varied, but the ionomer and elastomer content was fixed at 20 wt% in the blend. Extrusion temperature, elastomer/ionomer ratio, cation type and neutralization degree of the ionomer all had significant effects on impact toughness of the ternary blends. Super-toughness was successfully obtained for some blends compounded at 240„aC. It was found that zinc ionomer initiated and catalyzed the dynamic vulcanization of epoxy-containing elastomer, and it also promoted the reactive compatibilization at the interface between PLA and elastomer. TEM observation revealed that the °¥salami'-like phase structure was formed in the ternary systems. The effective reactive interfacial compatibilization at elevated blending temperatures was believed to be responsible for the significant increase in notched impact strength rather than enhanced crystallization of PLA matrix.

See more of this Session: Biobased Materials I
See more of this Group/Topical: Forest and Plant Bioproducts Division

(8b) PHA Bioplastic and Composites for Sustainable Residential Construction

Zachary Wright, Department of Chemical Engineering, Stanford University, Stanford, CA, Sarah Billington, Dept. of Civil & Environmental Engineering, Stanford University, Stanford, CA and Curtis W. Frank, Chemical Engineering, Stanford University, Stanford, CA

Polyhydoxyalkanoates (PHA) are a class of biorenewable polymers that are commercially underutilized, in part due to the lack of potential applications and high cost. Natural fiber/PHA composites use components that are biodegradable and combine the high tensile strength of the fibers with the stress transfer capabilities of the polymer matrix. These composite materials can be used as structural components in residential building materials. Natural fibers are typically hydrophilic, whereas PHAs are hydrophobic. Contrasting surface character potentially detracts from the overall mechanical strength of composite materials due to imperfect adhesion between the matrix and fibers at the interface. Understanding molecular interactions at the interface can lead to modification schemes that enhance the strength of the interface and thus the strength of the bulk composite. Two modifications have been investigated. First, wood flour has been treated with octadecyltrimethoxysilane to give the cellulosic materials a more hydrophobic character. For the second modification scheme a third component, maleated PHA, was added to the composite as a compatibilizer between the two phases. Molecular and microscale techniques are used to try and understand bulk mechanical behavior.

See more of this Session: Biobased Materials I
See more of this Group/Topical: Forest and Plant Bioproducts Division

(8c) New BioBased Carbon Nano Structures and Their Nanocomposites

Manju Misra, School of Engineering and the Department of Plant Agriculture, University of Guelph, Guelph, ON, Canada

There is a huge demand for polymer nanocomposites (with a polymer matrix and a filler with at least one dimension less than 100 nm) using carbon-based fillers because of its improved mechanical, thermal and electrical properties. Green nanocomposites; that are derived from bio-based polymers and bio-based nanoparticles tend to be viewed as the next generation of materials for innovative industrial uses.

See more of this Session: Biobased Materials I
See more of this Group/Topical: Forest and Plant Bioproducts Division

(8d) Mechanism Study of Starch Nanoparticle Formation

Delong Song, School of Chemical and Biomolecular Engineering, Georgia Institute of Technology, Atlanta, GA and **Yulin Deng**, School of chemical and biomolecular engineering, Georgia Institute of Technology, Atlanta, GA

Preparation and application of biomaterial based nanoparticles have been attracting more research interests. Nano-sized starch particles can be prepared by using different approaches, such as acid hydrolysis, inverse microemulsion polymerization, and mechanical treatment. Reactive extrusion method is a simple, efficient, and environmental friendly method for nanostarch preparation. In this study, the mechanism of starch nanoparticle formation during extrusion was investigated. The effects of extrusion conditions, including temperature (50~100oC), screw speed, residence time, starch moisture content and crosslinker addition, on particle size were studied. The results indicate that after extrusion at 100oC, the starch particles with a size of 320~400 nm were obtained. With the addition of appropriate crosslinkers, the starch particle size could be further reduced to around160 nm. The morphology, crystalline property, thermal stability and rheology properties of starch nano particles were also characterized.

See more of this Session: Biobased Materials I
See more of this Group/Topical: Forest and Plant Bioproducts Division

(8e) Efficient Conversion of Crop Stalk Into Succinic Acid by Actinobacillus Succinogenes

Jianmin Xing, Qiang Li and Maohua Yang, Institute of Process Engineering, Chinese Academy of Sciences, Beijing, China

Succinic acid is valued as one of the key platform chemicals in broad applications. Biosynthesis of succinic acid from renewable biomass resource was reported in this paper. Here batch approach was carried out to analyze influence of several carbon sources on succinic acid production from crop stalk wastes by Actinobacillus succinogenes BE-1. Crop stalk wastes, including corn stalk and cotton stalk, were converted into a carbohydrates-rich feedstock with subsequent hydrolysis by enzyme. In this strategy, the concentrations of glucose in the raw carbohydrate material could reach 65%~80% of the total reducing sugar. Batch fermentation was performed to utilize corn stalks and cotton stalks hydrolysates as the carbon sources, giving 17.8 g L-1 and 15.8 g L-1 succinic acid, respectively with 30 g L-1 of initial sugar. In particular, high conversion yield from cotton stalk was achieved with about 1.23 g succinic acid per g glucose consumed. The strategy enabled the exploration of crop stalk wastes for the economical and cost-effective production of succinic acid by A. succinogenes.

See more of this Session: Biobased Materials I
See more of this Group/Topical: Forest and Plant Bioproducts Division

(81a) A Novel Reactive Extrusion Process for Preparing Carboxymethyl Cellulose Ethers: Optimization of Reaction and Extrusion Conditions

Pratik N. Bhandari, Industrial Agricultural Products Center, University of Nebraska, Lincoln, NE and Milford A. Hanna, Industrial Agriculture Product Center, Biological Systems Engineering, University of Nebraska, Lincoln, NE

In a previous study, we had prepared carboxymethyl ethers of cellulose using reactive extrusion and studied the effects of water, ethanol and NaOH on the reaction efficiency. In this study, we explored the effect of a nitrogen atmosphere, during the reactive extrusion process, to improve the viscosity of the resulting carboxymethyl cellulose. We optimized the reaction and extrusion conditions for preparing carboxymethyl cellulose with high viscosity. The molecular properties of the carboxymethyl cellulose, prepared using reactive extrusion, were studied using scanning electron microscopy. The result of the molecular properties on the functional properties of carboxymethyl cellulose were studied and compared to that of commercially available products.

See more of this Session: Biobased Materials II
See more of this Group/Topical: Forest and Plant Bioproducts Division

(81b) Single Step Functionalisation of Cellulose to Produce All-Cellulose Nanocomposites

Koon-Yang Lee, Department of Chemical Engineering, Imperial College London, London, United Kingdom and Alexander Bismarck, Chemical Engineering, Imperial College London, London, United Kingdom

Recent interests in green nanocomposites and public demands for more environmental friendly materials have sparked the development of nanocomposites derived from renewable sources. One of the most studied materials in the field of green nanocomposites is cellulose. Cellulose is the most abundant organic material in the world. It can be produced by plant or bacteria from the Acetobacter species. However, cellulose is extremely hydrophilic in nature due to the presence of hydroxyl groups. This will result in poor interfacial adhesion between cellulose and hydrophobic biodegradable polymer matrices. Therefore, novel methods are required to improve the compatibility between the cellulose and the matrix.

In this work, we reactively compatibilise bacterial cellulose (BC) and microcrystalline cellulose (MCC) in a single step to produce all-cellulose nanocomposites through organic acid esterification reaction. Due to the differences in the crystal structure and the hydroxyl groups availability of the two cellulose, significant bulk modification of MCC will occur and MCC will be modified into a polymer whilst BC will essentially be only modified on the surface of the nanofibrils. As a result, all-cellulose nanocomposites reinforced by highly crystalline BC nanofibrils can be produced. By preserving the crystalline structure of BC, the Young's modulus of BC (~114 GPa) will not be affected much, leading to the production of green nanocomposites with improved properties. It was found that both the tensile strength and modulus of the nanocomposites improved by as much as 20% at 5 wt% BC loading in MCC esterified with hexanoic acid. We have also developed a method to determine the accessibility of hydroxyl groups on cellulosic materials in this work.

See more of this Session: Biobased Materials II
See more of this Group/Topical: Forest and Plant Bioproducts Division

(81c) Integrating Pulsed Pyrolysis/Laser Ablation with High Sensitivity Laser Ionization Mass Spectrometry to Measure Plant Cell Wall Composition of Single Cells

Calvin Mukarakate, Adam. M Scheer, David. J Robichaud, Robert. W Sykes, Mark. R Nimlos and Mark. F Davis, National bioenergy Center, National Renewable Energy Laboratory, Golden, CO

Biomass composition can have a significant impact on pretreatment and bioconversion processes. Understanding the chemical changes that occur at the cellular level in heterogeneous biomass materials can aid in identifying chemical reactions important for improving pretreatment processes and help engineer cell walls with traits more amenable to biomass conversion. This will require highly sensitive and selective experiments in which chemical measurements can be made on small groups of cells or even single cells. For example, laser ionization techniques combined with time-of-flight mass spectrometry (TOF-MS) may be useful. We have designed and developed a laser ablation/pulsed sample introduction/mass spectrometry platform that integrates pulsed pyrolysis with single photon ionization (SPI) and resonance-enhanced multiphoton ionization (REMPI) reflectron TOF-MS. We have used this experiment to measure cell wall chemical traits of biomass materials. In this study, the 3rd harmonic of a Nd:YAG laser (355 nm 3.50 eV) is used to pyrolyze the sample. The products are entrained and cooled in a He supersonic jet exiting the orifice of a solenoid pulse valve placed directly above the sample. The resulting molecular beam is then skimmed and introduced into an ionization region. The one color REMPI was used to improve the ionization efficiency of aromatic compounds, leading to highly selective detection of lignin and aromatic metabolites. Lignin pyrolysis products based on both guaiacyl and syringyl monomers have been assigned. The SPI (118.2 nm, 10.49 eV) is suitable for detecting all other compounds with ionization potentials below 10.49-eV. The recorded photoionization mass spectra are compared with that obtained from 22.5-eV electron-impact ionization in a molecular beam mass spectrometer. The mass spectra obtained using photoionization are characterized by high sensitivity, selectivity, high resolution and they display predominantly molecular ion peaks (soft ionization) compared with that of electron impact. The implication of using REMPI under supersonic jet conditions for measuring cell wall composition at sub-tissue level will also be discussed. Samples consisting of single types of cells are collected for pulsed introduction using laser microdissection.

See more of this Session: Biobased Materials II
See more of this Group/Topical: Forest and Plant Bioproducts Division

(81d) Flexible Extruded Sheets From Low Cost Soymeal

Murali M. Reddy, Bioproducts Discovery & Development Centre (BDDC), Department of Plant Agriculture, University of Guelph, Guelph, ON, Canada, **Amar K. Mohanty**, Bioproducts Discovery & Development Centre (BDDC), University of Guelph, Guelph, ON, Canada and Manju Misra, School of Engineering, University of Guelph, Guelph, ON, Canada

One of the biggest challenges of plastic menace arises from their one trip applications like shopping bags, packaging films etc. This problem can be better handled by developing affordable and compostable materials. Soy protein which is a natural biopolymer offers biodegradability and renewability is used in its crude form (soybean meal) to develop compostable sheets for packaging.

Soy proteins are stabilized and strengthened by hydrogen bonds, hydrophobic interactions and disulfide bonds and these provide strength and stiffness to proteins in the solid-state. However, this stability observed in proteins becomes hurdle during their processing in melt state. In our approach, we overcome this by dry processing of soybean meal with plasticizers and reducing agents to obtain bioplastic. FTIR confirmed that urea additive helped in destructurization of soybean meal by breaking hydrogen bonds and disulphide bridges in soy protein. Reactive extrusion of soybean bioplastic and flexible biodegradable polyester was carried out in a twin screw extruder to obtain thermoplastic soy blends. Tensile properties increased with increase in urea content and this was highest for 12.5 wt% urea in the soy bioplastic. This was due to chemical bonding of destructed soy meal and biodegradable polyester. Furthermore, cast film processing of thermoplastic soy blends were carried out in single screw microtruder. Cooling rate, throughput, roll speed have significant effect on the film properties. This work is financially supported by Hannam soybean utilization 2008 fund and the Ontario Ministry of Agriculture, Food and Rural Affairs (OMAFRA) 2009 New Directions & Alternative Renewable Fuels Research Program Project number SR9223.

See more of this Session: Biobased Materials II
See more of this Group/Topical: Forest and Plant Bioproducts Division

(81e) Characterization of Mechanical Properties of Jute Fiber Mats Reinforced Polyester Matrix Composites

Hiroyuki Hamada, Kyoto Institute of Technology, Kyoto, Japan

In advance of the enforcement of the recycling law in Japan, Japanese automobile makers have been started to develop easy-to-recycle vehicle structures and taking steps to reduce substances of environmental concern. Therefore, many efforts in this area focus on the use of renewable resources such as natural fibers to reinforce biodegradable polymer resin composite materials for a variety of structural applications in automotive industry. In this research, polymeric composite materials were fabricated from jute fibers mats reinforced unsaturated polyester resin. The reinforcing mats were prepared from jute fibers slivers (50%) and recycled jute fibers (50%), respectively. Evaluation of mechanical properties (flexural, impact, etc.) was carried out and compared with those of plain woven jute fibers reinforced unsaturated polyester matrix composites.

See more of this Session: Biobased Materials II
See more of this Group/Topical: Forest and Plant Bioproducts Division

(134a) Melt Spun Nanocomposite Fibres of Polylactide/Bacterial Cellulose Nanofibrils; Towards Composites with Anisotropic Properties and Nanophase Alignment

Jonny J. Blaker[1], Koon-Yang Lee[2] and Alexander Bismarck[1], (1)Chemical Engineering, Imperial College London, London, United Kingdom, (2)Department of Chemical Engineering, Imperial College London, London, United Kingdom

Conventionally, low-cost racemic polylactide (PDLLA) has been regarded as an inferior product over its semi-crystalline PLA counterpart, however it does has the advantage of being processed at temperatures sub 150°C, which is below the degradation point of many bio-based fillers, making it attractive as a matrix in composites. In an effort to enhance the properties of PDLLA we incorporate bacterial cellulose derived nano-fibrils, which have an elastic modulus reported at 114 GPa, via a twin-screw micro extruder and melt spin nanocomposite fibres. The hot extruded material is spun into fibres using a DSM Xplore (The Netherlands) fibre-spinning rig. These fibres are compared to those we produce based on a semi-crystalline PLLA matrix, produced at elevated temperatures (185°C). We also modify bacterial cellulose nano-fibrils via organic acid esterification (hexanoic and lauric acid) in an effort to improve the interface between the nano-phase and the matrix and investigate the effect of matrix and fibril modification on fibre properties. Neat PDLLA, PLLA and composite fibres filled with up to 5 wt.% bacterial cellulose nano-fibrils have been produced and characterized in terms their mechanical, and thermal properties, and the distribution of nano-fibrils within the fibre matrix assessed. By combining semi-crystalline PLA/bacterial cellulose nanocomposite fibres with amorphous PLA (processed at lower temperatures) nano-composites with anisotropic mechanical properties can be realized.

See more of this Session: Biobased Materials III
See more of this Group/Topical: Forest and Plant Bioproducts Division

(134b) Effects of Treatments On the Mechanical Properties of Cellulose Reinforced Composites From Recycled Jute Woven Cloth

Ying Yu, Yuqiu Yang and **Hiroyuki Hamada**, Kyoto Institute of Technology, Kyoto, Japan

Natural fiber reinforced composites were prepared from unsaturated polyester and recycle coffee bag (jute plain woven cloth). It is found that the higher moduli of composites were obtained by the adding of jute compared to the neat resin based on the results of tensile, flexure and Izod impact tests. However from the observation on the fracture surface of the tested specimens, it is found that the interface property between jute fiber and polymer seems weak. Therefore, at current study, in particularly, some treatments were carried out on the jute woven cloth before impregnation with polymer. The effects of the treatments on the mechanical properties of the composites were discussed compared to the natural ones.

See more of this Session: Biobased Materials III
See more of this Group/Topical: Forest and Plant Bioproducts Division

(134c) Electrospun Bionanofibers: Processing, Properties and Applications

Manju Misra, School of Engineering and the Department of Plant Agriculture, University of Guelph, Guelph, ON, Canada

A variety of bio-nanofibres can be made for applications in energy storage, healthcare, biotechnology, environmental engineering, and in defence and security. This paper will highlight the preparation, properties and applications of electrospun bio-nanocomposites.

See more of this Session: Biobased Materials III
See more of this Group/Topical: Forest and Plant Bioproducts Division

(134d) Energy Savings and Improved Mixing Performance of High Consistency Biosolids with Modified Impeller Technology

Wojciech Wyczalkowski, Philadelphia Mixing Solutions, Palmyra, PA and Marc Moseley, Pulp&Paper, Philadelphia Mixing Solutions, Palmyra, PA

Mixing of biosolids for generation of renewable energy is more challenging than mixing of paper pulp. This is due to the difficult rheology, large volumes of vessels, requirement for high fiber consistency and the need for energy conservation. Biosolids consist mostly of fibrous materials similar to paper pulp where the main component is cellulose. Recently tested in numerous paper mills, this full scale proven technology can be applied to biosolids mixing. At present, most cellulosic/paper pulp vessels are only partially mixed. Maximum consistency is restricted to less than 6.5% due to mixing limitations. This new impeller technology allows mixing of the entire vessel, significantly reduces fluctuations of biosolids consistency. The ability to uniformly mix biosolids throughout the entire vessel maximizes exposure of fibers to chemical/enzymatic treatments and provides high treatment efficiency at low mixing power. This technology also allows for uniform mixing of cellulosic pulp above 20% consistency. The paper describes the provincials of the new technology, case studies, measurement of mixing performances by Electrical Conducive Tomography and the method of measurement of biosolids properties.

See more of this Session: Biobased Materials III
See more of this Group/Topical: Forest and Plant Bioproducts Division

(136a) Selective Dehydration of Monosaccharides to 5-Hydroxymethylfurfural: Catalyst and Solvent Effects

Mark H. Tucker[1], Anthony Crisci[2], Susannah L. Scott[3] and James A. Dumesic[1], (1)Chemical and Biological Engineering, University of Wisconsin, Madison, Madison, WI, (2)Department of Chemistry, University of California, Santa Barbara, Santa Barbara, CA, (3)Departments of Chemical Engineering and Chemistry, University of California, Santa Barbara, Santa Barbara, CA

A key obstacle in the conversion of biomass to chemicals is the selective removal of excess oxygen functionality. To accomplish this we have investigated the dehydration of monosaccharides to 5-hydroxymethylfurfural (HMF) using solid acid catalysts and separation friendly solvents. HMF is a versatile platform chemical which has applications in a range of industries including uses for polymers, pharmaceuticals, and fuels. The production of HMF is limited due to degradation reactions of HMF and the monosaccharide starting material under acidic conditions. To limit these degradation reactions, previous research has employed difficult to separate solvents and homogeneous catalysts. In this work we have explored the use of solid acid catalysts including a set of mesoporous SBA-15 materials functionalized with modified sulfonic acid groups and surface modifiers. Catalysts were characterized using solid-state NMR, powder XRD, elemental analysis, and N2 adsorption/desorption. We have found that hydrophobic functional groups in close proximity to the acid site increase the selectivity to HMF. Furthermore, we have explored a range of solvents to exploit the properties of these novel catalysts.

See more of this Session: Catalytic Processing of Fossil and Biorenewable Feedstocks: Chemicals
See more of this Group/Topical: Catalysis and Reaction Engineering Division

(136b) Systematic Evaluation of Bio-Oil Hydrotreating Catalysts

Haijun Wan[1], Jackson W. Ford[2], Raghunath V. Chaudhari[1] and Bala Subramaniam[1], (1)Department of Chemical & Petroleum Engineering, Center for Environmentally Beneficial Catalysis, University of Kansas, Lawrence, KS, (2)Center for Environmentally Beneficial Catalysis and Department of Chemical & Petroleum Engineering, University of Kansas, Lawrence, KS

Though many studies of bio-oil upgrading with various noble and non-noble metal hydrotreating catalysts have been reported in the literature, there isn't a clear picture of the comparative performance of these catalysts for either common model bio-oil substrates or bio-oil samples. In this talk, we will present results of the systematic evaluation of the performances of various types of mono-metallic and multi-metallic catalysts with respect to hydrodeoxygenation of model bio-oil compounds such as aliphatic acids, furfurals and phenolic compounds, as well as bio-oil samples themselves. The evaluated catalysts include Pd/C, Ru/C, Pt/C mono-metallic catalysts and Pd-Re/C, Ru-Re/C multi-metallic catalysts, and the performance measures include hydrogen uptake rates, yields of liquid and gas samples along with their compositions, and catalyst stability to deactivation based on recycle studies and physical property measurements. The availability of such comprehensive information is essential for rational catalyst design/selection and process development. In addition to conventional slurry reactor studies, we will also present hydrodeoxygenation of some of these substrates using supercritical reaction media that we recently reported.

See more of this Session: Catalytic Processing of Fossil and Biorenewable Feedstocks: Chemicals
See more of this Group/Topical: Catalysis and Reaction Engineering Division

(136c) Transition Metal Catalyzed Oxidation of Lignin and Lignin Model Compounds in Room Temperature Ionic Liquids

Joseph Zakzeski, A. L. Jongerius, P. C. A. Bruijnincx and B. M. Weckhuysen, Inorganic Chemistry and Catalysis, Utrecht University, Utrecht, Netherlands

Lignocellulosic biomass is a renewable resource that has the potential to serve as a feedstock for the production of fuels, chemicals, and energy. Dissolution of wood and other lignocellulosic materials is a critical step for the valorization of biomass. Once dissolved, lignin is susceptible to a wide range of transformations to yield high value aromatic products. In this work, lignin was dissolved in 1-ethyl-3-methyl imidazolium diethyl phosphate (EMIM DEP) and catalytically oxidized using a CoCl2•6H2O catalyst precursor in the presence of NaOH and molecular oxygen. The system dynamics, including changes that occurred in the lignin as well as the formation of the active catalyst species, were monitored using in-situ ATR-IR, Raman, and UV-Visible spectroscopy. This spectroscopic analysis revealed increases in alcohol and aldehyde functionality in lignin during the course of the reaction and also provided information regarding the characteristics of the catalyst. In addition to lignin, reactions of several lignin model compounds that contain the salient features of lignin were conducted. These model compounds included those with phenolic, methoxy, and benzylic functionality and beta-O-4, 5-5', and phenylcoumaran linkages, which represent the most abundant linkages in the lignin polymer. Analysis of these compounds revealed that the catalyst system readily oxidized alcohol functionality to form aldehydes. For example, veratryl alcohol, which is a simple non-phenolic lignin model compound, was oxidized to form veratraldehyde and veratrylic acid with a maximum turnover frequency of 1440/h relative to 10-15/h for previously described systems. In addition to the oxidation of the propyl side-chain contained in cinnamyl alcohol to form cinnamaldehyde, the catalyst system was also capable of disrupting the carbon-carbon double bond to form benzoic acid. The role played by the components of the catalytic system, including the interaction of the diethylphosphate anion, hydroxide, substrate, and molecular oxygen with the cobalt to yield this high oxidative catalytic activity, will be discussed. We conclude with a discussion of the potential application of this catalyst system as a lignin pretreatment method in a biorefinery or, alternatively, as a step to functionalize the compounds formed after other lignin monomeric-compound forming pretreatment steps.

See more of this Session: Catalytic Processing of Fossil and Biorenewable Feedstocks: Chemicals
See more of this Group/Topical: Catalysis and Reaction Engineering Division

(136d) Sol-Gel and Solution Combustion Synthesized Ni/Al2O3 Catalysts for Aqueous-Phase Reforming of Ethanol

Banasri Roy, Chemical Engineering, New Mexico Tech, Socorro, NM and Corey Leclerc, Chemical Engineering, New Mexico Institute of Mining and Technology, Socorro, NM

Aqueous phase reforming (APR) is a single step and low temperature (≤500K) energy efficient process, which can produce hydrogen as alternative energy source from water-diluted oxygenated hydrocarbons directly from fermentation, eliminating the energy intensive distillation of ethanol/water mixtures. The typical operating pressure and low temperature for APR can be helpful for the separation of H2 and CO2 from other products that are volatile at atmospheric pressure. Additionally, APR is useful for producing fuel cell grade H2 with small amount of CO in a single chemical reactor as a consequence of the water–gas shift (WGS) reaction thermodynamically favored at lower temperature reaction condition. We have investigated the catalytic effect of alumina support based nano-scale nickel catalysts for aqueous-phase reforming of ethanol. The Ni/Al2O3 catalysts were prepared by a sol-gel (SG) method and a solution combustion synthesis (SCS) route. Compared to SG and commercial catalyst the SCS sample showed • Higher EtOH conversion, H2 production, and H2 selectivity • Lower CO production and CO selectivity In order to understand these catalytic activity results we ran a series of microstructural and temperature programmed characterizations. The findings of these experiments are as follows: • Compared to the commercial and SG powders the particle size of SCS Al2O3 support is much smaller (XRD) and surface area is much higher (BET). • Compared to the commercial sample Ni-Al2O3 interaction is much stronger for SCS and SG powders (TPR) • After being used in reactor the SCS catalyst showed less carbon deposition than that of the SG and commercial powders (TPO) • Bright field TEM showed nano structure and wide dispersion of Ni metals on SCS Al2O3 sample. The particle size analyzed from TEM images match well with the sizes calculated from the corresponding XRD profile. The SCS sample clearly has higher surface area and smaller particle size which lead to higher catalytic activity. In a separate experiment the effect of surface modification of SCS alumina support powder on aqueous-phase reforming of ethanol has been explored. The surface of the Al2O3 powder was modified by a nonthermal RF plasma treatment using nitrogen gas. The catalytic activity results showed that while nature, relative ratio and selectivity of the products both in gas and liquid effluents did not change, catalytic activity of the sample increased after plasma modification. Microstructural (XRD, surface area) analysis showed that phase content and surface area of unmodified and modified catalysts are very similar, H2-chesorption shows higher metal surface area, higher metal dispersion and lower active metal particle size for the modified sample compared to the unmodified sample. The temperature programmed reduction (TPR) analysis demonstrated stronger support-metal interaction and smaller NiO particles for the modified catalyst at lower heat treatment temperature. The temperature programmed desorption (TPD) of ammonia analysis showed stronger acidity for the modified support, which can explain better dispersion of the metal particles on the modified catalyst compared to the unmodified sample.

(136e) Electrochemical Conversion of Glycerol

Kanako Okada and Levi Thompson, Chemical Engineering, University of Michigan, Ann Arbor, MI

Biodiesel consumption has increased at a compounded annual growth rate of 180%, expanding from 91 million gallons in 2005 to 412 million gallons in 2008 [1]. This dramatic growth is fueled by regulatory incentives, increased demand for fleet use, and the desire to achieve energy independence through use of a renewable source. The production of glycerol (C3H8O3), the principal by-product of biodiesel and oleochemical production, is expected to increase significantly with the expansion of biofuels. Although glycerol is used in food and personal care products, supply currently outstrips demand so that up to 350,000 tons of glycerol is incinerated in the US annually [2]. The electrolytic conversion of glycerol into value added products has been reported [3]. This process was carried out at low temperatures using Pt/C electrocatalysts and resulted in a production of aldehydes and acids depending on the applied potential. Research described in this paper explored the use of an electrochemical reactor that resembled a proton exchange membrane (PEM) fuel cell for the oxidation of glycerol. Noble metals based anode and cathode (1 mg/cm2 loading) were used with Nafion® 117 to produce the membrane electrode assemblies (5 cm2). The activities and selectivities were evaluated at 50 °C and varying potentials with humidified N2 (100 mL/min) flowing to the cathode and glycerol (100 mL/min) flowing to the anode. The products were characterized using a gas chromatograph equipped with flame ionization and mass spectrometer detectors. The electrocatalysts were characterized using x-ray diffraction and BET surface area analysis. The selectivities and conversion varied significantly with the applied potential as shown in Figure 1, with high glyceraldehyde production rates at higher potentials. The conversions and reaction rates increased with potential. Those results provide key information to study the mechanism of electrochemical oxidation of glycerol in an electrochemical reactor and to optimize experimental conditions to achieve high selectivity. References [1] Renewable Energy Consumption and Electricity Preliminary Statistics 2008: Energy Information Administration, DOE/EIA (2009) [2] Zheng Y et al. Commodity Chemicals Derived from Glycerol, an Important Biorefinery Feedstock. Chem. Rev. (2008), 108, 5253–5277 [3] Roquet et al. Kinetics and Mechanisms of the Electrocatalytic Oxidation of Glycerol as Investigated by Chromatographic Analysis of the Reaction-Products - Potential and pH Effects. Electrochim Acta (1994) vol. 39 (16), 2387-2394

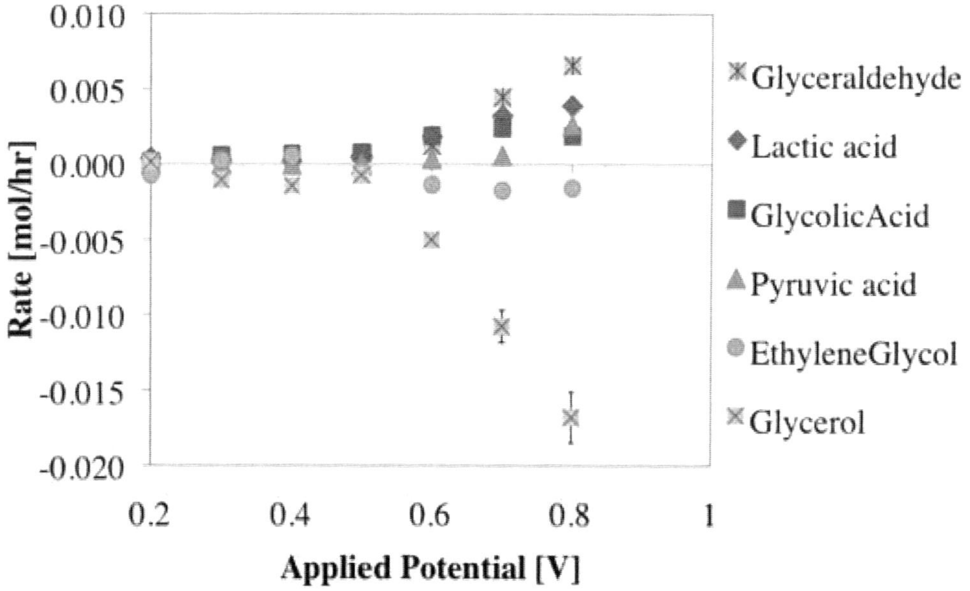

Figure 1. The effect of applied potential on the glycerol reaction rates.

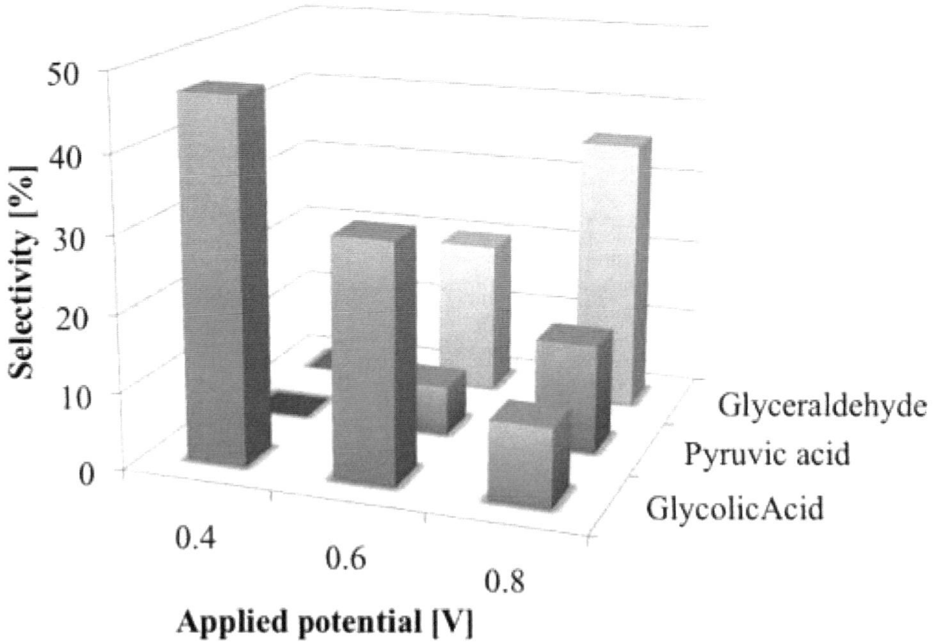

Figure 2. The effect of applied potential on the selectivities.

See more of this Session: Catalytic Processing of Fossil and Biorenewable Feedstocks: Chemicals
See more of this Group/Topical: Catalysis and Reaction Engineering Division

STUDY ON AN ENVIRONMENTALLY BENIGN PROCESS OF ETHYLENE GLYCOL PRODUCTION FROM SYNGAS

Jing Lv, Yujun Zhao, Baowei Wang, Shengping Wang, Yan Xu, Zhenhua Li, Xinbin Ma and Jinyu Han, Key Laboratory for Green Chemical Technology of MOE, School of Chemical Engineering & Technology, Tianjin University, Tianjin 300072, China

Introduction

CO coupling reaction to dialkyl oxalate and then hydrogenated to ethylene glycol runs as the following reactions:

$2CO+2RONO \rightarrow (ROOC)_2+2NO$

$(ROOC)_2+4H_2 \rightarrow (CH_2OH)_2+2ROH$

R can be methyl, ethyl; propyl etc. and RONO was prepared from methanol, ethanol, propanol correspondingly by the following reaction:

$2NO+0.5O_2+2ROH \rightarrow 2RONO+H_2O$

The total reaction for ethylene glycol production from syngas is:

$2CO+4H_2+0.5O_2 \rightarrow (CH_2OH)_2+H_2O$

In this process, dialkyl oxalate synthesis is the key step for ethylene glycol production. Also, dialkyl oxalate is the intermediate for production of oxalic acid used for pesticides, dyestuffs and drugs[1-4]. Since methyl nitrite and ethyl nitrite have low boiling point and are easily recycled as gaseous phase, the synthesis of dimethyl oxalate(DMO) and diethyl oxalate(DEO) attracts more attention. In this paper, the space-time yield, selectivity of dimethyl oxalate and diethyl oxalate and the conversion of CO were measured and compared in the same reaction condition over Pd-Fe/Al_2O_3 catalyst. The reaction was carried out in a fixed bed reactor (i.d. 20 x 200mm) by flowing the gas mixture of CO/C_2H_5ONO (or CH_3ONO)/N_2 (0.2/0.2/0.6) from the top of the reactor through 3mL catalyst bed. The inlet and outlet gases were analyzed by SP-2100 gas chromatograph and the liquid products were analyzed by SP-3420 gas chromatograph.

Results and discussion

The experiment was carried out at atmospheric pressure, residence time of 1.8s and temperature 100℃~140℃. The conversion of CO, space-time yield and selectivity of DEO were shown in Fig.1 to Fig. 3.

Conversion of carbon monoxide

From Fig. 1, it can be seen that with temperature increase, CO conversion increase first and then decrease. The optimum temperature was 130 ℃ for both DMO and DEO synthesis. Under the experimented temperature range, CO conversion for dimethyl oxalate synthesis is higher than that for diethyl oxalate synthesis, which indicated that methyl nitrite is more active than ethyl nitrite for CO coupling reaction.

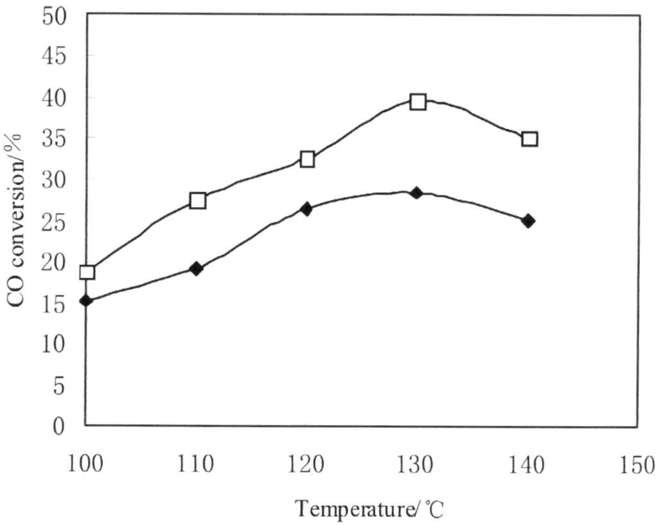

□ dimethyl oxalate, ◆ diethyl oxalate.

Fig.1 CO conversion vs.reaction temperature

Space-time yield of dialkyl oxalate

Fig.2 showed similar trend with Fig.1 and the space-time yield of dimethyl oxalate was higher than diethyl oxalate too, especially when the temperature was above 120℃. The optimum temperature for DMO and DEO synthesis was 120℃ and 130℃ respectively. Above the optimum temperature, STY of dialkyl oxalate decreased with temperature increase.

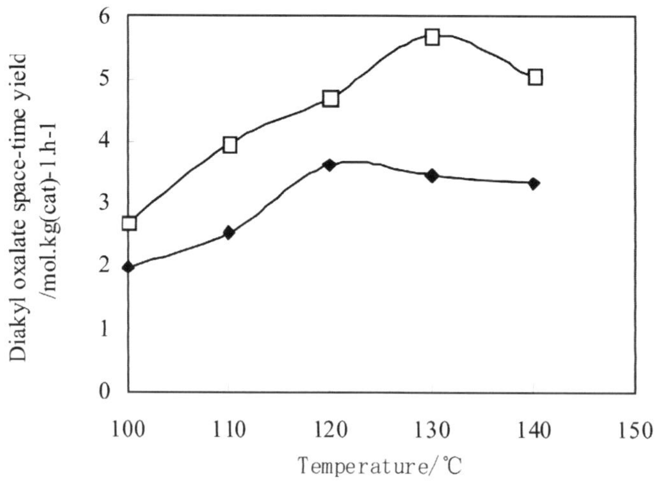

□ dimethyl oxalate, ◆diethyl oxalate.

Fig.2 STY of oxalete vs.reaction temperature

Selectivity of dialkyl oxalate

Fig.3 indicated that selectivity of dimethyl oxalate was very high and has little change with temperature but the selectivity of diethyl oxalate decreased with temperature increase. Also the selectivity of DMO is higher than DEO. This suggest that compared with coupling reaction of CO and ethyl nitrite, the coupling reaction of CO and methyl nitrite is more favorable considering the product separation and purification process.

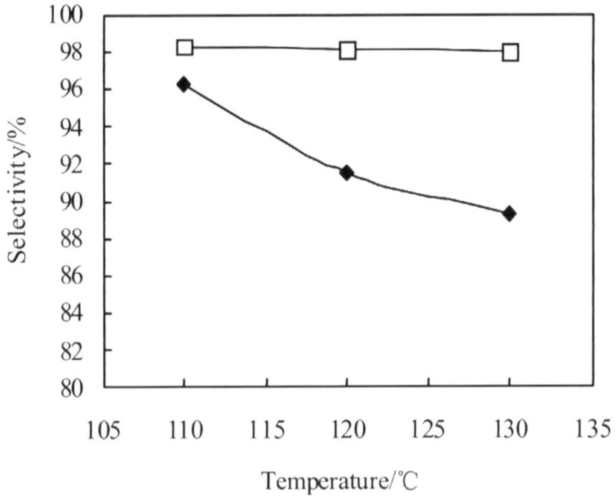

□ dimethyl oxalate, ♦ diethyl oxalate.

Fig.3 Selectivity of oxalate vs. reaction temperature

Conclusion

1. The optimum temperature for DMO and DEO was 130 ℃ and 120 ℃ respectively when the mole ratio methyl nitrite (or ethyl nitrite) and CO was 1:1.
2. In the same reaction condition, the space-time yield of DMO is higher than thant of DEO over Pd-Fe/Al_2O_3 catalysts because the decomposition conversion of methyl nitrite is lower than that of ethyl nitrite.

Acknowledgements:

We appreciate the financial support from Ministry of Science and Technology of China (No: 2006BAE02B03)

References:

1. Gen-hui Xu, Xinbin Ma, Fei He,Hongfang Characteristics of Carbon Monoxide Coupling Reaction. *Ind .Eng. Chem. Res*. 1995, 34(7):2379-2382
2. Jiang Xuanzhen, Su Yuehua, Lee Bor Jih, Chien Shuhua. A Study on the Synthesis of Diethyl Oxalate over Pd/α-Al_2O_3 Catalysts. *Applied Catalysis A: General*, 2001, 211:47-52
3. Wang Baowei, Xu Genhui. Journal of Fuel Chemistry and Technology, tudy of Catalyst of CO Coupling to Diethyl Oxalate in Gaseous Phase, *Journal of Fuel Chemistry and Technology*,1994, 22(1): 37-42
4. Meng Fandong, Xu Genhui. Carbon, Monoxide Coupling Reaction in the Presence of Hydrogen. *Journal of Natural Gas Chemistry,* 2002, 11: 57-62

AIChE 2010 Annual Meeting, November 7-12, Salt Lake City, UT.

Carboxylation of Glycerol in a Biodiesel Plant

Nghi Nguyen, Yaşar Demirel*
Department of Chemical and Biomolecular Engineering, University of Nebraska Lincoln, Lincoln, NE 68508

ABSTRACT

The US production of biodiesel is expected to grow with a target of 400 million gallons by the year 2012 and produce around 3.5 million gallons of crude bioglycerol. Refining crude bioglycerol is expensive, hence 50% of it is disposed of and the remaining is sold at a minimal price. Conversion of bioglycerol into value-added products has emerged in recent years as a result of bioglycerol's availability and for reducing the overall cost of biodiesel plants. Besides that reusing existing carbon dioxide as a feedstock for producing chemicals has received much attention recently. In a direct carboxylation process, glycerol reacts with carbon dioxide to produce bioglycerol carbonate and water. Bioglycerol carbonate is an intermediate chemical with many potential areas of application, such as reactive protic solvent, substitute for ethylene carbonate, propylene carbonate, cyclocarbonate derivatives, solvent for battery electrolyte, filming lubricants, filming plastifiers, ingredients for cosmetics, and monomers for polycarbonates. This study presents a conceptual process design for the production of glycerol carbonate using glycerol and carbon dioxide. The carboxylation reaction uses 1 mol% of *n*-dibutyltinoxide (n-Bu_2SnO) as catalyst and methanol as solvent at 3.5 MPa pressure and 80°C. A 35% conversion is assumed. The unreacted reactants and catalysts were recycled and 99.6 wt% pure glycerol carbonate is obtained in the process.

Keywords: Bioglycerol, carboxylation, carbon dioxide, bioglycerol carbonate, *n*-dibutyltinoxide

*Corresponding Author. Tel: 402 4722745, fax:402 4726989.
E-mail address: ydemirel2@unl.edu

1. Introduction

There are presently 173 companies with an annual production capacity of 2.69 billion gallons of biodiesel in the US [1]. However, this capacity is not fully utilized. The US production of biodiesel is expected to grow with a target of 400 million gallons by the year 2012. At this production level around 3.5 million gallons of crude bioglycerol is produced. On the other hand, 9 billion gallons of bioethanol production emitted around 25.9 million metric tons of CO_2 in 2008, in the US alone [2].

The production cost of biodiesel ranges from \$0.65-\$1.50 per gallon. However, as the cost of feedstock of oil and lipid exceeds \$0.3/lb, the price of biodiesel needs to be above \$2.8/gallon [3]. Besides, there is an inverse relationship between the production cost of biodiesel and the variations in the market value of bioglycerol. The production cost of biodiesel increases by \$0.008/gal for every \$0.01/lb reduction in glycerol selling price [4]. Therefore, economical utilization schemes of bioglycerol, such as carboxylation of bioglycerol to bioglycerol carbonate, can lead to more economical biodiesel production.

Conversion of bioglycerol into value-added chemicals has emerged in recent years as a result of rapid dropped in the market price of bioglycerol. Approximately one kilogram of bioglycerol is produced for every ten kilograms of biodiesel. Bioglycerol is also produced during soap manufacturing, fatty acid production, and fermentation of sugars directly or as a by-product of ethanol production from lignocellulose. Recently, selective glycerol-based catalytic processes have been proposed such as oxidation, hydrogenolysis, dehydration, esterification, and carboxylation, to produce polymers, ethers, and various fine chemical compounds including hydrogen and bioglycerol carbonate [5-7]. Among them, direct carboxylation of bioglycerol and carbon dioxide to yield glycerol carbonate and water has gained special attention due conversion of two by-products into a valuable product [8,9].

Bioglycerol carbonate is an intermediate chemical with many potential areas of application, such as, reactive protic solvent, a substitute for ethylene carbonate, propylene carbonate, cyclocarbonate derivatives, solvents for battery electrolyte, filming lubricants, filing plasticizers, agrosynthons, ingredients for cosmetics, and monomers for polymerization. It is also a novel component of gas separation membranes, coatings, paints and surfactants. It can act as a nonvolatile reactive solvent for several types of materials. In addition, it could serve as a source of new polymeric materials for the production of polycarbonates and polyurethanes.

Maximum conversion is around 7% for the direct carboxylation of bioglycerol in the presence of tin catalysts such as n-$Bu_2Sn(OMe)_2$[8].

$$\text{Bioglycerol} + CO_2 \xrightarrow{\text{catalyst}} \text{Bioglycerol carbonate} + \text{Water}$$

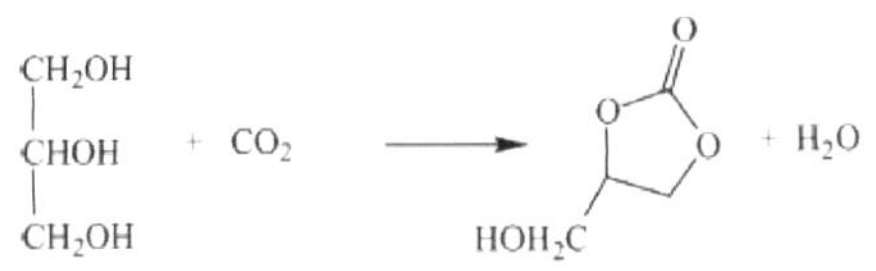

A conversion of 32.15 % can be achieved under supercritical CO_2 in the presence of zeolites and ion exchange resins [10]. However, excessive pressure results in higher equipment and operating costs. Finding a stable catalyst with a reasonable yield is crucial for reusing the catalyst and obtaining high conversion of bioglycerol [8-10]. A recent study showed that 35% conversion is possible at 80°C and 3.5 MPa using 1 mol% of n-dibutyltin(IV)oxide and methanol as solvent [9].

2. Process flow diagram

Aspen Plus version V7.2 is used for this simulation. Figure 1 shows the process flow diagram used for the production bioglycerol carbonate from pure carbon dioxide and bioglycerol feeds. Stream MAKEUP is the initial amount of catalyst (n-Bu_2SnO), which is recovered, treated and recycled to the reactor R101. Stream CO2 with a flow rate 15.811 kmol/hr is pressurized by compressor C101 to the reaction conditions before it is fed into reactor R101. Stream GLYCEROL with a flow rate 10.55 kmol/hr is preheated then sent into reactor R101. The carboxylation reaction takes place at 80°C and 3.5 MPa using 1 mol percent of n-dibutyltin(IV)oxide as catalyst. The conversion of bioglycerol is 35%. The reactor outlet, stream S5, contains the mixture of products, unreacted reactants, and catalyst. Most of the catalyst is separated in separator SEP101 and recycled.

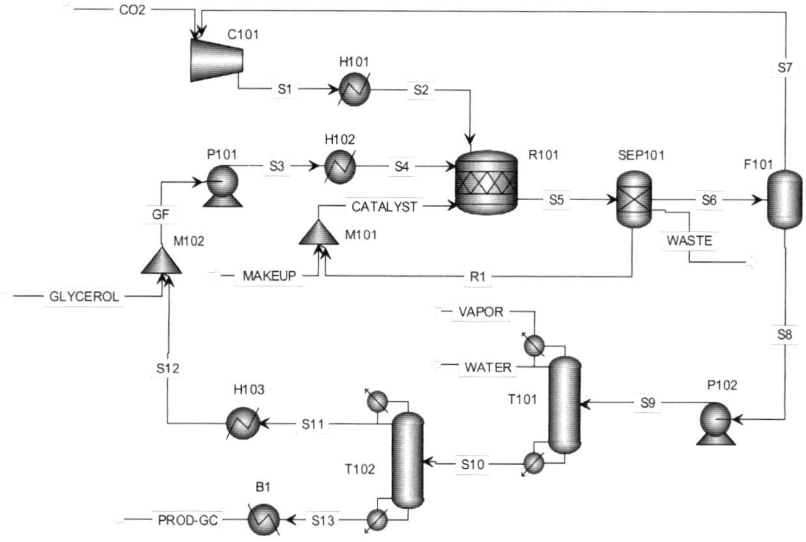

Figure 1. Process flow diagram for the production glycerol carbonates using bioglycerol and carbon dioxide.

Stream S6 is sent to flash column F101 to recover and recycle unused carbon dioxide. The top product, stream S7, contains 4.119 kmol/hr and 0.104 kmol/hr of carbon dioxide and water, respectively. Flash column F101 operates at 0.5 atm pressure. The bottom product, stream S8, contains bioglycerol, bioglycerol carbonate, and small amount of carbon dioxide and water. Stream S8 is pressurized to 1 atm and fed to column T101. Column T101 operates with 12 equilibrium stages with a kettle reboiler and a partial condenser. The duties of reboiler and condenser are 1095.35 kW and -537.37 kW, respectively. The feed, stream S9, enters above stage 7 of column T101. By-product water and small amount of carbon dioxide with flow rate of 15.88 kmol/hr emerges from the top in stream WATER, while stream S10 containing bioglycerol and bioglycerol carbonate and with flow rate of 30.05 kmol/hr is the bottom flow of column T101. Stream VAPOR contains mostly carbon dioxide.

The bottom flow of column T101 is introduced to stage 9 of column T102, which has 18 stages with a kettle reboiler and a total condenser. The distillate rate is 19.5 kmol/hr. The design spec sets the mass purity of bioglycerol in the distillate to 99.67 wt% by varying the molar reflux ratio. The molar reflux ratio 2.30 is obtained. The main product stream PROD-GC contains 99.63 %-wt of bioglycerol carbonate with the mass flow rate 1244.5 kg/hr (10.55 kmol/hr). The NRTL-RK model is used to estimate the phase-equilibrium and properties for the process.

3. Results and discussions

The process produces 1244.5 kg/hr glycerol carbonate from the feedstock of 699 kg/hr carbon dioxide and 972 kg/hr bioglycerol. The annual capacity of the product is 10752 ton/year based on 8640 hours of operation per year. Streams S7 and S11 are treated as tear streams. Two calculator blocks are employed to control the supply of carbon dioxide and bioglycerol.

Table 1

Some streams properties of the process

	GLYCEROL	CO2	S7	VAPOR	WATER	S10	S12	PROD-GC
Mass Flow kg/hr								
GLYCEROL	9.72E+02	0.00E+00	1.40E-04	2.15E-08	4.07E-02	1.80E+03	1.79E+03	4.57E+00
CO2	0.00E+00	6.99E+02	1.81E+02	2.22E+02	1.46E+01	1.14E-12	0.00E+00	0.00E+00
H2O	0.00E+00	0.00E+00	1.87E+00	9.19E+00	1.80E+02	1.37E-05	1.37E-05	2.41E-23
GC	0.00E+00	0.00E+00	1.06E-04	2.41E-11	4.76E-05	1.25E+03	5.91E+00	1.24E+03
CATALYST	0.00E+00	0.00E+00	0.00E+00	0.00E+00	0.00E+00	0.00E+00	0.00E+00	0.00E+00
Mass Frac								
GLYCEROL	1.00E+00	0.00E+00	7.65E-07	9.28E-11	2.09E-04	5.90E-01	9.97E-01	3.67E-03
CO2	0.00E+00	1.00E+00	9.90E-01	9.60E-01	7.51E-02	3.76E-16	0.00E+00	0.00E+00
H2O	0.00E+00	0.00E+00	1.02E-02	3.97E-02	9.25E-01	4.52E-09	7.65E-09	1.93E-26
GC	0.00E+00	0.00E+00	5.79E-07	1.04E-13	2.44E-07	4.10E-01	3.29E-03	9.96E-01
CATALYST	0.00E+00	0.00E+00	0.00E+00	0.00E+00	0.00E+00	0.00E+00	0.00E+00	0.00E+00
Total Flow kmol/hr	1.06E+01	1.59E+01	4.22E+00	5.56E+00	1.03E+01	3.00E+01	1.95E+01	1.05E+01
Total Flow kg/hr	9.72E+02	6.99E+02	1.83E+02	2.31E+02	1.95E+02	3.04E+03	1.80E+03	1.24E+03
Temperature C	2.50E+01	2.50E+01	3.50E+01	4.49E+01	4.49E+01	2.97E+02	2.50E+01	0.25E+02
Pressure bar	1.01E+00	1.01E+00	5.07E-01	1.01E+00	1.01E+00	1.01E+00	1.01E+00	1.01E+00
Vapor Frac	0.00E+00	1.00E+00	1.00E+00	1.00E+00	0.00E+00	0.00E+00	0.00E+00	0.00E+00
Liquid Frac	1.00E+00	0.00E+00	0.00E+00	0.00E+00	1.00E+00	1.00E+00	1.00E+00	1.00E+00
Enthalpy kcal/mol	-1.59E+02	-9.40E+01	-9.30E+01	-9.05E+01	-6.92E+01	-1.56E+02	-1.60E+02	-1.91E+02
Entropy cal/mol-K	-1.46E+02	6.70E-01	2.30E+00	8.20E-01	-3.80E+01	-2.74E+02	-1.47E+02	-6.08E+02

*GC: glycerol carbonate

Table 1 summarizes heat and mass balances as well as the properties of some of the input and output streams. Sensitivity analysis of the temperature in flash column F101 on the mass fraction of carbon dioxide shows that increasing the temperature yields higher total flow rate but decreases the mass fraction of carbon dioxide. High content of water in the reactor must be avoided because of inhibition of the forward reaction. The VAPOR stream contains 96-wt% of carbon dioxide, while WATER stream contains 92.5-wt% of water. However, the purity of those streams can be manipulated by varying the distillate vapor fraction.

4. Conclusions

Direct carboxylation of bioglycerol and carbon dioxide into bioglycerol carbonate is an environmentally friendly process as it helps to fix the carbon dioxide. It is also an economical process as it converts the glycerol into a value-added chemical product and lowers the unit cost of biodiesel production. Selection of a stable catalyst with a reasonable yield is crucial.

References

1. Biodiesel Organization: http://www.biodiesel.org. Accessed on Sep. 15, 2010.
2. Xu, Y., Isom, L., Hanna, M.A. (2010). Adding value to carbon dioxide from ethanol fermentations. Bioresource Technol., 101, 3311-3319.
3. Iowa State University: http://www.card.iastate.edu. Accesses on Sep. 15, 2010.
4. Zheng, Y., Chen, X., Shen, Y. (2008). Commodity chemicals derived from glycerol, an important biorefinery feedstock. Chem. Rev., 108, 5253-5277.
5. Guerrero-Perez, M.O., Rosas, J.M., Bedia, J., Rodriguez-Mirasol, J., Cordero, T. (2009). Recent inventions in glycerol transformations and processing. Recent Patents in Chem. Eng., 2, 11-21.
6. Zheng, Y., Chen, X., Shen, Y. (2008). Commodity chemicals derived from glycerol, an important biorefinery feedstock. Chem. Rev. 108, 5253-5277.
7. Zhou, C-H., Beltramini, J. N., Fan, Y-X., Lu, G. Q. (2008). Chemoselective catalytic conversion of glycerol as a biorenewable source to valuable commodity chemicals. Chem. Soc. Rev., 37, 527-549.
8. Aresta, M., Dibenedetto, A., Nocito, F., Pastore, C. (2006). A study on the carboxylation of glycerol to glycerol carbonate with carbon dioxide: The role of the catalyst, solvent and reaction conditions. J. Mol. Cat. A: Chemical, 257, 149-153.
9. George, J., Patel, Y., Pillai, S. M., Munshi, P. (2009). Methanol assisted selective formation of 1,2-glycerol carbonate from glycerol and carbon dioxide using *n*-Bu_2SnO as a catalyst. J. Mol. Cat. A: Chemical, 304, 1-7.
10. Vieville, C., Yoo, J. W., Pelet, S., Mouloungui, Z. (1998). Synthesis of glycerol carbonate by direct carbonatation of glycerol in supercritical CO_2 in the presence of zeolites and ion exchange resins. Catalysis Letters, 56, 245-247.

(199a) Poly-Pyrrole Loaded Fibers as New Adsorbent Materials

Samar Bose[1], **Bandaru Ramarao**[2], Raymond Francis[2] and Srikant Ramarao[2], (1)Department of Paper and Bioprocess Engineering, State University of New York, Syracuse, NY, (2)State University of New York, Syracuse, NY

Poly-Pyrrole coated silica granules form effective filtration material for separating humic acids from water. Besides being a conductive polymer, poly-pyrrole can adsorb various anionic species from aqueous solutions because of its cationicity. One limitation in the application of granular beds for filtration is the relatively high pressure drop per unit surface area available in the beds. Fibrous media provide an attractive option to operate at lower pressure drops and therefore achieve higher flow rates through the filters. This investigation was thus targeted at developing a new adsorbent material: recycled pulp fibers coated with poly-pyrrole which could be used as adsorbents for various materials from both aqueous and gas phases. We prepared poly-pyrrole laden pulp fibers by the in-situ polymerization of pyrrole under controlled acid conditions with FeCl3 as catalyst. Bleached and unbleached hardwood kraft pulps were treated with different amounts of poly-pyrrole. We found that the loading of poly-pyrrole was significantly larger on the unbleached pulps as compared to the bleached pulps. Measurements of the zeta potential of the pulps showed that they were cationic primarily due to the presence of the polypyrrole. Significant adsorption of ammonia was observed on the poly-pyrrole loaded pulps and particular effectiveness in controlling microbial activity was found with unbleached pulps with poly-pyrrole. These results indicate the effectiveness of poly-pyrrole loaded pulps as possible antimicrobial or antitoxic agents and in purification of water from different metal ions including arsenic, chromium and molybdenum.

See more of this Session: Biobased Materials IV
See more of this Group/Topical: Forest and Plant Bioproducts Division

(199b) Effect of Nanofiller Orientation On the Mechanical Properties of Cellulose Nanocrystal – Alginate Nanocomposite Fibers

Esteban E. Ureña-Benavides[1], Philip J. Brown[2] and Christopher L. Kitchens[1], (1)Chemical & Biomolecular Engineering, Clemson University, Clemson, SC, (2)Materials Science and Engineering, Clemson University, Clemson, SC

Alginate is a well known environmentally friendly material which has found many applications such as the preparation of dressings to treat exuding wounds, drug delivery, enzyme immobilization, etc.; however its use is limited due to low mechanical properties. Cellulose nanocrystals (CNC) isolated from cotton were added to calcium alginate fibers with the goal of improving its strength (tenacity), and modulus. The nanocrystals are elongated particles, with an average length of 130nm, an average width of 20.4nm, and an average height of 6.8nm; they were mixed with an aqueous sodium alginate dope solution, and wet spun into a $CaCl_2$ bath to form fibers. Addition of the nanocrystals reduced the tenacity and modulus of the material when the processing conditions were kept constant. This unusual behavior may be explained by the degree of orientation of the nanocrystals inside the fiber; wide angle x-ray diffraction (WAXD) experiments showed a decrease in nanocrystals alignment as the CNC concentration is increased. However a small concentration of CNC in the dope solution allowed to significantly increase the apparent jet stretch during processing. Characterization of the fibers produced at the maximum stretch showed an increase of 38% in tenacity, and 123% in modulus when the CNC concentration was 10%wt. WAXD showed that the degree of alignment of the CNC was recovered by increasing the apparent jet stretch. CNC have potential to become a green, biocompatible, and cost effective solution to reinforce alginate fibers, as well as other biopolymers.

See more of this Session: Biobased Materials IV
See more of this Group/Topical: Forest and Plant Bioproducts Division

(199c) Bioplastics in India

Manju Misra, School of Engineering and the Department of Plant Agriculture, University of Guelph, Guelph, ON, Canada

TBS

See more of this Session: Biobased Materials IV
See more of this Group/Topical: Forest and Plant Bioproducts Division

EVALUATION OF QUALITY OF AGRICULTURAL RESIDUE BASED COATED PAPERS USING PLACKETT BURMAN STATISTICAL DESIGN

DR. A. K. RAY * & SANJAY TYAGI**

** *Professor, Department of Paper Technology, IIT Roorkee (Saharanpur Campus), Saharanpur (INDIA)*

** *Research Scholar, Department of Paper Technology, IIT Roorkee (Saharanpur Campus), Saharanpur and Scientist B, Central Pulp & Paper Research Institute, Saharanpur (INDIA)*

ABSTRACT

Paper is pigment coated to produce a more uniform and more receptive surface for printing ink for better graphic reproduction. Large numbers of variables are involved in coating process which interacts with each other. In order to perform the coating experiments on indigenous base papers in a most systematic way with many variables, statistically designed experiments based on Plackett Burman design has been employed for examining the effects of different process variables. The variables considered were dosages of china clay, talc, natural and precipitated calcium carbonate, binder, total solids of coating color and thickener dosage (water retention chemicals). Based on the experiments the effects of above process variables on different properties of coated paper were identified.

Key words: coating, statistical design, precipitated calcium carbonate, brightness, talc, gloss

Introduction

Coating of paper is a layered construction which represents a multi-variable system and quantitative estimations on the effect of each parameter on the process are very difficult. The relative effect of so many variables on the properties of papers requires a systematic investigation, to obtain quantitative estimates of the effect of each of the parameter on the process, which is really a Herculean task. The complexity of the situation arises by the interaction of variables which cannot be separated from their main effect. In such an intricate situation statistical modeling can be of enormous use. It has been already probed that the statistical modeling can provide a means of reducing the enormous number of experiments which are normally required in traditional experimentation. In the traditional experiments, only one variable can be changed at a time keeping the others constant and thus may lead to erroneous conclusions. This can be avoided by various designs of experimental methodology, which not only avoid laborious experimentation, but also provide better quantification of parameter relationship. As a matter of fact, the prediction of different coating color formulation has been done using different statistical experimental design techniques such as full factorial design, McLean Anderson method, Scheffe method, Draper Lawrance, mixture design etc. by many researchers/ designers [1, 2, 3]. Of all the statistical design techniques, the most important being Plackett- Burman (P-B) design, generally favored to evaluate the effect of process variables [4].

Plackett- Burman Experimental Design:

The P-B design [5, 6] is a two level D-optimal saturated orthogonal designs (fractional factorial) constructed on the basis of fractional replicates of a full factorial design. Further, it is based on balanced incomplete blocks and the use of N experiments (N number of runs) to study (k = N- 1) process variables where N is a multiple of 4. For a design matrix to be orthogonal, the following conditions are necessary and sufficient:

1. The number of times each factor is fixed to each of its levels must be the same;

2. The number of times every two factors, each at any one of its levels are encountered must be the same; and

3. The number of observations must be divisible by square of the number of levels, defined as

$$N = nl^2 \qquad (1)$$

where n is integer. When the conditions stated above are available, to construct an orthogonal matrix (experimental design) requires combinatorial operations only.

If $N = nl^2$, then the number of factors for which one can compute their effects, given n is

$$k = (nl^2 - 1)/ (l - 1) \qquad (2)$$

or an integer part of the value given by equation (2).

If the number of levels for all factors is 2, then an optimum design is obtained by constructing N x N orthogonal matrix comprising +1's and -1's, where N is a multiple of 4, i.e. N = 4n. The maximum number of factors which can be included in the design is k = N -1.

Again, if the difference between the effects of each primary variable is not sufficient to permit confident identification of the important factors by inspection, an alternative is to construct normal or half – normal probability plots of the effect. In these plots the ordinate is the cumulative percentage probability P, with m data points taking values for constant spacing

$$P_i = 100(I - ½)/m, \text{ where } I = 1, 2, \ldots\ldots m \qquad (3)$$

over the cumulative distribution. The ordinate scale is adjusted to linearize the ideal sigmoid curve, so that if all the effects were normally distributed, they would plot in a straight line. Thus major deviations from this straight line are an indication that these effects are not easily explained as chance occurrences.

An incomplete factorial design which measures only the main effects (no interaction) is actually the P-B design where the numbers of experiments are reduced greatly but useful data is generated. Because these designs cannot be represented as cubes, they are sometimes called non geometric designs. Each variable is at one of two levels with high (+) and low (-). These levels were chosen far enough apart to expect a significant response in coating, but not so remote from normal conditions. (i.e. if the levels are set too close together or too far apart, they may give a response difference that is not significant, whereas the factor effect may be significant).

A major disadvantage of the factorial design is that an increase in number of factors is accompanied by a dramatic rise in the number of experiments. As an example, five factor two level unreplicated designs requires $2^5 = 32$ experiments and four factor 3 level design requires $3^4 = 81$ experiments. This difficulty can be overcome with reducing the number of experiments without a substantial loss of information by the use of fractional factorial design. P-B design is one of the classic examples for solving the above problem. In this a quarter fractional design of five factors at two levels has $(¼) \times 2^5 = 8$ experiments as in this present case. In fractional factorial designs with four or more factors, the higher interaction terms can be ignored so that only the main effects and two- factors interactions are evaluated.

P-B design is often called the ruggedness test and has application in validating methods that could be adopted as routine practice in laboratories. Factors which have a large effect on the results are thus identified and can be subjected to a more rigorous investigation before the method is validated. In collaborative trials this design is particularly useful in giving prior warning to experiments that require careful control.

Since the experimental design is only a fraction of complete factorial, each main effect is confounded with a large number of two factor, three factor, and high order interactions. However, since the chosen design has n as an integral power of two, it is possible to at least identifying the two factor compounding.

If it is not required to study as many as (n-1) variables in n experiments, the unassigned variables are dummies and no experimental condition is represented by the + or – in the design but the effects are calculated in the same way as the effects of the real variables. If there were no interaction and no experimental error, the effect shown by dummy variable would be zero. Non- zero results for included dummy variables can therefore be used to obtain an estimate of experimental variance. However, since it is impossible to determine whether the effect of a dummy variable is a measure of the uncertainty of the results or in fact due to one or more of the interactions with which it is confounded the values of calculating significance in this way is questionable.

Since the design matrix is orthogonal, this estimate of error mean square of the linear effects agrees with the estimates produced with the least squares. In addition the linear effects so obtained are unconfounded with one another and their estimates are unbiased. However, P-B design can be extended to the analysis of non- linear multivariate regression equation also with two significant two –factor interactions and more.

In the present investigation, an attempt has been made to determine the relative importance of seven selected factors or variables (viz. dosages of china clay, talc, natural and precipitated calcium carbonates, binder, total solids of coating color and thickner) by P-B design in coating of base paper provided by various mills and then to assess the most pertinent variables (factors) influencing the properties of coated paper. The coated paper properties studied were brightness, surface roughness, gloss, print gloss, printing picking velocity, print contact factor (m) and saturation ink density (D_∞). However, as the goal was to find out the relative effect of seven variables in terms of their preference for large scale experiment no attempt is made to develop linear multivariate regression equations with consequent evaluation of Student't' and Fisher 'F' test. This procedure of without considering development of multiple regression models is in agreement with Broderic et. al.[7], Cardwell [8] and Irvin et. al [9].

Also the above seven process variables were ranked relatively in accordance to their influence on different paper properties to assess the most pertinent variable influencing the properties of coated paper. An indication of the relative importance of these factors would assist in the design of effective laboratory scale simulations of the commercial processes.

Experimental Investigations

A design matrix based on above Plackett- Burman design for eight experiments for studying seven variables is given in Table 1. The assumption made was that within the restricted range of each variable, the response is entirely linear. Table 2 shows the combinations of process variables used for each coating experiment.

Table 1: Orthogonal design matrix of Plackett- Burman saturated design (N = 8; k = 7)

Exp. No.	Process variables						
	A	B	C	D	E	F	G
1	+	+	+	-	+	-	-
2	-	+	+	+	-	+	-
3	-	-	+	+	+	-	+
4	+	-	-	+	+	+	-
5	-	+	-	-	+	+	+
6	+	-	+	-	-	+	+
7	+	+	-	+	-	-	+
8	-	-	-	-	-	-	-

Table 2: Process variables used in the experiments

S. No.	**Process Variable**	**Process variable conditions**	
		Low Level	High Level
A	China clay (Parts)	25	77
B	Talc (Parts)	3	19
C	GCC (Parts)	6	56
D	PCC (Parts)	1	20
E	Binder (%)	10	18
F	Total Solids (%)	50	65
G	Thickener (%)	0.2	1

Thus for coating experiment 1, the high levels of process variables A, B, C and E were used and the low levels of variables D, F and G. The complete scheme is derived from this basic one line design which depends on the number of experiments. Subsequent lines are generated by cyclic rotation one space to the right (or left). The last line is always all low levels. When this is done, each variable will have an equal number of experiments at the high and low levels. Further examination shows that when variable A is at its high level, B is high twice and low twice. Similarly, when A is at its low level, B is high twice and low twice. The same is true for the other variables. Thus the net effect of changing the other variables cancels out when A is calculated according to the equation

$$E_A = [\sum R, \text{Response at } (+) / (n/2)] - [\sum R, \text{Response at } (-)/ (n/2)] \quad (4)$$

where E_A is the main effect of variable A and N is the number of results. The main effect of A is simply the difference between the mean values of the response at the high and low levels and the main effect of other variables can be calculated in the same way. Similarly, in this present case the effect of changing factor e. g. D from low level to high level (for N = 8) is given by

$$D = \tfrac{1}{4}(R_2 + R_3 + R_4 + R_7) - \tfrac{1}{4}(R_1 + R_5 + R_6 + R_8) \quad (5)$$

Coating Color Preparation

Different coating pigments were collected from mills and suppliers were evaluated for their particle size, particle size distribution and optical properties. The results are presented below.

Property	China clay	GCC	PCC	Talc
5%	<0.547μm	<0.605 μm	<1.55 μm	<3.185 μm
15%	<0.73 μm	<0.729 μm	<1.84 μm	<3.90 μm
35%	<1.19 μm	<0.896 μm	<2.26 μm	<4.81 μm
65%	<1.80 μm	<1.026 μm	<2.56 μm	<5.59 μm
Avg. Particle size (μm)	4.18	1.32	3.26	7.09
Specific surface area (m^2/cm^3)	3.12	5.15	2.15	0.97
Brightness (%)	81.45	91.68	96.54	86.86
Whiteness (%)	71.25	86.50	93.21	76.48

The coating color formulations were prepared according to experimental setup based on P-B statistical design. The coating color preparation process in brief is preceded in the order given below: The parenthesis indicates the type of mixing in either high speed or low speed mixer.

Pigments (China clay, GCC, PCC/Talc) → distilled water (high speed mixing) → dispersant (high speed mixing) → defoamer (high speed mixing) → latex (low speed mixing) → thickener (low speed mixing) → NaOH (low speed mixing) → distilled water. The total solids of coating slip were kept around 60% and the pH of coating slurry was maintained in the range of 8 to 9.

Coating Trials

A commercial base paper of 60 g/m^2 manufactured from the pulp of mixed hardwood and bamboo furnish in ratio 80:20 was taken for all coating trials. All coating chemicals i.e. pigment, latex, thickener, dispersant and deformer were procured from the coating plant itself. The viscosity of coating color prepared was measured using Brookfield viscometer. The water retention of coating slip was measured using AA-GWR gravimetric water retention meter (Finnish make model DT Paper Science, 1.5 bar, 120s, 5 μm filter). [10]

The coating trials were carried out on laboratory bar coater (Zethner ZAA- 2300) at the speed of 75 mm/s (4.5 m/min). The coated sheet was dried in laboratory oven at the drying temperature of 117°C for 30 seconds. The sheets were coated on one side with the coat weight of 20- 21 g/m^2. The coated paper was calendered in laboratory calender using soft nip only. Each paper was calendered at the linear load of 100 kN/m and temperature of 60ºC. The speed of calendered was kept at 20m/min.

RESULTS AND DISCUSSION

Effect of process variables on different characteristics of coated paper

Different sets of coating color based on P-B design was applied to base sheets and were evaluated for different coated paper characteristics. The results are given in (Table 3). The main effect of process variables on properties viz viscosity, water retention, brightness, surface roughness, gloss, print gloss, picking velocity, print density parameter (m, D_∞) were evaluated and ranked accordingly and given in (Table 4). The effect of main influential var The outcome of results from Plackett- Burman design is discussed below in the following paragraphs

Optical Characteristics

Brightness: Results from P-B design revealed that, out of the pigments (viz. china clay, talc, natural and precipitated calcium carbonate), china clay gave negative effect whereas other pigments gave positive effect on the brightness of coated paper. Binder and total solid of coating color showed negative effect whereas, the thickener, used to improve water retention gave positive effect. The main effect as a percent of mean reveals the highest effect by natural calcium carbonate (4%) followed by china clay (2.25%) and total solids (1.2%). Analysis of different optical properties of coating pigments also indicated that precipitated calcium carbonate used in above experiments had highest brightness and whiteness followed by ground calcium carbonate, talc and china clay which was the reason for the brightness value of coated paper.

Gloss: China clay gave positive effect on gloss whereas negative effect was observed in the case of talc, ground and precipitated calcium carbonates. Talc gave highest negative effect (21.6% of mean effect) followed by ground calcium carbonate (16.5%) and precipitated calcium carbonate (8.4%) indicating that introduction of these pigments with china clay will give coated paper with somewhat matt finish. The higher surface gloss in case of china clay (15.3%) is probably due to relatively lower void volume formation in the coating layer [8,9]. Binder level (11.8%) and total solid of coating color gave positive effect on gloss. The results are well in accordance to findings of Gron [11] and Hagymassy [12].

Surface Characteristics

Surface Roughness: China clay gave negative effect on surface roughness i.e. smoother surface is obtained when china clay is used as pigment in coating (15.2% of a mean effect). The positive effect of talc and precipitated calcium carbonate were comparable (3.0% and 4.2%) where as it was higher for ground calcium carbonate (25.5% of mean) indicating the ground calcium carbonate containing coated paper will be comparatively less smooth. Talc was able to enhance the evenness of paper surface; probably because the talc particles are platy shaped and possess high aspect ratio. Binder level gave positive effect of the order of 20.7% where as total solid and thickener gave negative effect of the order of 7.7% and 6.9% respectively.

Table 3: Coated Paper Characteristics obtained from experiments based on P-B design

Property	Brightness (%)	PPS Roughness (µm)	Gloss (%)	Print Gloss (%)	Picking velocity (m/s)	Density smoothness (m0 (μ^{-1})	Print density (D_∞)	Viscosity (mPas)	Water Retention (g/m2)
Exp. 1	80.0	1.77	41.4	73.5	225	0.34	2.42	50	2.24
Exp. 2	82.6	1.64	40.6	78.0	90	0.31	2.41	117	1.58
Exp. 3	82.4	1.92	38.2	75.8	240	0.30	2.43	57.4	1.48
Exp. 4	76.3	1.30	55.5	88.3	180	0.33	2.30	138	1.30
Exp. 5	77.5	1.41	40.0	84.8	276	0.30	2.38	450	1.38
Exp. 6	80.3	1.23	50.5	81.0	120	0.31	2.54	322	1.05
Exp. 7	78.4	1.07	44.0	78.7	93	0.24	2.91	402	0.95
Exp. 8	79.7	1.27	62.0	82.2	67	0.18	3.42	456	2.03
Mean	79.65	1.45	46.5	80.28	161.3	0.288	2.60	249.05	1.50

Table 4: Main effects of process variables on properties and relative ranking (based on P-B design)

Variables	Brightness (%)	PPS Roughness (µm)	Gloss (%)	Print Gloss (%)	Picking velocity (m/s)	Density smoothness m (μ^{-1})	Print density (D_∞)	Viscosity (mPas)	Water Retention (g/m2)
China clay	-1.80 **(2)**	-0.22 **(3)**	**7.15(3)**	0.27 **(6)**	40.25 **(3)**	0.03 **(2)**	-.011 **(5)**	-42.1 **(5)**	-0.23 **(4)**
Talc	0.5 **(6)**	0.04 **(7)**	-10.05 **(1)**	-3.07 **(3)**	12.75 **(6)**	0.01 **(3)**	-0.12 **(4)**	11.4 **(7)**	-0.07 **(7)**
GCC	3.3 **(1)**	0.37 **(1)**	-7.7**(2)**	-6.4 **(1)**	14..75 **(5)**	0.05 **(1)**	0.30 **(3)**	-344..9 **(1)**	0.17 **(6)**
PCC	0.55 **(5)**	0.06 **(6)**	-3.9 **(7)**	-0.2 **(7)**	-21.2 **(4)**	0.01 **(3)**	-0.18 **(6)**	-140.9 **(4)**	0.035 **(3)**
Binder	-1.2 **(3)**	0.300 **(2)**	5.5 **(5)**	0.63 **(4)**	137.8 **(1)**	0.05 **(1)**	-0.43 **(1)**	-150.4 **(3)**	0.20 **(5)**
Total solids	-0.95 **(4)**	-0.112 **(4)**	4.125 **(6)**	5.4 **(2)**	10.3 **(7)**	0.05 **(1)**	-038 **(2)**	15.4 **(6)**	-0.41 **(2)**
Thickener	0.1**(6)**	-0.11 **(5)**	-6.7 **(4)**	-0.4 **(5)**	41..7 **(2)**	0.01 **(3)**	-0.07 **(6)**	182.4 **(2)**	-0.63 **(1)**

Table 5: Main effect as percentage of mean (based on Plackett Burman Design)

Variables	Brightness (%)	PPS Roughness (µm)	Gloss (%)	Print Gloss (%)	Picking velocity (m/s)	Density smoothness m (μ^{-1})	Print density (D_∞)	Viscosity (mPas)	Water Retention (g/m2)
China clay	-2.25	-15.17	15.3	0.34	-24.9	10.3	-4.2	-17.0	-15.4
Talc	0.62	2.75	-21.6	-3.82	7.9	3.4	-4.6	4.5	-4.72
GCC	4.14	25.5	-16.5	-7.9	9.14	17.2	11.4	-138.5	11.0
PCC	0.70	4.22	-8.4	-0.25	-13.2	3.45	-6.9	-56.58	12.2
Binder	-1.5	20.7	11.8	0.78	85.4	17.2	-16.5	-60.4	13.0
Total solids	-1.2	-7.72	8.8	6.7	6.4	17.3	-14.6	6.18	-24.4
Thickener	0.12	-6.9	-14.4	-0.5	25.88	3.45	-2.72	73.2	-38.5

Printing Characteristics

Print Gloss: Print gloss was improved with the presence of china clay showing the positive effect. The highest negative effect is shown by ground calcium carbonate (7.9%) followed by talc (3.8%) and precipitated calcium carbonate (0.25%). Binder level and total solid gave positive effect (0.8% and 6.7%) where as thickener (0.5%) has shown negative effect.

Picking velocity: Presence of talc and ground calcium carbonate showed improvement in the picking velocity where as negative effect is shown by china clay and precipitated calcium carbonate. Binder level, total solid and thickener showed positive effect on the picking velocity. Out of all variables binder level shows maximum effect as a % of mean (84.5%) followed by thickener (25%) and PCC (13.8%).

Print density: The print density curve i.e. the optical density of print plotted against ink film thickness on the printing form gives indication about the ink requirement of paper in getting particular print density. In print density curve, the density D (i.e. the contrast with respect to unprinted paper) is plotted against the ink film thickness. Such density curve generally follows expression. (13)

$$D = D_{\infty} (1 - e^{-mx})$$

where D_{∞} is saturation density i.e. the density which is approached by theoretical curves if ink film thickness is increased indefinitely, m is the contract factor or density smoothness. It shows how quickly the saturation density is obtained and x is the ink layer thickness.
Ground calcium carbonate pigment has highest effect on contact factor i.e. m (i.e. 11.4%) followed by precipitated calcium carbonate (6.9%), talc, (4.6%) and china clay (4.2%). This indicated that ground calcium carbonate help in improving print density relatively higher than other pigments studied. It also gave positive effect on D_{∞}. The other pigments like china clay, talc and precipitated calcium carbonate had negative effect on D_{∞}. Binder level, total solids and thickener had positive effect on 'm' where as it was negative for D_{∞}.

Use of Talc in coating formulation

The introduction of talc should encouraged in the matt finish or semi matt finish coated papers to be printed by offset printing as the following features of offset process are especially demanding for talcs.

- Tacky ink emulsified by moisturizing solution.
- Hydrophilic moisturizing solution
- Offset blanket materials.

Talc helps to reduce the abrasivity of matt paper surface and lower the tendency for ink scuff or ink rub off. Ink scuff occurs when ink paper surface degrades the ink layer. With less abrasive surface, the ink layer better resists.

Coarse ground calcium carbonate generally used for matte papers create an abrasive surface against ink that can be controlled by using Talc in the recipe. Compared to 100% carbonate recipe, talc improves the ink holdout. Viscosity is positively affected by total solids and thickener. Ink setting/ ink hold out especially in offset printing is a function of ink penetration and/or ink absorption into the coating layer. The solvent of printing ink should obviously escape from the ink after printing. The more dried ink is retained on the outer surface of the paper, the higher will be the ink density or, accordingly lower will be the ink demand (14). If the pores of coating layer are too big, obviously the printing ink will penetrate easily into the paper and print gloss will be low. This is the case observed for ground calcium carbonate. To avoid this ground and precipitated calcium carbonate should preferably be used along with platy pigments like clay or talc

Effect on rheological properties of coating color

The effect on rheological properties of coating color made from different combinations of pigments and other additives were studied by measuring the viscosity and water retention value of different formulations. Results are given in (Table 3). Results shows that out of different pigments talc gave positive effect on coating color viscosity where as all the other pigment gave negative effect. The highest negative effect was shown by precipitated calcium carbonate indicating that introduction of excess precipitated calcium carbonate will change the rheological characteristics of coating color. This is probably due to high aspect ratio and narrow particle size distribution in the case of precipitated calcium carbonate. This may affect the ability to be pumped and also flow behavior in coating equipment.

The water retention value is negatively affected by china clay and talc indicating that these pigments helps in improving the water retained by coating layer where as water retention value (WRV) is positively affected by presence of ground and precipitated calcium carbonates. The positive water retention value means that less water is retained by coating color on the paper surface.

To get better coating, the pigment should be such that there is good water retention of coating color. If water is not retained somewhat, it will drawn in to paper instantly during application and a hard cake will build up on the metering device. Water retained is worst with ground and precipitated calcium carbonates than china clay probably due to reason that in the case of china clay and talc water simply has to go a longer way through tortuous platy packing where as in case of ground and precipitated calcium carbonates the particle size distribution is narrower so water has a shorter penetration path and tortuousity factor is less.

Conclusions

The Plackett- Burman design has served its purpose in ranking a large number of process variables according to their relative importance. The relative effect of different process variables as studied by P-B model indicates that

- China clay has negative effect on the brightness of coated papers as compared to other pigments like ground and precipitated calcium carbonate and talc.
- China clay gave positive effect on gloss while talc, ground and precipitated calcium carbonate gives negative effect. Ground calcium carbonate has highest negative effect on gloss. Binder level and total solids of coating color gave positive effect on gloss.
- Smoother surface is produced when China clay is used as a pigment followed by talc and precipitated calcium carbonate.
- China clay has more positive effect on print gloss while ground calcium carbonate has negative effect. Binder level and total solids shows positive effect while thickner gave negative effect on print gloss.

- China clay and precipitated calcium carbonate gave negative effect on picking velocity while talc and ground calcium carbonate shows an improvement in picking velocity.
- Ground calcium carbonate helps in improving the print density more than other pigments. It also has positive effect on D_∞ (saturation density). Binder level, total solids and thickner shows positive effect on m (contact factor or density smoothness) whereas negative for D_∞ (saturation density).
- Introduction of talc should be encouraged in matte finish or semi matte finish coated paper to be printed by offset printing.
- Water retention value (WRV) is negatively affected by china clay and talc indicating that these pigments help in improving the water retained by coating layer, whereas it is positively affected by ground and precipitated calcium carbonate.

References

1. R. Hari Krishnan A. Kumar A, A. Agarwal A and A. K. Ray, On the statistical prediction of performance of coating color formulations under laboratory air knife coater, *IPPTA Conventional issue,* 1995
2. A. K. Ray, Mitali Narang and Anshu Sharma, Statistically evaluating coating colour formulation for paper in a laboratory air knife coater, 4 P's Newsletter, 2001
3. A. K. Ray, Sanjay Tyagi, Aditya Tomar, Vikas Aggarwal and P. Monga, Predicting coating color formulation by Draper- Lawrence design with Titatimiun dioxide and zirconium sulphate, paper presented at conference on advances in Chemical Engineering (AChemE- 2009), Dept. of Chemical Engineering, Thapar University, Patiala, Punjab.
4. A. K. Ray, Dharm Dutt, C. H. Tyagi, Ankur Goel, A. Batra, G. Garg, Development of Coating colour formulation for duplex board on Plackett Burman statistical design, *Cellulose Chemistry and Technology*, **39**, 3 – 4, 265 – 275, May – August 2004.
5. Douglas C. Montgomery, "Design and analysis of experiments", Fifth edition, John Wiley & Sons (Asia) Pvt. Ltd., Singapore, 2007.
6. S. Akhnazarova and V. E. Kafarov, "Mascxperiment optimization in Chemistry and Chemical Engineering", MIR Publisher Mascow, 1982.
7. G. Broderick, J. L. Valade and J. Paris, High yield sulphite pulping based on Plackett-Burman design, *Pulp and Paper Canada,* **94**:9 (193) pp. 22 – 25.
8. R. D. Cardwell and S. B. Cundall, The effect of pulping process variables on pulp characteristics and physical properties, *Appita*, Vol. **29**, No.5 March 1976.
9. G. M. Irvin, N. B. Clark and C. Recupero, Extended delignification of mature and plantation eucalyptus wood. Part 2: The effect of chip impregnation factors, *Appita J.,* September 1996.
10. Stefan E. Sandas, Pekka J. Salminen and Dan E. Eklund, Measuring the water retention of coating colors, *Tappi Journa*l, December 1989. p. 207- 210.
11. Johan Gron and Peter Dahlvik, Influence of pigment particle characteristics on coating colour properties, *Paperi Ja Puu – Paper and Timber* Vol. **78**/No. 9/1996
12. J. Hagymassy and L. Haynes, Comparison of all clay minerals pigment system to clay/plastic pigment combination under varied finish conditions, *Tappi* , **60** (1977) 7, pp. 126 – 129

13. Y. V. Sood and S. R. D. Guha, Evaluation of printing characteristics of offset grade papers using IGT printability tester, *IPPTA* Vol 19, No. 4 Dec 1982.
14. Gill P. Drag, A. G. Hiorns, John D. Parsons and Les Coggon, Factors governing print performance in offset printing of matte papers, *Tappi Journal*, Vol 81, No. 11, November 1998.

HEAT TRANSFER MODEL IN CALENDER NIP – VERIFICATION FROM EXPERIMENTAL DATA ON AGRICULTURAL RESIDUE BASED COATED PAPER

Sanjay Tyagi [a] # and A. K. Ray[b]

[a] *Scientist C, Central Pulp & Paper Research Institute, P. B. No. 174, Paper Mill Road, Saharanpur (U. P.) (INDIA) and*
Research Scholar, IIT Roorkee (Saharanpur Campus), Department of Paper Technology Saharanpur(U.P.), INDIA

[b] *Professor, IIT Roorkee, (Saharanpur Campus), Department of Paper Technology, Saharanpur, (INDIA) (Email: amiyakumarray@gmail.com)*

#corresponding author (email: sanjaykhatuli1@rediffmail.com; styagi77@yahoo.co.in
Phone no. 919412479054; Fax no. 91132 - 2471057)

Abstract

The existing models for heat transfer during calendering were mainly due to Keller [1994], Hestmo and Lamvik [2002], which were applied only for calendering of newsprint in hardnip machine calenders. The experimental data of newsprint pilot machine were used for simulation. However, the models were never applied for soft nip calender to check their validity. Besides, the derivations of models from basic unsteady heat transfer equation were not shown. In present day, demand for high quality papers both surface size uncoated and pigment coated requires modification or extension of above models. Therefore, the present investigation deals with simulation of models with data obtained from laboratory soft nip calender for uncoated/coated paper. The basic heat transfer equation was used along with functional dependence of thermal conductivity, thermal diffusivity, specific heat, density etc. The extended models have been used for simulation and compared with those of Keller's for constant value of coefficient for coated/uncoated papers. This model is integrated with nip mechanics model of Hertz.
Keywords: calender rolls, heat transfer, coated paper, pressure, temperature

1.0: Introduction

Calendering is one of the last operations in dry end of paper machine where paper is conveyed through the contact line, the nip between two rotating parallel rolls. Due to line load in the nip, the compression of paper web takes place which reduces paper thickness and impart smoothness and gloss (in case of coated papers) to its surface. During the calendering process, the general aspect of heat transfer in calendering is that heat is transferred to calender rolls from the heated web leaving the dryer section, heat generation during the compression of paper in the nip and supply of heat from hot roll to the web to raise the temperature of paper. The above fact is well known to everybody, but the amount of heat transfer in rolling nip or in wrap of paper on calender roll is not well known because of the strong time dependence due to the very short periods of heat transfer process. Due to a very short period at calender nip, web is neither heated nor cooled to an equilibrium state. It is always in a transient state wherein the temperature distribution through its thickness is changing with time. The heat transfer between two materials depends on a heat transfer coefficient and a contact resistance. The contact resistance depends on the smoothness of the surfaces so that the resistance increases with increasing roughness. The pressure between surfaces affects the contact resistance and the value of contact resistance is estimated to be zero when the pressure exceeds 700kPa [1]. In dry paper heat transfer occurs mainly through conduction. As a porous material paper has a poor ability to transfer heat when dry. With increasing moisture content in paper, the importance of convection as a heat transfer mechanism increases. The heat conduction of paper depends both on the paper density and on the moisture content. With increasing density and moisture content the amount of air in the paper structure decreases and heat transfer improves. Calendering improves the heat transfer due to the increase in density [2]. Increasing the filler content increases the heat transfer for two reasons. Firstly, pigments have higher heat conductivity than fibers and secondly the density of the paper increases with the increasing amount of filler. The thermal conductivity of mineral pigments is 10 to 20 times higher than that of the fibers [3].

Kartovaara [2] has shown that with newsprint the temperature in the middle of the 50 g/m^2 paper starts to increase after only 2ms of contact time (calender surface temperature 120°C). With increasing moisture content the temperature increase starts even sooner (0.5 ms). Keller [4] in his work studied the heat transfer in calender nip for coated papers and found that in coating the amount of mineral pigment in the paper structure increases. This means that heat transfer through the coating occurs faster than through the fiber matrix and the temperature of the whole coated paper structure can already increases in the nip to a value very near to the surface temperature of the thermo roll. On the other hand, for uncoated paper it has been calculated [5] that the temperature gradient at the paper surface in the calender nip is very high and the temperature of the paper was approximately 35°C lower than the surface of the hot roll even at low running speed (250m/min). Guerin et. al. [6] studied the influence of some parameters on heat transfer during calendering for woodfree coated papers.

In gradient calendering the basic idea is to use the temperature and/or moisture gradient so that the calendering effect can be limited to the surface layer of the paper only. The aim is to concentrate the smoothing action of calendering on the surface of the paper and hence minimize the negative structural changes, such as density increase and thickness reduction of the middle part of the paper structure. Kerekes et al [7] showed the effect of high calendering temperature on the properties of newsprint. The actual gradient calendering concept was presented by Crotogino [8] in hard nip temperature gradient calendering of newsprint. Hestmo and Lamvik [9] studied the temperature gradient calendering and derived an empirical expression for the heat transfer coefficient as a function of line load and physical properties of paper. They found that for high speed calenders, the heat transferred mainly by conduction to the fibers in contact with the calender roll. The fibers in the surface of the paper are heated nearly to the temperature of the calender roll.

The nip width calculation is an important factor for calculation of heat transfer in the nip. Differential equations addressing the heat transfer problem can be written and solved analytically for different initial and boundary conditions. In the present investigation an attempt has been made to validate the heat transfer model due to Keller [4] and Hestmo and Lamvik [9] for both coated and uncoated paper made from agro residue pulps blended with softwood in the ratio of 75: 25 calendered hard nips and soft nip using laboratory calender. A systematic procedure interlinking the aforesaid model of nip mechanics was adopted for simulating the experimental data generated on laboratory calender under the varying conditions of calender load and temperature.

2.0: Experimental Investigation

2.1: Base Paper: The base paper used in this study was collected from the paper mill in reel form and having the furnish composition of 75% agricultural residue (bagasse/wheat straw) and 25% of soft wood chemical pulp.

2.2: Coating formulation and coating trials:

The paper coating trials were conducted on DT pilot coater at the speed of 30m/min. The distance between the applicator roll and blade was kept at 250mm. The metering blade measured 84mm x 315 mm was used in coating trials. The blade angle was kept at 31°. The blade load was adjusted by pneumatic cylinders at the pressure of 1.5 bar. During the run only one top side of 75 g/m^2 base paper was coated by the coat weight of approx 15 g/m^2, thus giving the total weight of 90 g/m^2.. The coating formulation consisted of 100 parts pigment (fine ground calcium carbonate and fine clay in ratio of 80:20), 12 parts SB latex, 1 part CMC , 0.12 part OBA and 0.5 part dispersant). For drying of coated paper two IRT- Monocassettes of infrared units were used to give a temperature of 60 – 70°C. The temperature of air drying unit was kept at 50 °C.

2.3: Calendering Conditions

Both uncoated and coated sheet of paper were calendered in laboratory calender (DT Paper Science, Finland) using soft nip condition of calendering. For both conditions, different set of linear load were tried (viz. 100kN/m, 120 kN/m, 150kN/m, 170kN/m and 200kN/m). For each set of load, three different calendering temperatures of 20° C (or room temperature), 70 °C and 120 °C were used. All above experiments were also tried at the higher calender speed of 25m/min.

2.4: Evaluation of coated papers

Calendered coated/uncoated sheets were conditioned at a temperature of 27± 1 °C and a relative humidity of 65 ± 2% prior to testing. All testing experiments were carried out according to relevant ISO standard test methods.

3.0: Validation and simulation for heat transfer models due to Keller [4, 10] and Hestmo and Lamvik [9] with the experimental data

3.1: Derivation of Conduction Heat Equation for a rectangular slab:

The unsteady state Fourier heat conduction equation in rectangular slab can be applied to paper machine calender without much error, though the letter is a cylindrical system. Details of derivation of equation for a rectangle slab is given elsewhere [1, 11, 12]

The system equation, however, involves constant physical (density), physico-thermal (specific heat, coefficient for thermal expansion) and transport properties (thermal conductivity and thermal diffusivity). In fact, these properties are constant for homogeneous isotropic solids but vary widely with the roll composition, even for machine calender with different type of thickness of chilled iron layer.

$$\frac{\partial T}{\partial t} = \alpha \left[\left(\frac{\partial^2 T}{\partial x^2} \right)_{yz} + \left(\frac{\partial^2 T}{\partial y^2} \right)_{zx} + \left(\frac{\partial^2 T}{\partial z^2} \right)_{xy} \right] + \frac{q_v}{c_p \rho} \qquad (1)$$

The equation can be further approximated with one dimensional form and after dropping the heat generation term q_v=0 (which is true for any system under investigation) can be written as

$$\frac{\partial T}{\partial t} = \alpha \frac{\partial^2 T}{\partial x^2} \qquad (2)$$

The above partial differential equation is parabolic in nature and can be solved analytically and numerically by different methods. [1, 11]
The problem can be solved in the case of hot calender using various kinds of initial and boundary conditions which in turn depend on design and operational conditions of various calenders.

PDE: $\frac{\partial T}{\partial t} = \acute{\alpha} \frac{\partial^2 T}{\partial x^2}$

I. C. :	T (0, x) = Ti	(3)
B. C. 1:	T(t, 0) = Ts	(4)
B. C. 2	T(t, ∞) = Ti	(5)

The B.C. 1 implies that since the thermal conductivity of the solid is finite the temperature at 'infinite' distance from the surface remains equal to the initial temperature for all finite time. Let us introduce a dimensionless temperature and a modified time variable

$T = \frac{T - Ti}{Ts - T_i}$ and T= ά t

The DE, the I.C. and the B. C's get transformed to the following forms

PDE : $$\frac{\partial \bar{T}}{\partial t} = ά \frac{\partial^2 \bar{T}}{\partial x^2} \quad (6)$$

I. C. : T (0,x) = 0 (7)

B. C. 1: T (τ, 0) = 1 (8)

B. C. 2 : T (τ, ∞) = 0 (9)

The detail of the derivation is given elsewhere [1, 11, and 12] for conduction of semi – infinite solids.

Based on the above boundary condition for dimensionless transient heat conduction equation for semi- infinite solid, one can set the following équation

$$T = (T - Ti)/ (Ts - Ti) = 1- \frac{2}{\sqrt{\pi}} \int_0^{\eta} e^{-\eta^2} d\eta \quad (10)$$

where $\eta = x/ 2(ά\, t)^{1/2}$ dimensionless parameter

The function as shown in the second part of Eq. (10) is known as the Gauss error integral or probability integral. Eq. (10) represents the temperature distribution in calenders. This can be suitably modified for simulating both machine as well as soft calendering operation which involves heat transfer between paper and calender roll. These are discussed as follows.

3.2: Analysis of unsteady state conduction heat transfer model to hot calendering operation

The above heat transfer model equation based on semi-infinite solids are extended to hot calendering based on the guidelines of Kerekes [7], Keller[4], and Hestmo and Lamvik [9] as under:

Eq. (10) indicates that at any time after the surface temperature is changed there will be some change in temperature at all points in the solid, even points far removed from the hot surface. The actual change at such distant points however is negligibly small. Beyond a certain distance from the hot surface not enough heat can penetrate to affect the temperature significantly. This penetration distance x, is arbitrary defined as that distance from the surface at which the temperature change is 1% of the initial change in surface temperature, i.e. $(T - T_a)/ (T_s - T_a) = 0.01$ or $(T_s - T)/ (T_s - T_a) = 0.99$.

3.3: Development of simplified equation for industrial practice on hot calendering:

To find the total heat transferred to a semi infinite solid in a given time, it is necessary to find the temperature gradient and heat flux at the hot surface as a function of time The temperature gradient at the surface can be estimated based on Holman [1] is found by differentiating eq. (10) to give

$$\left(\frac{\partial T}{\partial x}\right)_{x=0} = (T_s - T_a)/ (\pi ά t)^{1/2} \quad (11)$$

The heat flow rate at the surface is therefore,

$$\left(\frac{q}{A}\right)_{x=0} = -k\left(\frac{\partial T}{\partial x}\right)_{x=0} = k\,(T_s - T_a)/\,(\pi\acute{\alpha}t)^{1/2} \qquad (12)$$

After substitution of dQ/dt for q, equation (12) can be integrated to give the total quantity of heat transferred per unit area Q_t/A in time t_T as follows

$$\left(\frac{Q_T}{A}\right) = \left(\frac{k(Ts-Ta)}{\sqrt{\pi\alpha}}\right)\int dt/\sqrt{t} = 2k(Ts-Ta)\sqrt{t_T/\pi\alpha} \qquad (13)$$

Using $2k/\sqrt{(\pi\acute{\alpha})}$ as constant K, one can write

$$Q_T = K.\Delta T.\,(t)^{1/2} \qquad (14)$$

Where $\alpha = k/\rho C_p$ and $\Delta T = T_{roll} - T_{sheet}$

$$K = 2(k\rho\, C_p/\pi)^{1/2} = 1.128\,(k\rho C_p)^{1/2} \qquad (15)$$

Now rewriting $Q_T = 1.128\,[k.\rho.\,Cp.]^{1/2}\,\Delta T\,(t))^{1/2}$

And putting the value of dwell time t = W/S

$$Q_T = 1.128\,[k.\rho.\,Cp.]^{1/2}\,\Delta T\,(W/S))^{1/2} \qquad (16)$$

But $(k.\rho.\,C_p)^{1/2} = [k\,(\rho).\,\rho\,(P, T, S).\,C_p\,(T)]^{1/2}$

Therefore

$$Q_T = 1.128\,[k\,(\rho).\,\rho\,(P,T, S).\,C_p\,(T)]^{1/2}\,\Delta T * (W/S))^{1/2} \qquad (17)$$

The values of k, ρ and C_p depend on temperature, pressure, speed and type of paper to be calendered. The values of different thermal properties as given by different authors are given in reference [12]. The above equation (Eq. 16) can be generalized as Eq. (17). The parameter under parenthesis, i.e. square root of the product k (ρ). ρ (P,T,S). C_p (T) without their parametric variation has been defined as temperature penetration number [9].

Keller [4] has assumed constant K of the order of 699.50 for 65 g/m^2 uncoated paper and 1350.60 g/m^2 for coated paper

Thus the equations have been simplified as

$$Q = 699.50\,\Delta T\,(t)^{1/2} \qquad (18)$$

For coated paper,

$$Q = 1350.60\,\Delta T\,(t)^{1/2} \qquad (19)$$

Determining Temperature Penetration Number has been used a practical means for estimating heat transfer to paper in a hypothetical calendering operation. The nip dwell time would be calculated from roll radius and material properties and machine speed using Hertz equation.

3.4: Parameter estimation of nip mechanics model

In soft nip calendering where one hard nip and one soft nip rolls are in contact of paper, the Hertz model can be simplified for two roll theory of different radius, modulus of elasticity and Poisson ratio

$$w = (4FD_E/\pi E^*)^{1/2} \qquad (20)$$

For contacting bodies 1 & 2

$$1/D_E = 1/D_1 + 1/D_2 = (D_1 + D_2)/\,D_1D_2$$

$$D_E = D_1D_2/(D_1 + D_2)$$

$$1/E^* = [(1 - v_1^2)/E_1 + (1 - v_2^2)/E_2]$$

$$E^* = [(E_1E_2)/\,E_2\,(1 - v_1^2) + E_1(\,1- v_2^2)] \qquad (21)$$

Putting the values of D_E and E* in Hertz equation, we got nip width (w)

$w^2 = [(16PD_1D_2/ \pi (D_1 + D_2)) (1 - v_1{}^2)/E_1 + (1 - v_2{}^2)/E_2)]$
(32)

Now average nip pressure (σ) can be given by

$\sigma = P_{max}/ w$

$= 1.87\ E/D_E [1 + 0.207\ (w/h)^2 - 3.75 \times 10^{-3}\ (w/h)^4]$ (23)

where h = thickness of cover material

The Hertz formula is often quoted although it is not accurate for soft rolls

$$W^2 = 16\ LR_1R_2/ \pi(R_1 + R_2) [(1 - \sigma_1{}^2/ E_1) + (1 - \sigma_2{}^2/ E_2)] \quad (24)$$

where L is the line load and R_1, R_2 is the radius of two rolls. In industry, it is preferred to report the effects of calendering in terms of average nip pressure, that is line load/nip width which should overcome the problem of relating the results of one pair of rolls to another of different size or even of different stiffness. The above equations were used for the calculation of different parameters like average nip pressure, nip width and equivalent modulus of elasticity of soft nip calenders for different cases of coating formulations.

3.5: Simulation of Heat Transfer Models:

Based on equations of nip mechanics (i.e. equations 21, 22 and 24) and heat transfer equations (18) for uncoated papers and equation (19) for coated papers, heat transfer rate through uncoated/coated paper was found out at the varying conditions of linear load, (from 100 kN/m to 120 kN/m), calendering temperature (20°C, 70°C, 120°C) and calender speed (20 m/min, 25m/min) for 75 g/m^2 uncoated base paper made from the furnish of agri-residue and soft wood pulp. For coated paper, same paper was coated to 90 g/m^2 with GCC and china clay in ratio 80/20.

4.0: Results and Discussion

4.1: Simulation results for heat transfer rate for uncoated/coated paper calendered at soft nip calendering condition with calender speed of 25m/min and 20m/min.

Both uncoated paper and coated papers were calendered in laboratory soft nip calender under the varying conditions of load and calender temperature. Nip width and nip residence times were calculated from nip mechanic equations and heat transfer rate was calculated from equations (18) and (19). Results are given in Table 1 to 4. Higher calender speed of 25 m/min was also tried. The results of heat transfer rate for coated and uncoated papers at different nip residence time (due to varying load) and at different set of calendering temperature for calender speed of 20m/min and 25m/min are depicted in figures 1 and 2.

From the results of first trial at the calender speed of 20m/min, it is obvious that for uncoated paper, as the nip load of calender increases from 100kN/m to 200kN/m, nip width increases by 29% resulting an increase in nip residence time of 2.19 ms. Increasing nip load also increases the heat transfer rate across the nip due to the densification of paper matrix with increasing load (Table 1). When load increases from 100 kN/m to 200kN/m, heat transfer rate across the nip increased by 15.9%.

Table 1: Heat transfer rate for uncoated paper, calendered softnip at the calendar speed of 20 m/min at different conditions of liner load and roll temperature

Load (kN/m)	Nip width (mm)	Nip residence time (ms)	Δ T (°C)	Heat transfer rate (Q) (KJ/m^2)
100	1.76	5.28	20	1.017
			70	3.560
			120	6.103
120	1.93	5.79	20	1.064
			70	3.762
			120	6.387
150	2.15	6.47	20	1.125
			70	3.940
			120	6.755
170	2.29	6.89	20	1.161
			70	4.065
			120	6.969
200	2.489	7.47	20	1.209
			70	4.233
			120	7.258

The calendering temperature also has profound effect on heat transfer rate at calender nip. For uncoated paper, at the same nip load, increasing the calendering temperature from 20°C (room temperature) to 120°C, heat transfer rate changes by 83.3%. Thus calendering temperature has more pronounced effect on heat transfer rate as compared to nip load.

Coated paper also exhibits the same trend of increasing nip width and nip residence time with increasing nip load as the case with uncoated paper. The results are given in Table 2. Compared to uncoated paper, heat transfer rate in case of coated paper increases by 48%. This may be explained by both relative density of two sheets and difference in their material properties. (i.e. fiber v/s mineral pigments).

In another trial, higher calendering speed of 25m/min is used for the calendering of both uncoated and coated papers. Results are given in Table 3 & 4 respectively. It is obvious that at higher speed nip residence time in calender nip reduces by 1.05ms. This reduces the heat transfer. At the same nip load and calender temperature, heat transfer rate decreased by 10.5% for both uncoated and coated paper. The same is true for different calendaring temperatures. The above results are in good agreement with the findings of Lloyd et. al. [13]

Table 2: Heat transfer rate for uncoated paper, calendered softnip at the calender speed of 25 m/min at different conditions of liner load and roll temperature

Load (kN/m)	Nip width (mm)	Nip residence time (ms)	ΔT (°C)	Heat transfer rate (Q) (KJ/m^2)
100	1.76	4.23	20	0.910
			70	3.185
			120	5.460
120	1.93	4.63	20	0.952
			70	3.334
			120	5.715
150	2.15	5.18	20	1.007
			70	3.525
			120	6.043
170	2.29	5.51	20	1.039
			70	3.637
			120	6.235
200	2.489	5.985	20	1.082
			70	3.788
			120	6.493

Table 3: Heat transfer rate for coated paper, calendered softnip at the calender speed of 20 m/min at different conditions of liner load and roll temperature

Load (kN/m)	Nip width (mm)	Nip residence time (ms)	ΔT (°C)	Heat transfer rate (Q) (KJ/m2)
100	1.76	5.28	20	1.964
			70	6.874
			120	11.784
120	1.93	5.79	20	2.055
			70	7.194
			120	12.334
150	2.15	6.47	20	2.173
			70	7.607
			120	13.041
170	2.29	6.89	20	2.242
			70	7.849
			120	13.456
200	2.489	7.47	20	2.335
			70	8.174
			120	14.014

Table 4 : Heat transfer rate for coated paper (coated with china Clay /GCC in ratio 80: 20) calendered softnip at the calender speed of 25 m/min at different conditions of liner load and roll temperature

Taking $Q = 1350.6\ \Delta T\ (t)^{1/2}$

Load (kN/m)	Nip width (mm)	Nip residence time (ms)	ΔT (°C)	Heat transfer rate (Q) (KJ/m2)
100	1.76	4.23	20	1.757
			70	6.150
			120	10.543
120	1.93	4.63	20	1.839
			70	6.437
			120	11.030
150	2.15	5.18	20	1.944
			70	6.806
			120	11.668
170	2.29	5.51	20	2.006
			70	7.020
			120	12.039
200	2.489	5.985	20	2.089
			70	7.314
			120	12.538

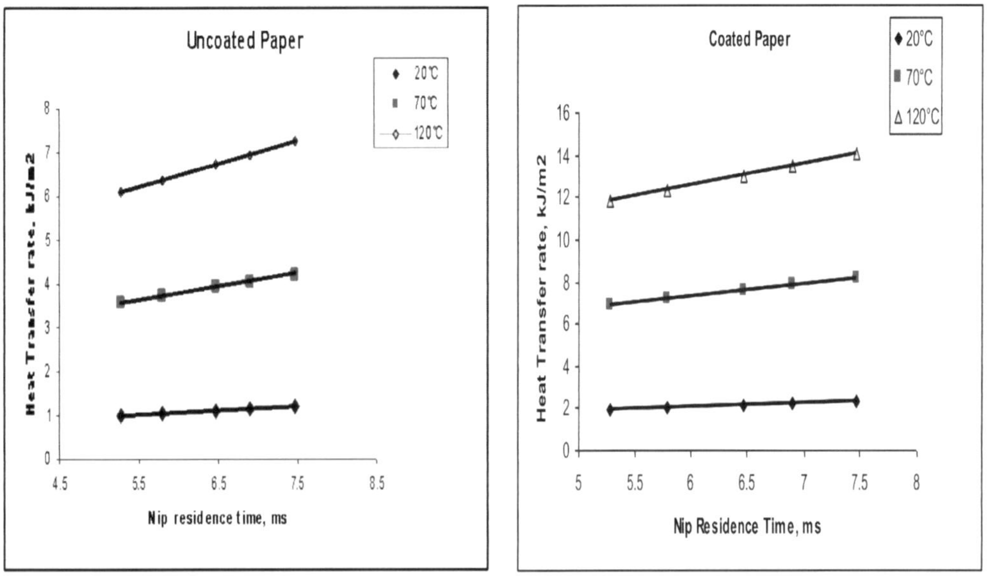

Fig 1: Comparison of heat transfer rate for uncoated and coated paper samples during soft nip calendering (calendering speed 20m/min) at different calendering temperature

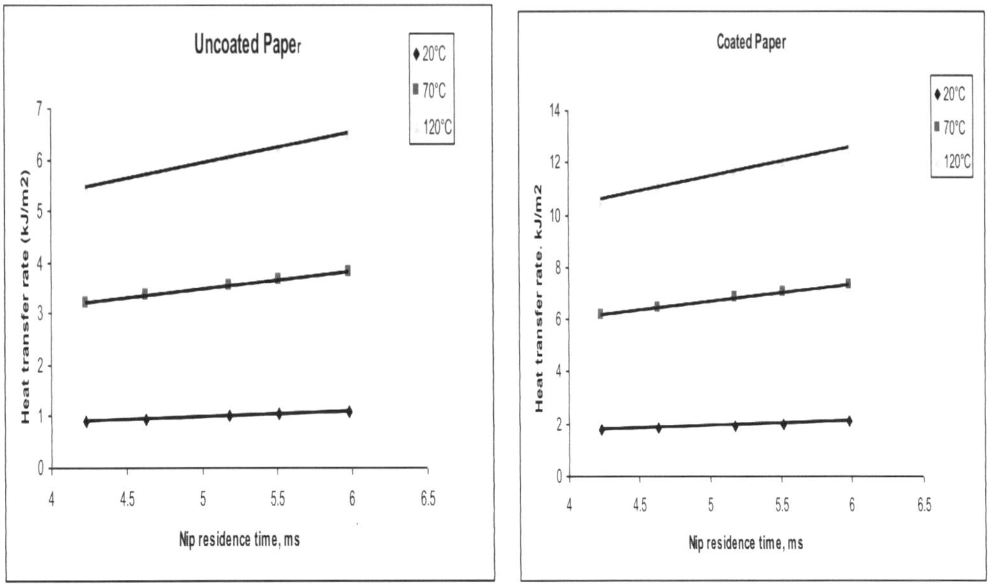

Fig 2: Comparison of heat transfer rate for uncoated and coated paper samples during soft nip calendering (calendering speed 25m/min) at different calendering temperature

5.0: Conclusions:

The Keller's model is well correlated with the experimental data generated from the soft nip calendaring of uncoated and coated papers. The heat conduction characteristics of paper change considerably in calender nip due to web compression. In the initial stage of compression, due to low intensity of nip load and temperature, heat transfer value is low limited by surface roughness of the paper web. At higher nip load and high calender temperature, heat transfer rate increases reflecting the web densification.

Increasing nip load influences heat conduction since paper densification also depends on nip pressure. A low nip pressure transfers less heat than a high nip pressure at a given dwell time and roll temperature interface.

Heat transfer to coated paper is much greater than uncoated paper calendered under same conditions of softnip calendering. This may be explained by both relative density of two sheets and the difference in their material properties (i.e. fiber v/s mineral pigments).

Under the constant condition of linear load and temperature, with an increase in calender speed, heat transfer rate decreases (for both uncoated and coated papers). A change of speed from 20 m/min to 25 m/min decreases the nip residence time of coated paper in order of 19.8% resulting in reduced heat transfer rate by 10.5%. The same is true for different calendering temperatures.

Nomenclature

A = surface area, m^2
C_p = specific heat of solid, J/(kg K)
k = thermal conductivity, W/(m K)
t = time, s, h ; T = Temperature, K or °C
T_s = surface temperature, K or °C
ΔT = temperature difference between two bodies, K or °C
w = nip width, mm
x = distance (from leading edge of plate) longitudinal coordinate, m
y = transverse coordinate, m; z = axial position of cylinder, m
W = half contact length (m),
D_E = relative roll radius
E^* = composite elastic modulus of the contacting bodies.

where D_E = equivalent diameter of two rolls, D_1, D_2 = diameters of two rolls or twice the curvature radii of rolls, E = equivalent elastic modulus of two rolls, E_1, E_2, and v_1, v_2 = Elastic modulus and Poisson ratio of two rolls.

Greek Nomenclature

α = thermal diffusivity, m^2/s
β = coefficient of volumetric expansion
η= similarity variable
δ = fluid film thickness, m
ρ= density of solid, kg/m^3

References

[1]. J. P. Holmen, Heat Transfer, (McGraw – Hill , New York), 1976.

[2]. I. Kartovaara, R. Rajala, M. Luukkala, K. Sipi, Conduction of heat in paper. Papermaking raw material transactions of 8th fundamental research symposium 1985, 381- 412.

[3]. I. Incropera, D. DeWitt, Fundamental of Heat and Mass transfer, 6th edition, John Wiley & Sons, New York, 1990.

[4]. S. Keller, Heat Transfer in Calender Nip, Journal of Pulp and Paper Science 20 (1) (1994), J33 – J37.

[5]. J. E. Olsen, On the significance of physical properties in heat transfer during calendering The Nordic Conference in calendering. Paper 4, Trondheim, Norway, (2000) pp. 14 – 20.

[6]. David Guerin, Veronique Morin and Peter Svenka, Influence of some parameters on Heat Transfer Calendering: Case of a wood free Coated Paper, Ippta Convention Issue, (2003) pp. 17-22

[7]. R. Kerekes, I. Pye, Newsprint calendering: an experimental comparison of temperature and loading effects. Pulp and paper Canada 75 (11) (1974), T379-T386.

[8]. R. Crotogino, Temperature – gradient calendering Tappi J. 65 (10) (1982) pp. 97 – 101.

[9]. R. H. Hestmo and M. Lamvik, Heat transfer during calendering of Paper, Journal of Pulp and Paper Science: Vol. 28, No. 4, (April 2002) pp. 128 – 135.

[10]. S. F. Keller, Measurement of the pressure time profile in a rolling calender nip. J. Pulp Paper Sci. 18(1) (1992) pp.144-148.

[11]. Binay K. Dutta, Heat transfer – Principles and Applications (Prentic-Hall of India Private Limited), New Delhi, 2001.

[12]. Sanjay Tyagi, Response of coating and calendering on quality of papers made from indigenous raw materials, Ph. D. Thesis, submitted to Dept. of Paper Technology, IIT Roorkee (Dec 2009).

[13]. M. D. Lloyd and Alan R. Dickson Effect of calendering speed on sheet structure and paper properties, Appita Journal, Vol. 53 No. 5 September 2000. p. 487- 492.

(276a) Biomass Feedstock Market Supporting the Emerging Biorefinery Industry: How Will It Develop?

Yogendra Shastri[1], Ming-Che Hu[2], Alan Hansen[1], Luis Rodriguez[1] and K.C. Ting[1], (1)Department of Agricultural and Biological Engineering, University of Illinois at Urbana-Champaign, Energy Biosciences Institute, Urbana, IL, (2)Department of Agricultural and Biological Engineering, University of Illinois at Urbana-Champaign, Energy Biosciences Instiute, Urbana, IL

An efficient and sustainable biomass feedstock production system is critical for the success of lignocellulosic biomass based bioenergy sector. However, there is currently a lack of understanding regarding the development of the feedstock value chain as well as the possible configuration of the mature value-chain. One of primary reasons for this is the involvement of multiple stakeholders, such as farmers, refinery, transportation companies, storage elevators, farm consultants and equipment renters that must participate and yet compete for a fraction of the profit. Moreover, the individual stakeholder behavior in reality is based on factual details as well as other aspects like experience and personal preferences. This adds to the complexity and makes the analysis difficult. The objective of this work is to take a complex adaptive systems approach to analyze the development and functioning of the feedstock market in the presence of such complexity. The power of this approach lies in its ability to simulate emergent system behavior through the modeling of key features and interrelations. It uses the agent-based modeling philosophy to model multiple stakeholders such as farmers and refinery in a typical feedstock production system. Each stakeholder is modeled as a class of independent agents using an object-oriented approach. The behavior and decision making of each class of agents is modeled with a set of rules. These rules are based not only on the economic rationale such as profit maximization, but also on agent-specific mental models and cognitive processes such as risk aversion. This enables the model to replicate decision making process that happen in reality to a certain extent. Key interactions between different classes of agents are modeled to mimic contractual negotiations and agreements in a real system. The role of the government, assumed to be very important to stimulate growth in this field, is also modeled in the form of policies that affect the interaction and decision making process for each agent. The model will be used to carry out simulation studies over multiple years to predict different possible development paths for the biomass feedstock market. The impact of agent diversity as well as policy interventions will be quantified. This should, subsequently, help in the identification of the most desirable incentives and policies to ensure a stable feedstock production system in the long run.

See more of this Session: Forest Products Biorefinery Feedstock and Logistics
See more of this Group/Topical: Forest and Plant Bioproducts Division

(276b) Woody Biomass and Mill Waste Utilization Opportunities in Alabama: Transportation Cost Minimization, Optimum Facility Location, Economic Feasibility and Impact

Burak Aksoy[1], Harry T. Cullinan[1], David Webster[2], Kevin Gue[3], Mario R. Eden[4], Norman E. Sammons Jr.[5] and Sujith Sukumaran[6], (1)Auburn University, Chemical Engineering Department, Alabama Center for Paper and Bioresource Engineering, Auburn University, AL, (2)Ark Resources, LLC, Birmingham, AL, (3)Industrial and System Engineering, Auburn University, Auburn, AL, (4) Department of Chemical Engineering, Auburn University, Auburn University, AL, (5)Intramicron, Inc., Auburn, AL, (6)Industrial and Systems Engineering, Auburn University, Auburn, AL

Four biorefinery technologies were studied for feedstock allocation, optimum facility location, economic feasibility, and their economic impacts on Alabama. The studied technologies are: 1) circulated fluidized bed gasification of woody biomass for Fischer-Tropsch (F-T) fuels and power production (CFB); 2) simultaneous saccharification and fermentation (SSF) of paper sludge for ethanol production; 3) direct spouted bed gasification with air and steam of woody biomass for power (DSB); and 4) direct combustion of woody biomass for power production (DC). The optimum biorefinery locations for all four processes were southwestern Alabama. The average transportation distance for woody biomass was about 32 miles (51.5 kilometers) to the biorefinery plants; however, longer transportation distances were required for paper sludge where the feedstock is provided by the pulp and paper plants in the region. The SSF process along with the DSB gasification process were the only economically feasible biorefinery technologies with an over 80% and 22.5% internal rate of return, respectively. Economic impacts of all the studied biorefinery technologies were comparable to each other.

See more of this Session: Forest Products Biorefinery Feedstock and Logistics
See more of this Group/Topical: Forest and Plant Bioproducts Division

(276c) Mixing and Conveying High Solids Biomass Using Rheological Modifiers

J. R. Samaniuk[1], D. J. Klingenberg[1], T. W. Root[1] and C. Tim Scott[2], (1)Chemical and Biological Engineering, University of Wisconsin, Madison, WI, (2)USDA Forest Products Laboratory, Madison, WI

Rising global demand for energy combined with an increasing appreciation for the environmental consequences of petroleum-based fuel use is motivating research and development of more carbon-neutral, widely accessible sources of renewable energy. Particularly critical to the transportation industry is an abundant source of renewable liquid fuels to counter rising petroleum costs. A likely feedstock in the production of fuel ethanol is lignocellulosic biomass. Many processes designed to convert lignocellulosic materials to liquid fuel require operations in which biomass slurries are heated (e.g. the NREL process that employs hydrolysis of cellulose). Reducing the amount of water present in those slurries lowers operating costs associated with heating and makes such processes more economical. Processing biomass at high solids concentration leads to a number of difficulties that are not present in more dilute systems. Concentrated biomass acts as a Bingham fluid with a yield stress that increases rapidly with solids concentration. This behavior makes pumping and transporting the process slurry difficult if not impossible. In addition, the mixing required to obtain adequate rates of mass transfer when introducing reactants, enzymes and reagents into the slurry is severely inhibited. Controlling the rheology of biomass will help improve the economics of high solids processing. Prior work on fiber suspensions and recent preliminary work on corn stover slurries have shown that controlling the magnitude of yield stresses present in concentrated biomass slurries is possible using small amounts of high molecular weight polymer additives. We examined the rheological behavior of biomass with polymer additives as a function of process conditions using torque rheometry. Temperature, pH, solids concentration and biomass type (e.g. corn stover vs. switchgrass) were considered as process variables. Water soluble polymers reduced yield stresses in a variety of lignocellulosic materials. Biomass pH and solids concentration significantly affect polymer efficacy. Unusual behavior in the form of a non-monotonic increase of stress with strain rate was observed which can be modified with the addition of polymer. We will also describe more recent experiments in which those rheological modifiers are exploited to facilitate extrusion of biomass at high solids concentrations.

See more of this Session: Forest Products Biorefinery Feedstock and Logistics
See more of this Group/Topical: Forest and Plant Bioproducts Division

EFFECT OF STEAM EXPOSION ON WOOD PELLET QUALITY

Pak Sui Lam[1], *Shahab Sokhansanj*[2], *Xiaotao Bi*[3] *and C.J. Lim*[3]

(1)Chemical and Biological Engineering, University of British Columbia, Vancouver, BC, Canada,
(2)Oak Ridge National Laboratory, Oak Ridge, TN,
(3)Department of Chemical and Biological Engineering, University of British Columbia, Vancouver, BC, Canada

Introduction

Wood pellet is a green renewable fuel for heat and power generation. It is a compressed cylindrical pellet with 6.4 mm diameter and 25.4 mm length that can be made by densification of sawdust, shavings or ground wood chips. This enhances the energy density of a low bulk wood residue for energy production. The manufacturing process of wood pellets includes drying, size reduction, pelletizing, cooling and packaging (Mani et al., 2006a). Its uniform shape is favorable for transportation and feeding for combustion. Due to its low emissions during combustion and credits/subsidies given to the users of wood pellets for energy production, there is an increasing demand of wood pellets as a renewable fuel. This leads to growing opportunities in export and import of wood pellets across the countries. In 2006, Canada produced around 1.2 million tons of wood pellets and a high proportion of them were sent to Europe through Rotterdam harbor in the Netherlands (Peksa et al., 2007).

The handling and storage ability of wood pellets limits its use in energy production applications. Wood pellets are easy to disintegrate into fines which cause blockage of the conveyor or hopper during processing. This also leads to potential dust explosion. Fines inhalation is also an occupational health problem to the site workers (Aleksandra et al., 2006). All of these are related to the lack of natural binders of wood fibers. In practice, most pellets from agricultural crops and hardwood are difficult to be densified without the addition of any expensive binders (Sokhansanj et al., 2005).

Hygroscopicity of wood pellet also leads to disintegration under humid conditions. High surface area of fines favors the susceptibility of the microbials attack during storage (Lehtikangas, 2000). The anaerobic condition leads to local heat generation and toxic gas emissions, e.g. terpenes (Rupar et al., 2005). Local heat generated ignites the volatiles from the pellets and causes fire hazard. The off-gases accumulations inside the storage silos are toxic to the site workers. High moisture content reduces the pellet durability and combustion efficiency at the power plant.

The wood products become more dimensionally stable and less hygroscopic after steam treatment (Angles et al., 2001; Bruno et al., 2007). Angles et al. (2001) reported that the physical and mechanical properties of the steam exploded wood panel improved as the severity of the pre-treatment increased up to a point where mechanical properties deteriorated. Poplar pellet treated with steam explosion possessed higher breaking forces than untreated poplar pellet (Shaw et al., 2009). The increase in total lignin content and restructure of lignin after steam explosion may both contribute to the increase in breaking strength of wood pellets (Shaw et al., 2009). Lignin droplets acted as a self-binding agent between fibers upon densification. Nielsen et al. (2010) reported that the extractives seemed to prevent close contact between bonding sites of the lignocellulose particles and thereby decreased pellet strength. The removal of extractives significantly increased pellet strength. On the other hand, extractives acted as plasticizers and lubricants and thereby decreased the energy requirements for the pelletizing process.

The objective of this research is to study the effect of steam exploded wood powder at different severity on physical, chemical and mechanical properties of steam exploded wood pellet. Mechanical strength and moisture sorption behavior of steam exploded wood pellets were characterized.

Materials and Methods

Sample preparation

The effects of steam explosion conditions (temperature and time) and particle size on pellets properties were investigated. The detailed steam explosion procedures were described in a previous work (Lam et al., 2010). Four different severities of 3.54, 3.94, 4.23 and 4.53 were used to steam treat the (Douglas Fir) wood samples. The moisture content of each samples were conditioned by drying at 50 ^{o}C in a convection oven from average 33% to around 10% moisture content (wet basis) prior to pelletization. The samples from each steam treatment were stored in a sealed plastic container at 6 ^{o}C and randomly selected for pelletization.

Pelletization

An MTI (Measurement Technology Inc., Roswell, GA) equipment was used to exert a pressure force on the wood powder inside the piston-cylinder unit to make pellets (Figure 1). The pellets were made by a piston-cylinder unit (Figure 2). There are 3 sections of the whole system. The piston has 90 mm length, 6.35 mm diameter, with a flat end bottom. The top cylinder part is 6.35 mm inside diameter and 69.85 mm length, and is used for guiding the piston movement. The outside of the compression zone is wrapped by an electric heating tape and insulated to simulate the normal compression temperature in the pellet mill.

Table 1 lists the samples particle size and moisture content before pelletization and their respective preset load used in this study. About 0.7 – 0.8 g biomass was placed in the cylinder and allowed to be heated to 70 ^{o}C for 3 minutes before compression. The MTI was preset to exert a vertical force of 4000 N (128 MPa). The movement of piston was set at 6.67 mm/min during compression. The samples were compressed up to a specified preset load and held for 30 s to arrest the spring back effect. The compression data on stress natural strain and strain rate during stress relaxation were logged into the computer. The pellet was collected by extruding it out using the piston at 13.34 mm/min. The extrusion data on stress natural strain were also logged into the computer. The pellet was cooled to room temperature and stored inside a sealed glass bottle for further characterization. Each compression was repeated fifteen times. Extrusion data were logged for five times.

Pellet density and specific density

The mass, length and diameter of each pellet were measured immediately after its removal from the die. The pellets were stored inside a sealed glass bottle for one week. After one week, length and diameter were measured to evaluate the percentage expansion in diametrical and longitudinal direction.

Specific density was determined by measuring the total pore volume by the adsorption of nitrogen into void spaces of the biomass powders with a Quantachrome Multipycnometer

(Quantachrome, Boyton Beach, FL, USA). The measurement was done by measuring the pressure difference for a known quantity of pressurized nitrogen flowing from a reference volume into a sample cell loaded with samples. Particle volume measurements were repeated five times for each sample, for the determination of an average, and a reference volume of 29.42 cm^3 was used in all measurements. Equation (1) was used to obtain the true volume of the biomass grinds deduced from the ideal gas law.

$$V_p = V_C - V_R\left[\frac{P_1}{P_2} - 1\right] \tag{1}$$

where V_p= true volume of biomass grinds (m^3), V_c= volume of sample cell (m^3), V_R= reference volume (m^3), P_1= pressure reading after pressurizing the reference volume (Pa) and P_2 = pressure reading after including V_c(Pa). The particle density was determined by the ratio of the mass of the biomass grinds and the true volume using Equation (2):

$$\rho_p = \frac{m}{V_p} \tag{2}$$

where m = mass of the sample (kg) , V_p= true volume of biomass grinds (m^3)

Interparticle porosity (ε_o) provides packing information of the biomass grinds inside a known container and is determined by equation 3:

$$\varepsilon_o = 1 - \frac{\rho_b}{\rho_p} \tag{3}$$

where ε_o is the porosity, ρ_b = bulk density of biomass grinds (kg/m^3), ρ_p = specific density of biomass grinds (kg/m^3)

Moisture content, ash content, elemental analysis and heating value

Moisture contents of the samples before and after pelletization were analyzed according to ASABE S358.2 (2010). Around 0.5 g of sample before grinding and around 1 g of pellets were used for moisture content measurement. Triplicate samples were oven dried at 103 °C for 24 hours. Moisture contents of each sample were determined by the weight difference and reported in wet basis.

The heating calorific values of the pellets were determined by the oxygen bomb calorimeter (Parr 6100) from two replicates.

The ash content was determined according to the Determination of Ash in Biomass from NREL (NREL/TP-510-42622, 2005) as the percentage of residue remaining after dry oxidation (oxidation at 550 °C to 600 °C). All results are reported relative to the 103 °C oven dried weight. Ash content analysis was made in three replicates and an average value was calculated.

Chemical composition characterization

Extractive contents were analyzed through modified acetone extraction (TAPPI T280 pm-99, 1999). About 3 g of samples were put inside extraction thimbles and extracted by acetone in a Soxhlet apparatus for 4 – 6 hours in order to achieve Soxhlet siphons 4 times per hour for 5 hours, totaling 20 cycles. The acetone in the extractive-acetone mixture was separated by a TurboVap evaporator (Zymark, Hopkinton, MA, USA) under nitrogen, and finally the weight of extractives was measured after freeze drying. All measurements were made in duplicates.

Klason lignin contents of extractive-free samples were measured according to the modified version of the Determination of Structural Carbohydrates and Lignin in Biomass from NREL (NREL/TP-510-42618, 2005). Two-step acid hydrolysis was used to fractionate the sample into forms that were more easily quantified. The lignin was fractionated into acid-insoluble and acid soluble material. The acid-insoluble material may also include ash and protein, which must be accounted for during gravimetric analysis. The acid soluble lignin was measured by an Ultraspec 1000 UV-Vis spectroscopy (Pharmacia Biotech, Piscataway, NJ, USA). All measurements were made in triplicates, and an average value was calculated.

Stress relaxation and hardness test

The relaxation data for each sample were analyzed by the method of Peleg, which was applied to food materials (Peleg, 1980) and biomass grinds (Mani et al., 2006b). The force relaxation curves of pellets were normalized and linearized as a straight line.

$$\frac{\sigma_0 t}{\sigma_0 - \sigma(t)} = k_1 + k_2 t \qquad (5)$$

where σ_0 is the initial stress (MPa), $\sigma(t)$ the stress after time t at relaxation (MPa), t the time (s) and k_1, k_2 are the empirical constants.

The slope k_2 of the straight line must be greater than one, and from a rheological point of view, the slope can be considered as an index of how solid the compacted specimen is on a short time scale. Any large value of greater than one is an indication of the existence of stresses that will eventually remain unrelaxed (Moreyra and Peleg, 1980).

Moreyra and Peleg (1980) and Scoville and Peleg (1981) stated that the asymptotic modulus (E_A) can be an empirical index of solidity which is the ability of the compressed powder to sustain unrelaxed stresses; E_A is defined as

$$E_A = \frac{\sigma_0}{\varepsilon}\left(1 - \frac{1}{k_2}\right) \qquad (6)$$

where E_A is the asymptotic modulus (MPa) and ε is the strain (dimensionless or m m^{-1}).
Eq. (5) was fitted to the stress relaxation data to estimate k_1 and k_2. The estimated k_2 value was then used in Eq. (6) to calculate E_A for each sample.

The maximum breaking load, hardness modulus and Meyer hardness of the pellets were calculated and average values from five replicates were reported. The hardness modulus was determined from the linear slope in the initial portion of the force-deformation diagram. The Meyer hardness (H_M) is the applied force divided by the projected contact area and it was calculated according to the following equation (Tabil et al., 2002):

$$H_M = \frac{F}{\pi\left(Dh - h^2\right)} \qquad (7)$$

where H_M is the Meyer hardness (N/mm^2) and h is the indentation depth (mm).

Moisture sorption rate test

The moisture sorption rate of different pellets was measured in a humidity chamber (ESPEC CORP, LHU-113, Japan) at 30 °C at 90% relative humidity for 5 hours. Before the test, all the samples were first dried in the convection oven at 103 °C for 24 hours to make them fully dried. Two pellets at different treatment conditions were kept inside a petri glass dish. In total, three replicates were measured for the untreated sample and the sample treated at different steam explosion condition. The samples inside the Petri dish were covered with a glass cap to prevent moisture loss when removed from the humidity chamber and weighted in an ALC-80.4 analytical balance with 0.1 mg accuracy (Acculab, Edgewood, NY). The weight of the samples were measured for every 10 minutes for the first hour and followed by every 30 minutes for the consecutive 4 hours. The moisture sorption rate constant was calculated by curve fitting on the data points using the Eq. (8) (ASABE S448.1, 2001).

$$\frac{M - M_e}{M_i - M_e} = e^{-kt} \quad (8)$$

where M is the instantaneous moisture content (%, decimal dry basis), M_e is the equilibrium moisture content = 0 (%, decimal dry basis), and t is the time (min). The moisture ratio against time was plotted to obtain the moisture sorption rate kinetic constant (k). Three replicates of measurement were conducted.

Scanning electron microscope of cross section of pellet

The binding characteristics of the wood pellet were studied by analyzing the cross section of the wood pellet by field emission scanning electron microscope (FESEM). The wood pellet was immersed in liquid nitrogen for 10 minute and fractured into two portions in the frozen condition. The surface created after the fracture was mounted on the top of the specimen stubs and coated with gold under vacuum. All photographs were taken at 10 to 20 kV accelerating voltage by using a field emission scanning electron microscope Hitachi S4700 (Hitachi, Japan) in UBC bioimaging facility.

Results and Discussions

Pellets compression and stress relaxation

Figure 3 shows the typical force and displacement relationship of compression test of different samples with a preset load of 4000N. The displacement of the piston with higher temperature and longer treatment time of steam explosion was shorter than the control sample because particles were initially packed closer inside the die when loaded. The bulk density of steam exploded D.F. powder treated at 220 °C and 10 minute was 2.3 times of the untreated samples (Lam et al., 2010). For the untreated sample, the powder underwent elastic compression until a bioyield point was reached. Beyond the bioyield point the force increased rapidly as a result of the elimination of air voids and compression of the solid wood. For the steam exploded samples, the bioyield point existed at a lower displacement compared to the control sample. This is contributed to the higher initial bulk densities of the steam exploded samples than the control sample. The high initial bulk densities were brought by partial disruption of the cell wall structure by explosion effect and particle size reduction as discussed in Lam et al. (2010). There seems to be a plastic deformation between 2 stages of densification process for steam exploded samples beyond the elastic region. The restructure of lignin on the cellulose fiber surface underwent plastic deformation and allows packing closely to form solid bridges.

Table 2 lists the measured preset load, compression and extrusion energy required for the untreated and steam exploded samples from the compression test. The compression energy of the untreated sample was 22.3 J. The compression energy of steam exploded samples increased with the severity of the steam explosion from 25 J to 40.4 J. The hemicellulose acted as a viscoelastic interface between cellulose fibers and lignin in the cell wall. After steam explosion, most hemicelluloses are hydrolyzed into simple monosugars and interfacial bonding between cellulose and lignin is destroyed

(Ramos et al., 2003). Therefore, the resulting material becomes brittle and has higher elastic modulus in elastic region. The steam exploded wood treated at a severity of 3.9 was reported to be brittle, having microcrystalline cellulose with recondensed lignin droplets on fiber surface by SEM photos (Angles et al., 2001).

Asymptotic modulus is defined as the ability of compressed powder to sustain unrelaxed stresses, i.e., an indication of the pellet solidity since it reflects the stresses that can be supported without dissipation through plastic flow of the solid matrix and reorientation of the interparticle bridges (Moreyra et al., 1980). Mani et al. (2006b) and Shaw et al. (2009) used this technique to characterize the biomass compression behavior. From our experiment, the asymptotic modulus of the control sample was 1331 MPa, which was the highest among all samples. The asymptotic modulus of samples after steam explosion decreased with increasing the severity from 1251 to 1174 MPa. This showed that the rigidity of the sample decreased with increasing the severity of steam explosion. The steam explosion bursted the cell wall structure of the wood sample. The modulus decreases at a specific strain due to miniature cracks developed. Another reason is the modification of the cellulose and lignin structure after steam explosion. The new lignin-cellulose fiber structure after steam explosion was reported to have lower glass transition temperature (T_g) compared to the lignin in the control sample with the same moisture content (Startsev et al., 2000). This allows easier reorientation of the interparticle bridges and plastic flow within the matrix under compression above the T_g.

Physical properties of the pellets

Figure 4 shows the physical appearance of the wood pellets. The color of the pellets is getting darker with increasing the treatment time or severity. Table 3 lists the specific, initial and relaxed densities of the pellets made at different conditions. The solid densities of the control and steam exploded pellets measured with pycnometer were in the range of 1.42 – 1.43 g/cm^3. There was no significant difference in specific density between the control and the steam exploded pellets. The initial pellet density measured from mass and volume of the control sample was 1.09 g/cm^3. The initial pellet density of steam exploded pellet increased to 1.12 g/cm for the samples treated at 200 ^oC and 5 minutes. With increasing the severity of the treatment, the initial pellet density of steam exploded pellets decreased from 1.12 g/cm^3 to 1.07 g/cm^3. The initial pellet density of pellets increased as the geometric mean diameter of the steam exploded powder before pelletization increased (Table 1).

Similar trend was observed for the relaxed pellet density for all samples measured after 7 days of pellet production. The relaxed pellet density of the untreated pellets decreased from 1.09 g/cm^3 to 1.07 g/cm^3 due to a 0.56% volumetric expansion. In particular, the diametric expansion of 0.29% was higher compared to the longitudinal expansion of 0.06%. In contrast, the steam exploded pellets showed an increase of volumetric contraction after relaxation with increasing the steam explosion conditions. Steam treatment prior to compression can increase the compressibility of wood and significantly reduce the buildup of internal stresses during compression (Hsu et al., 1988). This helps to eliminate the springback of compressed wood by irreversible swelling due to its hygroscopic nature. The reduction in longitudinal direction between -0.24% and -0.47% was higher than diametric direction between -0.12% - -0.24% for steam exploded pellets. This is in a good agreement with the previous result from Shaw et al. (2009).

Moisture content, ash content and heating value

Table 5 lists the fuel properties of the untreated and steam exploded pellets. The moisture content of the untreated pellet is 6.9%, the highest among all the samples. The steam exploded pellets at all conditions are around 6% except the most severely treated pellets at 220 °C for 10 minute has the lowest moisture content of 5.3%. The ash content of the control pellet is 0.27 % by weight. There is a slight increase to 0.32% for pellets made from the materials treated at 200 °C regardless of treatment time. The ash content of the samples treated at 220 °C increased further to around 0.41 – 0.52 % by weight.

The wet basis high heating value of the steam exploded pellets increased with the severity of the steam explosion treatment from 18.82 MJ/kg for untreated pellets to 20.09 MJ/kg for most severely steam treated pellets. Meanwhile, the percentage of carbon increased from 48.44% to 53.09% and the percentage of hydrogen decreased from 6.23% to 5.91% from elemental analysis. This increase of C/H ratio from 7.78% to 8.98% was due to the removal of –OH group of hemicelluose after the hydrothermal treatment (Negro et al., 2003).

Binding characteristics of pellets

The binding characteristics of the wood pellets made from untreated and steam exploded at 220 °C and 5 minute under 4000 N force was investigated from the cross section of the pellets under scanning electron microscope. Figure 5 shows the cross section of the untreated and steam exploded pellets at a low magnification of 30. For the untreated pellet, each particle stacked on each other after compression. Clear boundaries and gaps between individual particles were observed. In contrast, the particles in the steam exploded pellet were more closely packed and with no gaps in between. The dark regions between individual particles boundaries indicated that there may be a glue effect brought by pseudolignin binding.

The pseudolignin formed on the wood fiber surface was partially formed by lignin flow. Figure 6 shows that some of the lignin droplets with 500 nm diameter were deposited on the particle surface. The lignin was softened and flowed at the compression temperature after steam explosion treatment. Donohoe et al. (2008) provided experimental evidences supporting the idea that thermochemical pretreatments reaching temperatures above the range for lignin phase transition cause lignins to coalesce into larger molten bodies that migrate within and out of the cell wall, and can redeposit on the surface of plant cell walls. The pseudolignin formed solid bridges after compression and cooling. The binding was contributed by the solid bridge formed by the lignin.

Hardness test

Figure 7 shows a force displacement graph of the hardness test of different pellets. The elastic modulus, maximum breaking force and the area under the curve (i.e., the energy absorbed to break the pellet) was higher for steam treated samples than the untreated one. Table 7 summarizes the hardness test results of the control and steam exploded pellets. The maximum force required to break the control pellet is 18 N. The maximum breaking force required for steam exploded pellets increased from 26 N to 59 N with increasing treatment severity from 200 °C for 5 min to 220 °C for 5 min. This indicates 3.3 times of forces are required to break the pellets. The increase in total lignin content after steam explosion from 30% to 43% (Table 6) may contribute to the increase in breaking strength. The restructure of lignin droplets on the wood fiber surface (Figure 6) after steam explosion caused higher availability of lignin for cross-linking and binding. Lignin acts as a stiffener of the cellulose microfibrils/fibrils (Sweet et al., 1999) and an increased cross linking of this polymer appears to prevent or limit movement perpendicular to the grain (which occurs during compressive loading parallel to the grain). Furthermore, lignin is the main component of the middle lamella (Fengel et al., 1984) and an increased cross linking of the lignin polymer network improves the strength of the middle lamella which subsequently affects the strength properties of the cell wall. According to Ramos et al. (2003) steam explosion is an acidic autohydrolysis reaction which depolymerizes hemicellulose into monosugars, e.g. xylose. Additional bonding strength of steam exploded pellets may be caused by the formation of highly branched polysaccharides by reverse reactions of monosugars during densification.

Meyer hardness is the indentation hardness to measure how resistant the solid matter is to various kinds of permanent shape change when a force is applied. The average Meyer hardness increased from 1.6 N/mm^2 for control pellets to 6.6 N/mm^2 for the steam exploded pellets treated at 220 °C for 5 min. Similar trends hold for hardness modulus increase from 62 MPa for control pellet to 248 MPa for steam exploded pellets treated at 220°C for 5 min. The control pellet with the lowest hardness modulus may indicate that the pellets have a greater pore space which reduces the resistance of the pellets to deformation. This is in a good agreement with the observations from the cross section SEM photo of the control pellet (Figure 5). The increase in hardness modulus and Meyer hardness can be explained by the hemicelluose solubilization and the resulting crystalline cellulose fiber turns brittle after hydrothermal treatment. From x-ray diffraction result, the cellulose crystallinity increased from 49.5% for the control sample to a maximum of 67.5% for steam exploded pellets treated at 220 °C for 5 min.

The maximum breaking force, hardness modulus and Meyer hardness of the pellets treated at 220 °C for 10 min decreased from the maximum value at 220 °C and 5 min due to the destruction of cellulose crystalline structure by prolonged steam treatment (Bhuiyan et al., 2000). The average maximum breaking force of the pellets treated at 220 °C and 10 minutes decreased from the maximum of 59 N to 44 N. The average Meyer hardness of steam exploded pellets treated at 220 °C and 10 min decreased from 6.6 N/mm^2 to 5.6 N/mm^2. The crystallinity of cellulose of steam exploded pellet decreased to 61% for the most severely treated sample.

Moisture sorption of pellets

The moisture sorption rates of different pellets are shown in Figure 8. The control sample exhibits the most hygroscopic behavior, in which the moisture content is around 10.3% after 5 hours exposed to RH = 90% at 30 °C. The moisture content of the steam exploded pellets treated at 200 °C for both 5 min and 10 min was close at 7.7% after 5 hours exposure in the humidity chamber at RH = 90% at 30 °C. The moisture content of the steam exploded pellets treated at 220 °C for both 5 min and 10 min was around 6.0% after 5 hours exposure in the humidity chamber at RH = 90% at 30 °C. From t-test, the treatment time seems to have little effect on moisture sorption of the pellets with 5% confidence interval. The moisture sorption rate constant of different samples were listed in Table 8.

Conclusions

In summary, the effect of steam exploded wood powder at different severity on physical, chemical and mechanical properties of steam exploded wood pellets was investigated. It was found that steam exploded wood powders required a higher compression energy of 40.4 J compared to 22.3 J for control samples from the force displacement test. This was contributed by hemicellulose solubilization, increase in cellulose crystallinity and lignin redistribution after steam explosion and thereby the resulting fiber was more brittle. The asymptotic modulus of steam exploded pellets decreased with increasing the severity of steam explosion. This indicated that the rigidity of the sample decreased with increasing the treatment severity the wood sample. From hardness test, the maximum breaking force and Meyer hardness of the steam exploded pellets treated at 220 °C for 5 minutes was improved by 3.3 times and 6.6 times, respectively, compared to the control pellets.

Steam exploded pellets showed an increase of volumetric contraction from 0.03 to 0.79% after relaxation with increasing the severity of steam explosion. This is opposite to the volumetric expansion of 0.56% after densification for the control pellets. Steam treatment prior to densification can help to reduce the building up of internal stresses during compression and reduces the spring back of compressed wood by irreversible swelling.

Moisture sorption rate of steam exploded pellets was slower than the untreated pellets. The equilibrium moisture content of the steam exploded pellets treated at 220 °C for 5 min and control pellets after 5 hours exposure in the humidity chamber at RH = 90% at 30°C was 6.0% and 10.3%, respectively. The reduction in moisture sorption is contributed by polymerization of condensed substances that results in high hydrophobicity.

All of these improved properties of woody biomass brought by steam explosion pretreatment could ease the handling, storage and transportation of densified pellets.

Tables

Table 1. Physical properties of steam exploded wood powder before pelletization (n = 3)

Treatment	**Severity**	**Geometric mean diameter (mm)**		**Moisture content (%) w.b.**	
		Average	SD	Average	SD
Untreated		0.42	0.03	8.7	3.0
200 °C, 5 min	3.64	0.45	0.01	9.4	0.2
200 °C, 10 min	3.94	0.40	0.02	10.0	0.2
220 °C, 5 min	4.23	0.38	0.02	11.8	0.5
220 °C, 10 min	4.53	0.35	0.04	8.6	0.1

Table 2. Compression, stress relaxation and extrusion data of different wood pellets (n=15)

Treatment	**Preset load (N)**		**Compression Energy (J)**		**Extrusion Energy (mJ)**[1]		**Asymptotic modulus (Ea) (MPa)**	
	Average	SD	Average	SD	Average	SD	Average	SD
Untreated	4290	70	22.3	1.3	48.9	3.3	1331	30
200 °C, 5 min	4190	145	25.0	1.0	117.7	29.9	1251	16
200 °C, 10 min	4120	39	31.8	1.6	117.3	13.8	1244	18
220 °C, 5 min	4297	54	38.8	4.4	177.5	45.4	1193	13
220 °C, 10 min	4079	27	40.4	1.6	235.4	2.8	1174	7

[1]: n = 5

Table 3. Solid, initial and relaxed densities of the untreated and steam exploded wood pellets

Treatment	Solid density[1] (g/cm^3)		Initial pellet density[2] (g/cm^3)		Relaxed pellet density[2] (g/cm^3)	
	Average	SD	Average	SD	Average	SD
Untreated	1.4303	0.003	1.0921	0.039	1.0723	0.020
200 °C, 5 min	1.4220	0.005	1.1236	0.046	1.1084	0.012
200 °C, 10 min	1.4297	0.003	1.0851	0.012	1.0857	0.011
220 °C, 5 min	1.4208	0.002	1.0813	0.039	1.0795	0.040
220 °C, 10 min	1.4243	0.002	1.0673	0.028	1.0555	0.034

[1]: n = 5
[2]: n = 15

Table 4. Dimensions expansion of wood pellets

Treatment	Longitudinal expansion[1] (%)		Diametric expansion[1] (%)		Volumetric expansion[1] (%)	
	Average	SD	Average	SD	Average	SD
Untreated	0.06	0.58	0.29	0.38	0.56	0.89
200 °C, 5 min	-0.24	0.57	0.13	0.39	0.03	0.96
200 °C, 10 min	-0.3	0.53	-0.12	0.38	-0.46	0.79
220 °C, 5 min	-0.42	0.38	-0.24	0.32	-0.79	0.64
220 °C, 10 min	-0.47	0.28	-0.14	0.26	-0.67	0.6

[1]: n = 15

Table 5. Fuel properties of wood pellets

Treatment	Moisture content[1] (%)		Heating value[2] (MJ/kg) (w.b.)		Ash content[1] (wt. %)		Elemental analysis			
	Average	SD	Average	SD	Average	SD	C (%)	H (%)	N (%)	C/H ratio
Untreated	6.89	0.12	18.82	0.30	0.27	0.02	48.44	6.23	0.22	7.78
200 °C, 5 min	5.96	0.61	18.94	0.04	0.32	0.01	49.14	6.08	0.17	8.08
200 °C, 10 min	6.03	0.26	19.15	0.04	0.32	0.01	50.46	6.10	0.00	8.27
220 °C, 5 min	6.03	0.22	19.50	0.05	0.52	0.09	52.42	5.95	0.18	8.81
220 °C, 10 min	5.25	0.12	20.09	0.01	0.41	0.02	53.09	5.91	0.17	8.98

[1]: n=3
[2]: n=2

Table 6. Chemical compositions of wood pellets (n=2)

Treatment	Extractives		Acid insoluble lignin (%)		Acid soluble lignin (%)		Total lignin (%)
	Average	SD	Average	SD	Average	SD	
Untreated	1.42	0.01	28.49	3.75	1.60	0.16	30.09
200 °C, 5 min	2.22	0.43	24.23	0.87	1.65	0.00	25.88
200 °C, 10 min	3.23	0.17	28.23	2.37	1.42	0.08	29.65
220 °C, 5 min	8.84	0.17	36.26	0.15	1.37	0.05	37.62
220 °C, 10 min	7.36	10.06	42.14	1.54	1.09	0.07	43.23

Table 7. Hardness test results of wood pellets

Treatment	Maximum breaking force[1] (N)		Hardness modulus[1] (MPa)		Meyer hardness[1] (N/mm^2)		Cellulose crystallinity[2] (%)	
	Average	SD	Average	SD	Average	SD	Average	SD
Untreated	18.0	4.0	62	14	1.6	0.2	49.5	0.7
200 °C, 5 min	25.7	2.1	99	7	2.1	0.3	51.5	0.7
200 °C, 10 min	34.5	11.9	112	14	2.3	0.9	57.5	2.1
220 °C, 5 min	59.3	11.1	248	72	6.6	2.1	67.5	0.7
220 °C, 10 min	44.2	7.6	210	38	5.6	1.8	61.0	1.4

[1]: n=5

[2]: n=2

Table 8. Fitted moisture sorption rate kinetics parameters of wood pellet

Treatment	Moisture sorption rate constant K (min^{-1})	Goodness of fit R^2
Untreated	0.0152	0.9804
200 °C, 5 min	0.0125	0.9811
200 °C, 10 min	0.0135	0.9728
220 °C, 5 min	0.0129	0.9713
220 °C, 10 min	0.0128	0.9660

Figures

Figure 1. The MTI machine with computer control system and temperature controller.

Figure 2. The piston – cylinder unit for pellet production.

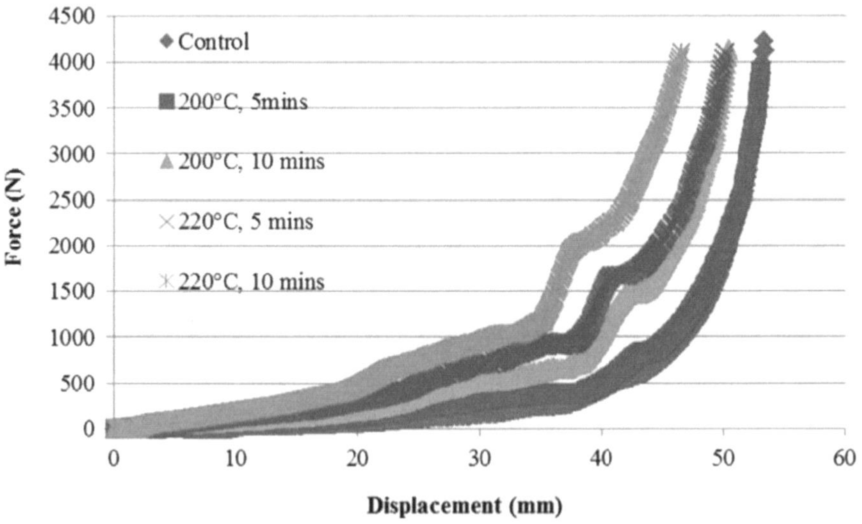

Figure 3. A typical force v.s. displacement graph for producing pellets

Figure 4. Physical appearance of wood pellets treated at different steam explosion conditions. From left to right: (a.) Control Douglas Fir, (b.) 200 °C, 5 min, (c.) 200 °C, 10 min, (d.) 220 °C, 5 min, (e.) 220 °C, 10 min

Figure 5. Cross section SEM photos of untreated and steam exploded (severity = 4.23) pellets at low magnification (x30).

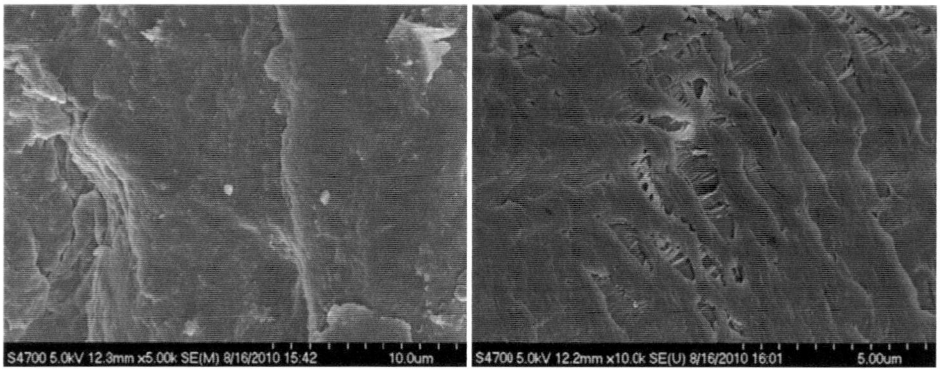

Figure 6. Cross section SEM photos of steam exploded (severity = 4.23) pellets under high magnification (x5000 and x10000, respectively).

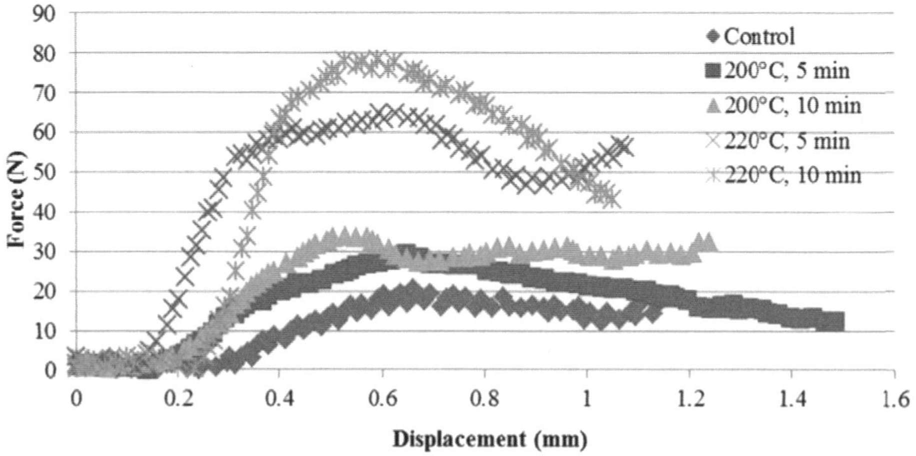

Figure 7. Force displacement graph of hardness test of different pellets.

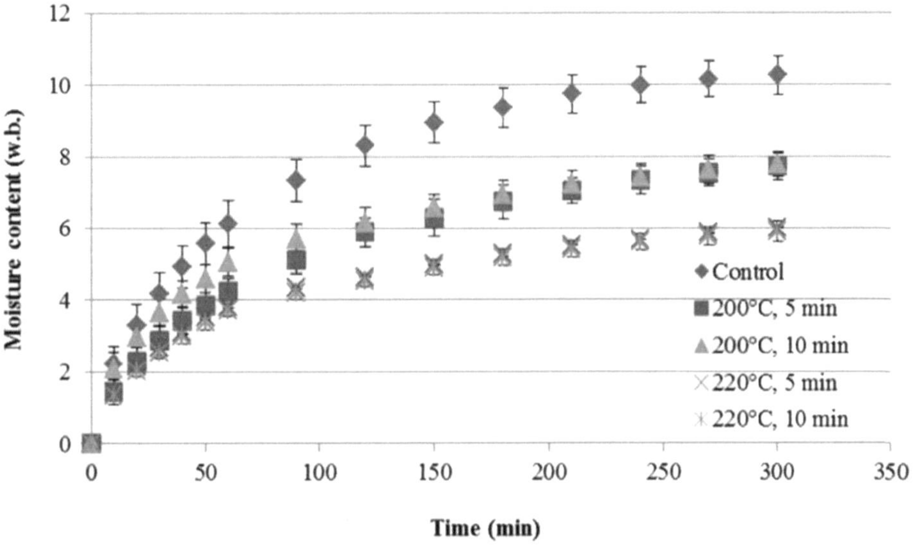

Figure 8. Moisture sorption behavior of control and steam exploded pellets treated at different temperature and time.

References

1. Aleksandra S., A. M. Madsen, L. Martensson, D. Pomorska, L. Larsson (2006),"Assessment of microbial exposure risks from handling of biofuel wood chips and straw – effect of outdoor storage",*Annals of Agricultural and Environmental Medicine*, 13:139-145.
2. Angles M. N., F. Ferrando, X. Farriol, J. Salvado. (2001), "Suitability of steam exploded residual softwood for the production of binderless panels. Effect of the pre-treatment severity and lignin addition", *Biomass and Bioenergy* 21(3): 211-224
3. ASABE Standards 47th ed., (2010), S358.2 - Moisture measurement forages. American Society of Agricultural Engineers, St. Joseph, MI. pp.595
4. ASABE Standards 47th ed., (2006), S448.1- Thin layer drying of agricultural crops. American Society of Agricultural Engineers, St. Joseph, MI.
5. Bhuiyan M. T. R., N. Hirai, N. Sobue. (2000),"Changes of crystallinity in wood cellulose by heat treatment under dried and moist conditions",*Journal of Wood Science*, 46: 431-436
6. Bruno E. A., V. Marques, I. Domingos, H. Pereira. (2007),"Influence of steam heating on the properties of pine (Pinuspinaster) and eucalypt (Eucalyptus globulus) wood",*Wood Science and Technology*, 41:193-207
7. Donohoe B. S., S. R. Decker, M. P. Tucker, M. E. Himmel, T. B. Vinzant. (2008),"Visualizing lignin coalescence and migration through maize cell walls following thermochemical pretreatment",*Biotechnology and Bioengineering*. 101(5): 913-925.
8. Fengel D., Wegener G., (1984),"Wood: Chemistry, Ultrastructure,Reactions", Walter de Gruyter and Co. Berlin, New York
9. Hsu, W.E, W. Schwald, J. Schwald, J. A. Shields. (1988), "Chemical and physical changes required for producing dimensionally stable wood-based composites", *Wood Science and Technology*. 22: 281–289

10. Lam P. S., S. H. Larsson, S. Sokhansanj, X. Bi, C. J. Lim. (2010), "Effect of temperature, time, particle size and moisture content on physical properties of steam exploded woody biomass", ASABE 2010 annual international meeting, paper no: 1009796, Pittsburgh, U.S.A.ASABE: St. Joseph, MI
11. Lehtikangas P. (2000), "Storage effects on pelletised sawdust, logging residues and bark", *Biomass and Bioenergy*. 19(5):287-293
12. Mani S., S. Sokhansanj S., X. Bi, A. Turhollow (2006a), "Economics of producing fuel pellets from biomass",*Applied Engineering in Agriculture,* 22(3): 421-426
13. Mani S., L.G. Tabil and S. Sokhansanj. (2006b),"Effects of compressive force, particle size and moisture content on mechanical properties of biomass pellets from grasses", *Biomass and Bioenergy* 30:648–654.
14. Moreyra R. and M. Peleg. (1980),"Compressive deformation patterns of selected food powders", *Journal of Food Science*, 45:864–868
15. Negro M. J., P. Manzanares, J. M. Oliva, I. Ballesteros, M. Ballesteros. (2003), "Changes in various physical/chemical parameters of Pinuspinaster wood after steam explosion pretreatment", *Biomass and Bioenergy*. 25(3): 301-308
16. Nielsen N. P. K., D. J. Gardner, C. Felby(2010),"Effect of extractives and storage on the pelletizing process of sawdust",*Fuel,* 89:94-98
17. NREL Standard, (2005), "Determination of Ash in Biomass", NREL/TP-510-42622
18. NREL Standard, (2005), "Determination of Structural Carbohydrates and Lignin in Biomass", NREL/TP-510-42618
19. Peksa B. M., P. Dolzan., J. Heinimo., M. Junginger., T. Ranta., A. Walter, (2007), "Global Wood Pellets Markets and Industry: Policy Drivers, Market Status and Raw Material Potential",*IEA Bioenergy Task 40 technical report.*

20. Peleg M. (1980),"Linearization of relaxation and creep curves of solid biological materials",*Journal of Rheology,* 24 (4):451–463.

21. Ramos L. P.,(2003),"The chemistry involved in the steam treatment of lignocellulosic materials",*Quimica Nova.* 26(6):863-871

22. Rupar K., M. Sanati. (2005), "The release of terpenes during storage of biomass", *Biomass and Bioenergy,* 28(1):29-34.

23. Scoville E. and M. Peleg, (1981), "Evaluation of the effects of liquid bridges on the bulk properties of model powders", *Journal of Food Science* 46(1): 174–177.

24. Shaw M.D., C. Karunakaran., L.G. Tabil. (2009), "Physicochemical characteristics of densified untreated and steam exploded poplar wood and wheat straw grinds", *Biosystems Engineering* 103:198-207

25. Sokhansanj S., Mani S., Bi X., Zaini P., Tabil L. G.(2005),"BinderlessPelletization of biomass", Paper No. 056061. ASABE: St. Joseph, MI.

26. Sweet M.S., Winandy J.E.,(1999), "The influence of degree of polymerization (DP) of cellulose and hemicellulose on the strength loss offire-retardant-treated wood", *Holzforschung* 53 (3):311–317

27. Startsev V., Salin B, N., (2000), "Polycondensation of the components of the lignin-carbohydrate complex of steam exploded wood", *Doklady Chemical Technology*, 373-375 (0), 30-33.

28. TAPPI T280 pm-99, (1999), "Acetone Extractives of Wood and Pulp", Technical Association of the Pulp and Paper Industry

29. Tabil L.G., Sokhansanj S., Crerar W. J., Patil R. T., Khoshtaghaza M. H., and Opoku A., (2002),"Physical characterization of alfalfa cubes: I. Hardness",*Canadian Biosystems Engineering* 44(3):55-63

ECO- FRIENDLY UTILIZATION OF HAZARDOUS PINE NEEDLE WASTE FOR THE PRODUCTION OF PAPER

Majani Das, Vivek Kumar, A.K Ray,
Indian Institute of Technology Roorkee, Roorkee, India

ABSTRACT

Ever increasing demand of fibrous raw material for pulp production has affected the natural eco-system resulting environmental degradation due to deforestation in countries like India. It is therefore, an imperative necessity that utmost attention should be given to utilization of forest wastes, agricultural residues/wastes, and non-wood fibrous materials for the production of high yield pulp using latest technology, which will help in reducing the environmental pollution. Pine needles, leaves of pine, having 68.5% hollocellulose, 4.56% extractives and 31.0% lignin can be a good raw material for paper making. The commercial utilization of dry pine needle shall check pollution on one hand and safety of the Forest wealth on the other. Since the raw material is very new and no comparison is available to evaluate the results, in this paper an attempt has been made to evaluate the potential of pine needle for pulp production. For this a statistical model has been developed to predict the yield and kappa number through statistical experimental design.

The best pulping results were obtained at 18% chemical dose, 160°C cooking temperature and time of 4 hours. The screen yield was in the range of 28.02% to 46.65% which is comparable to other non wood based raw materials. Based on pulping experimental results, multivariable linear regression equation is developed to examine the effect of variables on pulp properties, like kappa number and yield.

1.0 Introduction

In the state of Uttaranchal, hills forest category area consists of around 55 to 70 % of total area with cultivated land of the order of 13 to 15%. Various plant fibers are available in Uttarakhand like industrial hemp, pine needles, agricultural residues, bhimal (Grevia optiva) and kandali. For obvious reasons the habitants of this area in the state are exclusively dependent on forests for mere survival. As per the statistics of State Forest Department 0.3 million hectares of land is under pure pine forest category which is the source of pine needle. The pine needles are a fire hazard to the forest and an irritant to the villagers, as these do not allow the grass to grow underneath for the cattle. Also its slow rate of decomposition hinders regeneration as they make thick layer and prevent the seeds to reach the soil. The fall of these needles is about 3 tonnes per hectare per season, depending upon the density of the forest.

Although several usage of pine needles have been reported such as fiber board, packing box, extracting essential oil, producing ligno-sulfonates and making pine wool etc. but still pine needles have not been utilized for any industrial scale [1].

Pine needle can be utilized in three ways. It can be utilized for manufacture of wide range of fancy and decorative handmade paper such as handmade drawing paper, envelopes, letter pads, paper stationery articles, paper bags, filter paper, electrical insulating paper, sanitary tissue, paper and paperboard and extracting oil using sub-critical and super-critical temperatures in a high pressure extractor or autoclave.

Chawla et.al [1-4] prepared various semi chemical (disk-refined cold-soda) pulps and thermo mechanical pulps from Indian slash pine needles for the preparation of fiberboard on hand mold. Using various dosages of lime yield ranging between 41% - 63% has been obtained. Uniform sheets were prepared successfully from the pulps, but the pine needle pulps required blending with other fibers. Fiberboard properties were considerably improved by heat tempering or oil tempering. Attempts have also been made to carry out sulphate pulping of pine needles and reported that pulp obtained is of poor quality with only 30% yield.

Majani Das [5, 6] attempted exploratory trials in the laboratory for the pulping of pine needles using various pulping methods such as soda, kraft and sulphite at a temperature of 170 °C with varying time of pulping.

The objectives of the above study was to upgrade and standardize the existing pulping technology of pine needles to manufacture hand made products and some high value paper products like sanitary pads and tissue paper etc. Further investigations were made to develop handmade paper from blends of pine needles and kenaf fiber to get stronger pulps and also to develop high absorbency and other suitable paper products from pine needle pulp. Results on the basis of screened yield and kappa number obtained from soda and kraft pulping showed that yield from soda pulping is very low. In kraft process yield improves slightly of the order of 29.5%. Though the pulp quality improves significantly, pulp yield is also too low for commercial application of pine needles [3, 4].

The present work first focuses on the results of exploratory studies as comparative one to examine which process has the maximum potential for pulping of pine needles. After getting the direction from first phase of trials, further experimental investigation has been carried out based on full factorial design [7]. In this design of experiments, numbers of levels chosen were three and number of factors two. The third phase of experiments is aimed to enhance the strength of pine needle pulp by blending with kenaf pulp. Finally, to explore potential further if any valuable chemical can be extracted from pine needle or not which can accrue economy in conjugation with paper products.

2.0 Experimental Work

The dry pine needles, collected from the forests of Uttaranchal Hills were firstly washed and air dried, then chopped to the required size (i.e. 38.1mm in length). The materials were mixed uniformly and packed in several packets for various pulping experiments. Moisture content of the material was estimated to be 11%.The physico-chemical characteristics of the pine needles and the standard methods used for characterization are given elsewhere [5, 6].

2.1 Preparation of Handmade Paper

Pulping of the packed (300 g) material were done in three different bombs (each of 100 g capacity) placed inside a 10 litre capacity electrically heated laboratory stainless steel rotary digester (Weverk Make). The cooking liquor used for sulphite pulping was prepared in two different methods, first with NaOH & Na_2SO_3 both and next with only Na_2SO_3. Finally results were compared only for 100% Na_2SO_3. In chemi-mechanical pulping cooked pulp was given mechanical treatment in the refiner. The refined pulp was further beaten in a Valley Beater upto the desired freeness level. Table 1 summarizes the conditions of pulping.

Table 1: Various Conditions of pulping

Type of Pulping		Chemical, %	Temp range, °C	Bath Ratio	Time at temp, min.
Chemical Pulping	Soda	8 – 20	150 -170	1:4	60-120
	Kraft	8 – 20 (20% sulphidity)	150 -170	1:4	60-120
	Alkaline Sulphite Pulping	4- 18 %	120 – 160	1:4	60-120
Chemi-Mechanical Pulping (Alkaline Sulphite Pulping)		8- 16%	120 – 160	1:4	180-240

Handmade sheets were prepared on the British sheet former for all the above soda, kraft and Na- sulphite pulps as per the standard procedure followed by wet pressing and drying. The dried sheets were subjected to testing for various paper properties.

Further investigation on chemi-mechanical pulping has shown yield upto 40% (screened) and kappa number upto 22 at sodium sulphite dose as high as 18%. Looking at the suitability of alkaline sulphite pulping, further studies are being made on the pulping at lower dosage of chemical and at milder conditions. Pulping at 8% and 12% sulphite chemical at 160° C showed yield around 50 % (but at higher kappa number).

The preliminary investigations carried out on 26°SR pulp for the properties of sulphite pulp of dry pine needles are shown in Table 2. The tear and burst strengths are found quite satisfactory though tensile strength and folding endurance are on the lower side.

Table 2: Properties of Sulphite pulp of Dry Pine Needles

Beating time (min)	Freeness (°SR)	Drainage time (sec)	Tear Index (mNm^2/g)	Burst Index ($kPam^2/g$)	Tensile Index (Nm/g)	Folding endurance (no.)
30	40	10	2.956	1.635	18.75	4

2.2 Preparation of Chemical bleached Pulp with experimental design

For chemical pulping similar procedure followed but according to statistical full factorial design of experiments [7]. In the design of experiments, for statistical experimental design the number of levels of chemical dosages considered were 12%, 16%, 18% whereas number of factors were two, temperature and chemical dose with 9 numbers of experiments conducted. The effect of two factors on the Yield and Kappa number at the different levels mentioned above was studied. In the dimensionless coordinates system, the upper and lower levels were at +1 and -1 with X_0, a dummy variable, has been taken as 1. Respectively, the coordinates of the center point of the design is zero and coincide with the origin of coordinate. The design of matrix is shown elsewhere [5-7]. Based on the suggested procedure of full factorial design a suitable regression equation of the form has been assumed as $Y=b_0+b_1X_1+b_2X_2+b_{12}X_1X_2$

The matrix developed was solved using MATLAB for each of the property and the corresponding coefficients were calculated. After doing the experiment final regression equations for each property were obtained. These all regression coefficients and interaction coefficients were checked for significance using student's t-test and test of adequacy of fit with Fisher "F" test.

2.2.1 Bleaching and strength property evaluation

The pulp obtained were subjected to multistage bleaching with sequences CEHH, CEHD, CEDPD and the properties of bleached pulp were evaluated in the laboratory.

2.3 Effect of blending with kenaf pulp: The experiments were also conducted to examine the effect of blending of kenaf with the pine needle unbleached pulp on the strength properties.

3.0 Results and Discussion

Analyzing thoroughly and considering all the aspects including the physical appearance of pulp, it can be inferred that the suitable conditions for pine needles pulping are 18% chemical dose and 160°C. At lower chemical dose the product obtained is pine wool and therefore cannot be used for paper making.

Effect of pulping variables:

3.1. Effect of temperature on yield and kappa number

It is evident from the graph that the screen yield increases with increase in chemical concentration while pulp kappa number decreases. It is due to the reason that the sulphite in cooking liquor greatly accelerates delignification rate. The total yield decreases with increasing sodium sulphite concentration. The effect of temperature on screened yield, kappa number and total yield is shown in figures 1, 2 and 3 respectively.

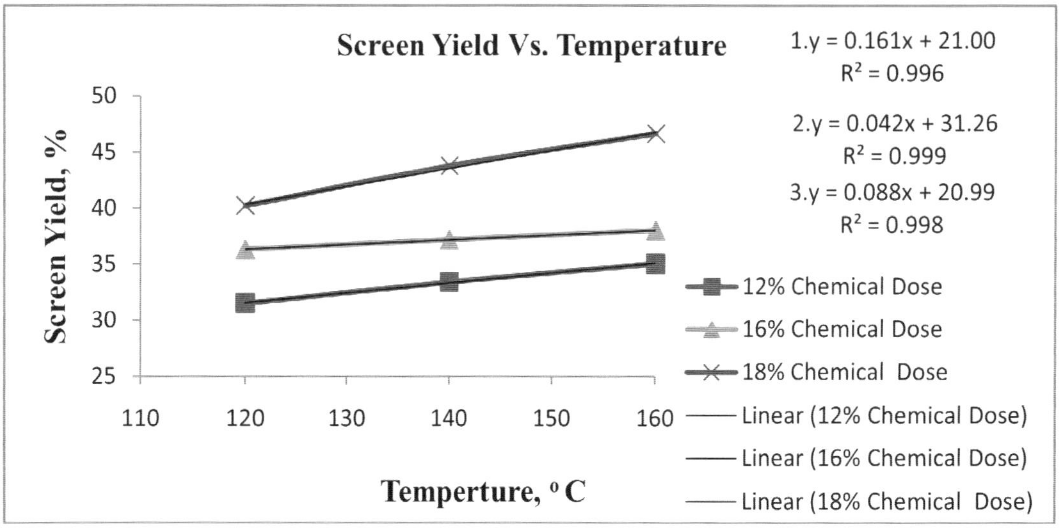

Fig 1: Graph showing relation between temperature and yield after 4 hours cooking at different chemical concentration.

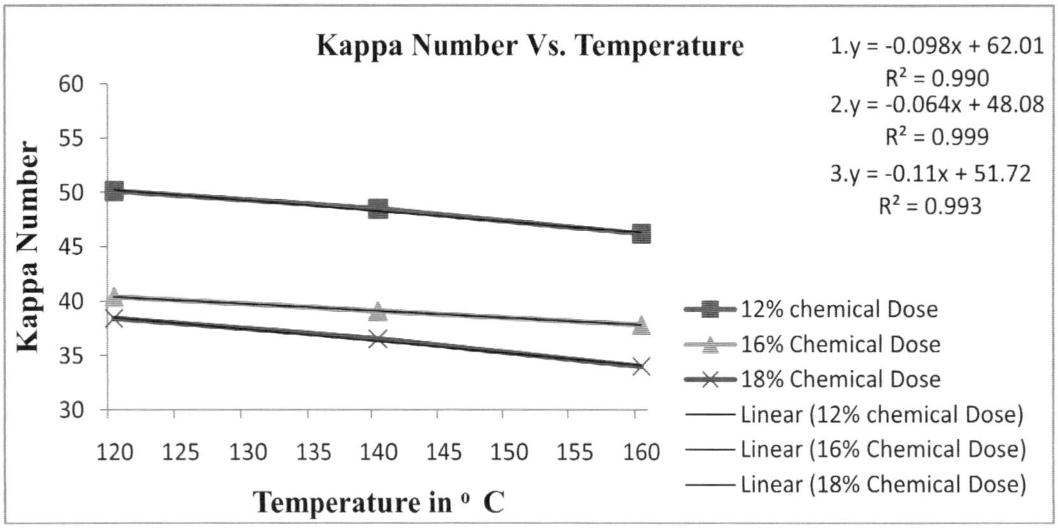

Fig 2: Graph showing the relation between pulp kappa number and temperature after 4 hours cooking at different chemical concentration.

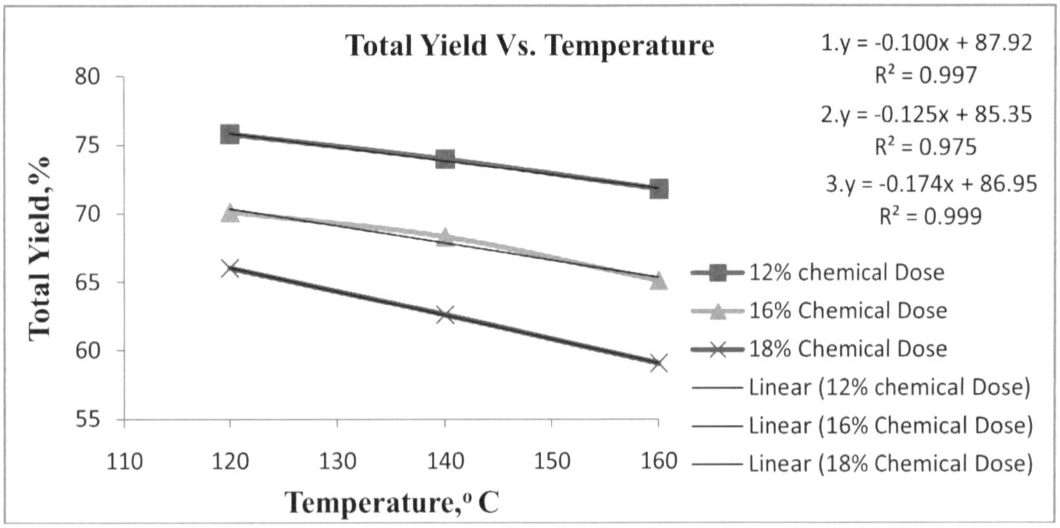

Fig 3: Graph showing the relation between total yields and pulping temperature after 4 hours cooking at different chemical concentration.

Best pulping results were obtained at 4.0 hours, 160^{0}C and 18% sulphite dose. At these conditions the overall yield was 59.06% and kappa number was lowest at 34.02. The screen yield was in the range of 28.02% to 46.65% which is comparable to yield from other non wood based raw materials. Both yield and kappa number show linear relationship with pulping temperature.

3.2 Effect of chemical dose on yield and kappa number

The results of pulping experiments of pine needles at constant time but varying doses of sodium sulphite (i.e. 12%, 16% and 18% as Na_2O) indicate that the amount of sodium sulphite charge has a considerable influence on the rate of pulping.

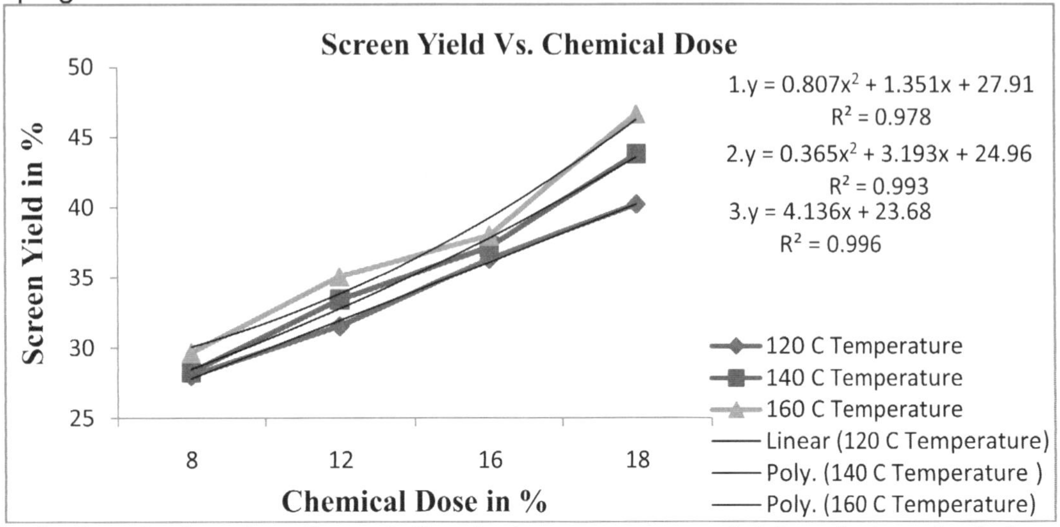

Fig 4: Graph showing the relation between screen yield and chemical dose after 4 hours cooking at different temperature.

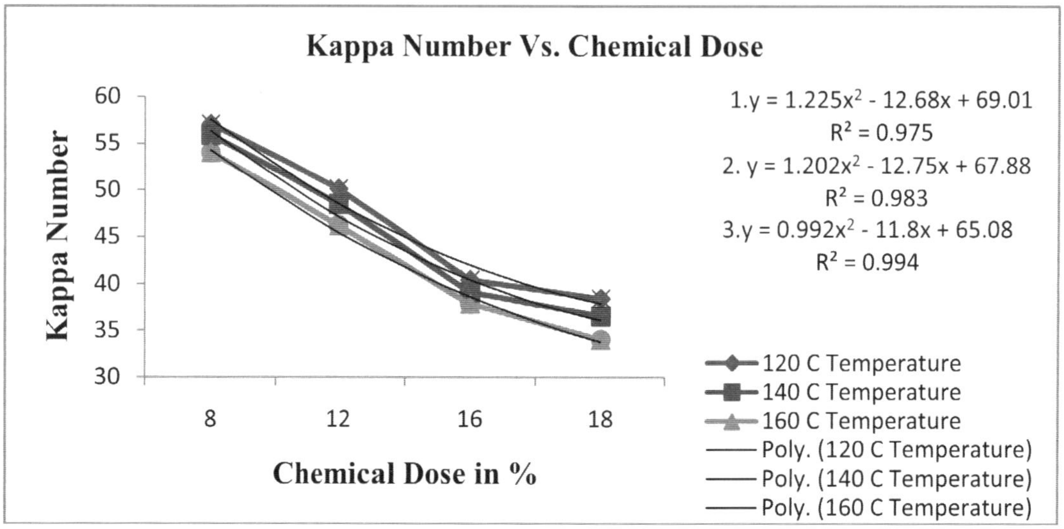

Fig 5: Graph showing the relation between kappa number and chemical dose after 4 hours cooking at different temperature.

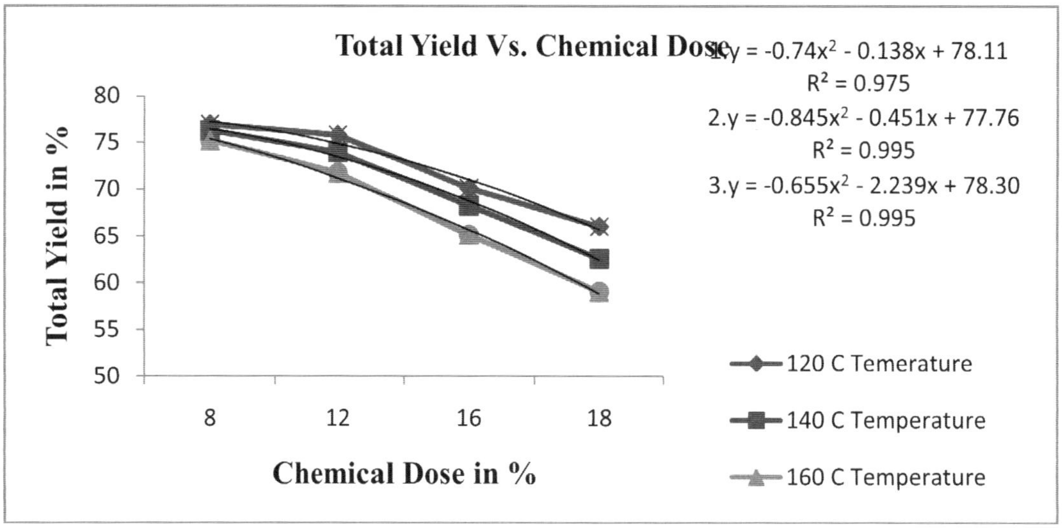

Fig 6: Graph showing the relation between total yield and chemical dose after 4 hours cooking at different temperature.

Figure 4 shows that screen yield varies linearly at lower temperature but as a polynomial function of chemical dose. This behaviour is due to better cooking of pine needles at higher chemical dose and temperature, which increases the screen yield and reduces the rejects.

Figure 5 and figure 6 indicate that the rate of delignification reaction shows an indirectly proportional relationship of kappa number and total yield with chemical dose. On increasing the chemical dose from 12% to 18% the kappa number decreases from 57.03 to 34.02. Figures 5 and 6 clearly indicate a sharp decrease in pulp yield and Kappa number with increasing sodium sulphite dose.

3.3 Regression equations

The regression equations were obtained on the basis of experimental data collected applying full-factorial design with three levels and two factors after significance test of all coefficients through students t-test. The student's t-test results for yield are found out as: $t_0 = 85.2783$, $t_1 = 11.4378$, $t_2 = 4.3581$, $t_{12} = 0.7251$. The student's t-test results for kappa number are: $t_0 = 92.501$, $t_1 = 13.412$, $t_2 = 4.066$, $t_{12} = 0.12$

The regression equations for yield and kappa number are as under:

For Yield: $Y_{Yield} = 38.0206+5.1X_1+1.9433X_2$

For kappa number: $Y_{Kappa\ Number} = 41.2556-5.9817X_1-1.8133X_2$

where: X_1=Sodium sulphite concentration in % ,X_2=Temperature in ^{O}C

3.3.1 Relationship between predicted kappa number and actual kappa number

The relations between actual values of yield and kappa number with the data obtained from respective models were attempted. It is found that, a linear relationship exists between the actual and predicted kappa number. The relationships are shown by the following graphs

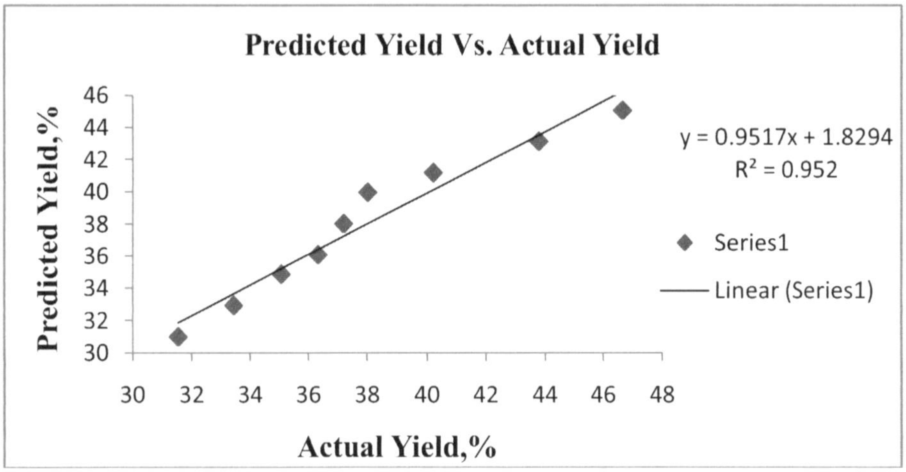

Fig 7: Graph showing the relationship between the predicted yield and actual yield

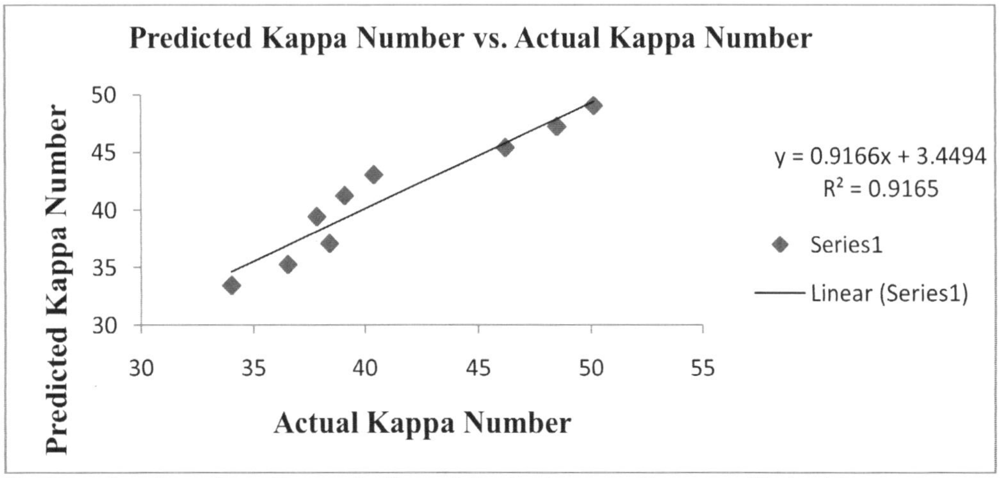

Fig 8: Graph showing the relationship between the predicted kappa number and actual kappa number

3.4. Bleaching of Pine Needle Pulp:

Based on the standard conditions of various bleaching agents in three sequences chosen namely, CEHH, CEHD, CEDPD, strength properties, brightness and yield values are plotted in figure 9.

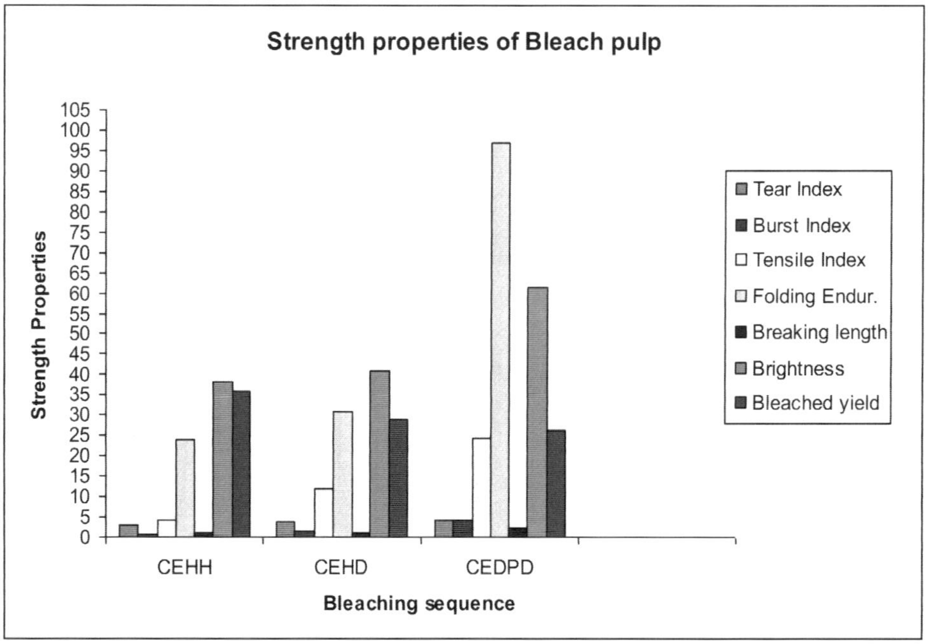

Fig.9 Strength Properties of Bleach Pulp

3.4.1 The effect of various bleaching sequences on strength properties and brightness:

It is evident from the graph that the bleached yield decreases with the increasing ClO_2 treatment. The bleach pulp yields were found 36% and 26.21 % for CEHH and 26.21% CEDPD respectively. The brightness increases with the increase of dioxide bleaching stages. The maximum brightness of 61.50% ISO is achieved with the bleaching sequence CEDPD. This is not very promising results. Tensile strength increases gradually when an H-stage is replaced by a D-stage but increases sharply when both of the H-stages are replaced by a DPD-sequence. It is clear that tensile index vs a vis breaking length are not up to the mark though tear, burst indices and double fold are acceptable. Thus bleaching of pine needle pulp is not recommended and the pulp may be used for kraft and board making.

3.5. Effect of blending of kenaf pulp on strength properties

The blending of kenaf pulp with pulp obtained from pine needles and the properties at 40°SR indicated that all the strength properties improve on blending with kenaf pulp. The strength properties of blended pulp were shown in Figures 10-13 for tear, burst, tensile and folding endurance respectively. It is evident from the figures that the best results were obtained with 30% kenaf pulp blending. It was also observed that after blending with long fiber pulp the formation improves at lower GSM. After blending with kenaf pulp the properties become compatible with that of the kraft paper.

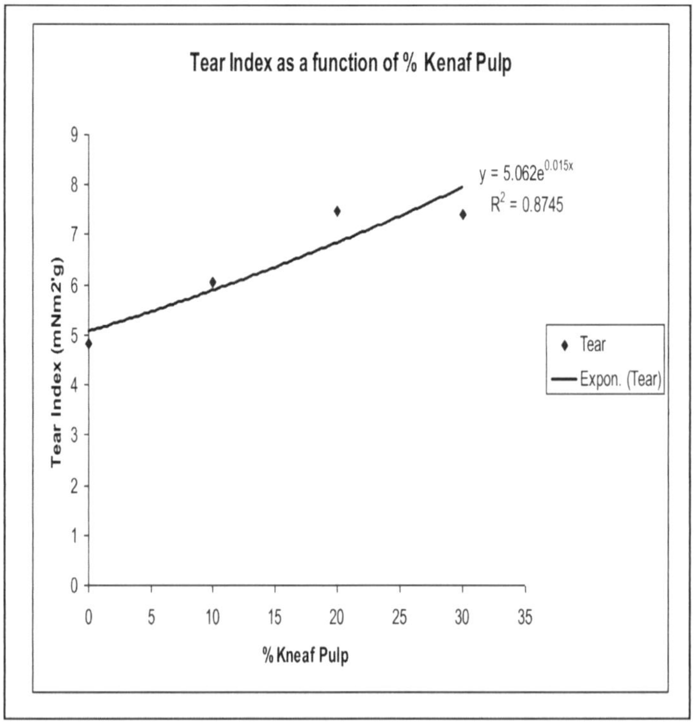

Fig.10 Tear Index as a Function of % Kenaf Pulp

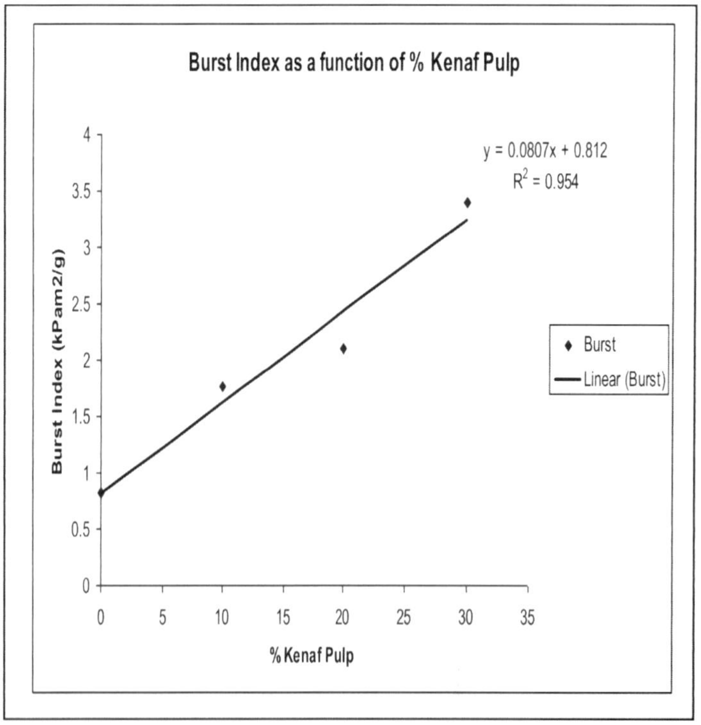

Fig.11 Burst Index as a Function of % Kenaf Pulp

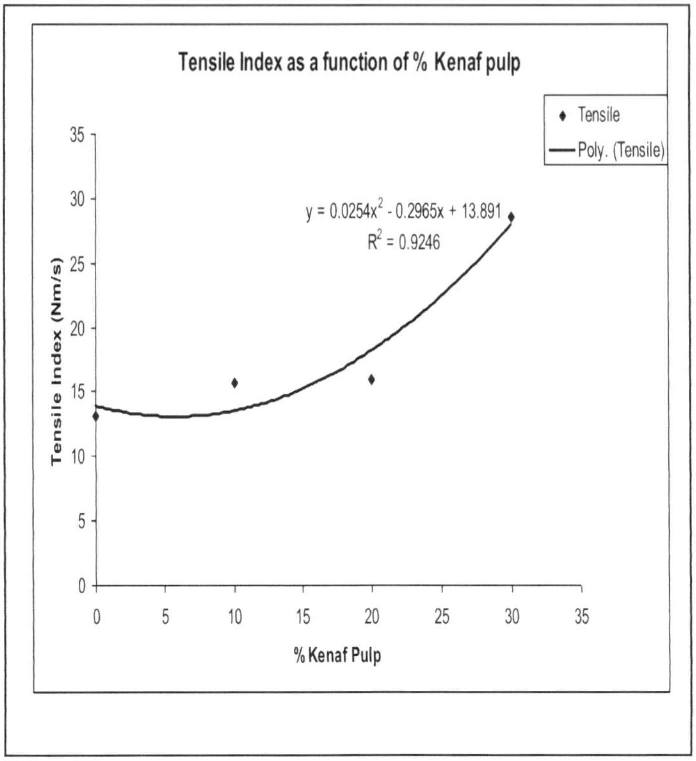

Fig.12 Tensile Index as a Function of % Kenaf Pulp

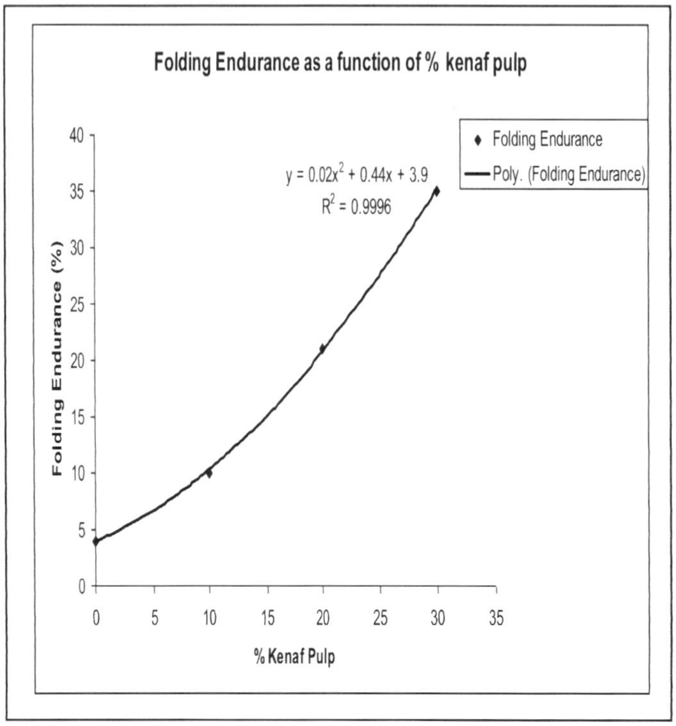

Fig.13 Folding Endurance as a Function of % Kenaf Pulp

3.6. Supercritical extraction of Pine Needles

Steam distillation the pine needles generally give only low yield of oil varying in the range between 0.08 -0.5% and therefore is not economical. Hence, in this investigation attempts have been made to explore production of essential oils by different means of extraction- supercritical and subcritical process in a high-pressure extractor/autoclave using CO_2. Both the processes gave low yield of the order of 4-5% but 8-10 times higher than those obtained by steam distillation methods. The product extracted a pale yellow coloured volatile oil possessing a strong balsamic odour.

The yield values and the chromatographic analysis are shown in table 3:

Table 3. Experimental results of Extraction of Dry Pine Needles.

Extraction process	Yield (%)
Super-critical	5
Sub-critical	4.33

From the results of GC run it can be concluded that the number of major components determined is 45 for sub-critical extraction and 49 for super-critical extraction. To identify these components the oil should be run in GCMS. Therefore, this study indicates that pine needle has a good potential for papermaking along with oil extraction being a good alternative but supercritical and sub-critical extraction of dry pine needle alone appears to be uneconomical at this low value of yield of oil.

4.0 Conclusion

Attempt to prepare pine needle pulp based on soda and kraft pulping processes produce low yield pulp. The alkaline sulphite process however has produced increased pulp yield. The unbleached pulps of dry pine needles showed good strength properties especially tear and burst. It was also observed that after blending long fiber pulp like rag pulp the formation improves even at lower. The results have shown that pine needle can be used as raw material for handmade paper making in Uttarakhand.

More extensive trials for chemical pulp based on statistical design of experiments indicated that the best pulping results could be obtained at 18% chemical dose, 160°C cooking temperature and time of the order of 4 hours. The screen yield was in the range of 28.02% to 46.65% which is comparable to other non wood based raw materials. Multivariable regression equations are developed to examine the effect of variables on pulp properties. Linear regression equations for both yield as well as kappa number as function of sodium sulphite dosage and temperature were developed. After blending with kenaf pulp the properties become compatible with that of the kraft paper.

Integrating recovery of essential oils by extraction into a program of forest thinning and biomass utilization may be an attractive option for improving forest and economic health in the region. Supercritical and sub-critical extraction of dry pine needle alone appears to be uneconomical at this low value of yield of oil.

5.0 References

1. Chawla, J. S., and Negi, J. S,; Fiberboards from Pinus Caribaea Needles Source, Holzforschung und Holzverwertung, v 32, n 5, 121-126, Oct., 1980.
2. Chawla, J. S; Studies on Pine Needle (Leaves) Fiber: Pinus Roxburghii, IPPTA, v 14, n 3, 205-208, July/Sept., 1977.
3. Chawla J.S., "Studies On Pine Needle (Leaves) Fibre : Pinus Roxburghii, Report of Regional Research laboratory, Jammu-Tawi, India
4. Negi, J. S.; Chawla, J. S, Indian Pulp and Paper, v 36, n 2, 9, 11-13, Aug. 1981.
5. Das Majani,Vivek Kumar and A.K.Ray" A statistical approach to evaluate the potential of pine needles for pulp production,Proceedings of the 23rd National Convention of Chemical Engineers on Recent trends in Chemical Engineering held on October 5-7,2007.,pp.312-320.
6. A Project Report on "Utilization of Dry Pine Needles" Department of Paper Technology, IIT Roorkee, September, 2006.
7. Drapper, N. And smith, H. "Applied Regression Analysis" Second addition John Wiley and Sons, New York, (1966)

AIChE 2010 Annual Meeting, November 7-12, Salt Lake City, UT.

Bioethanol Dehydration in Thermally Integrated Extractive Distillation Columns

Nghi Nguyen, Jiwon Sim, Yaşar Demirel*
Department of Chemical and Biomolecular Engineering, University of Nebraska Lincoln, Lincoln, NE 68508

ABSTRACT

Bioethanol plants produce dilute mixture of around 10-wt% of ethanol in water. Ethanol is distillated to 96%, and most commonly, sent to a molecular sieve, which absorbs water. The molecular sieve is regenerated by heating to remove water. As the capacity of ethanol production increases molecular sieve becomes thermally inefficient. Beside that the distillation of ethanol could consume up to 50% of the overall energy used in a typical grain alcohol plant. Ethanol purification by distillation requires extractive distillation with an entrainer, such as pentane, benzene, diethyl ether, ethylene glycol, toluene, cyclohexane, methoxy-ethanol, ethyl tert-butyl ether, and bioglycerol. Entrainer breaks the azeotrope which forms between ethanol and water at 95.6 wt% ethanol (at atmospheric pressure). The study analyzes the potential of using thermally integrated columns to save energy for the purification of ethanol from water using ethylene glycol as entrainer.

Keyword: Ethanol, azeotrope, thermally integrated column, ethylene glycol

*Corresponding Author. Tel: 402 472 2745, fax:402 472 6989.
E-mail address: ydemirel2@unl.edu

1. Introduction

A reasonable blend of anhydrous ethanol into gasoline can be used as fuels without any modification of engine necessary [1,2]. Usually, dilute ethanol-water mixture (around 10-wt% or less ethanol) is obtained from the fermentation process of a bioethanol plant [3]. However, separation of ethanol-water is a major drawback due high energy requirement because of the formation of azeotrope around 96-wt% of ethanol at atmospheric pressure. Several methods have been introduced to produce fuel grade ethanol, such as using molecular sieves, pressure-swing distillation, and homogeneous and heterogeneous azeotropic distillations using suitable entrainers. Molecular sieve is most frequently used but becomes thermally inefficient as the capacity of ethanol production increases. Pressure-swing distillation requires two distillation columns operate under a different pressure to cross the azeotrope. Hence, this research will focus on the possibility of using thermally integrated distillation columns to separate ethanol from water by homogeneous azeotropic distillation columns [4-6].

Conventional extractive distillation of ethanol-water mixture with an entrainer requires at least two distillation columns, extractive column and entrainer recovery column. The first column uses entrainer to break the azeotrope to separate ethanol with high purity while the second column recovers the entrainer from water and recycles. There are various entrainers available for the separation of water-ethanol mixture by homogeneous azeotropic distillation, such as glycerol and ethylene glycol [7,8]. In this simulation, ethylene glycol is used to break the azeotrope between ethanol and water. This separation unit can consume over 50% of the total energy used in a typical bioethanol plant [1,2].

Thermally integrated distillation column sequences eliminate either the reboiler or/and the condenser by interconnecting vapor and liquid streams from one column to another. Three well-known configurations are the side rectifier, side stripper, and Petlyuk column. Requirement of a solvent feed to the top and alcohol/water mixture feed to the bottom of the column limits the separation sequences. Petlyuk configuration can be the most energy efficient but only for a narrow range of operating conditions [4]. Thermally coupled side-stripper configuration is allowable but cannot separate ethylene glycol and water to high purity due to the contamination of ethylene glycol throughout the column. Therefore, in this study we explore thermally coupled side-rectifier configuration.

2. Homogeneous Azeotropic Distillation

Figure 1 shows the residue curve map for the mixture water-ethanol-ethylene glycol system calculated by using the NRTL-RK model. The line between the feed point (F) and pure ethylene glycol (EG) represents the addition of pure ethylene glycol into the feed of dilute ethanol water mixture. The ethylene glycol to feed molar ratio is 1:10. The line between pure ethanol and the bottom flow WAT-EG is the distillation line for column T101 (see Figure 2) producing pure ethanol as the distillate. The bottom flow of WAT-EG with around 90-mol% of water is sent to column T102, shown in Figure 2, to recover the entrainer.

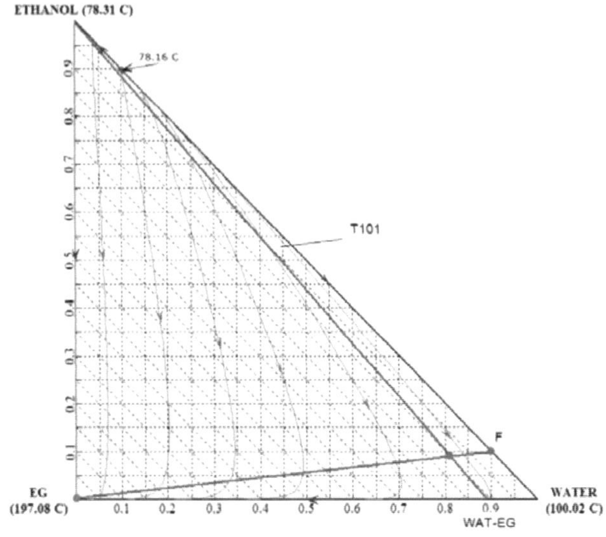

Figure 1. Residue curve map for water-ethanol-ethylene glycol system and distillation line for column T101. The FEG is the mixing line of ethylene glycol with the feed of 10-mol% ethanol.

Figure 2 shows the process flow diagrams of the conventional and thermally integrated azeotropic distillations for the separation of a dilute ethanol-water mixture using ethylene glycol as entrainer. In both designs, columns T101 and T102 operate with 16 and 6 total number of stages, respectively. Stream MAKEUP combined with stream S4 in mixer M101 is fed to column T101 at stage 3. A calculator block sets the sum of ethylene glycol flow in stream S4 and MAKEUP to equal to 10 kmol/hr. Stream WATETH containing 10-mol% ethanol and 90-mol% water is preheated in heat exchanger HX101 by the product stream S3 to 59.2 °C, and it enters column T101 at stage 12. The product stream ETHANOL contains 10 kmol/hr of at least 99.4-wt% of ethanol while the stream WATER contains 90 kmol/hr of 99.6-wt% of water. Stream WAT-EG, LIQUID, and S3 were treated as tear streams in this simulation.

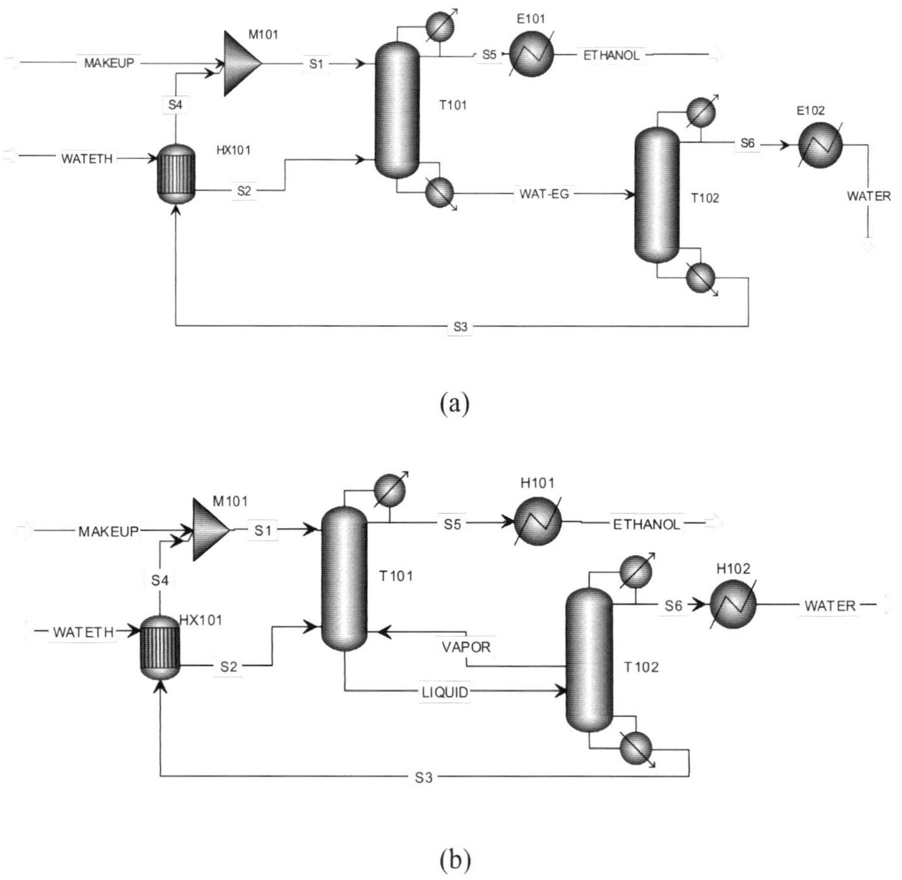

Figure 2. Process flow diagram for: (a) conventional design; (b) thermally integrated design.

3. Results and discussions

The properties of inlet and outlet streams of the conventional and thermally integrated designs are summarized in Table 1. The conditions of the feed streams are identical except that the flow rate of MAKEUP stream depends on the flow rate of EG in stream S4. EG loss is slightly higher in the thermally integrated design, hence requires higher flow rate for stream MAKEUP (see Figure 2).

In the thermally integrated design, the vapor and liquid streams interchange between stage 16 of column T101 and stage 4 of column T102. The vapor flow rate is set to 35 kmol/hr which supplies adequate energy for the separation of water-ethanol mixture. The reboiler is no longer needed for column T101.

Table 1

Inlet and outlet streams properties of the conventional and thermally integrated design

	Conventional Design				Thermally Integrated Design			
	MAKEUP	WATETH	ETHANOL	WATER	MAKEUP	WATETH	ETHANOL	WATER
Mass Flow kg/hr								
WATER	0.00E+00	1.62E+03	2.14E+00	1.62E+03	0.00E+00	1.62E+03	2.35E+00	1.62E+03
ETHANOL	0.00E+00	4.61E+02	4.55E+02	5.58E+00	0.00E+00	4.61E+02	4.55E+02	6.11E+00
EG*	9.88E-01	0.00E+00	1.40E-01	8.47E-01	9.98E-01	0.00E+00	1.51E-01	8.49E-01
Mass Frac								
WATER	0.00E+00	7.79E-01	4.68E-03	9.96E-01	0.00E+00	7.79E-01	5.13E-03	9.96E-01
ETHANOL	0.00E+00	2.21E-01	9.95E-01	3.43E-03	0.00E+00	2.21E-01	9.95E-01	3.76E-03
EG	1.00E+00	0.00E+00	3.07E-04	5.21E-04	1.00E+00	0.00E+00	3.30E-04	5.22E-04
Total Flow kmol/hr	1.59E-02	1.00E+02	1.00E+01	9.00E+01	1.61E-02	1.00E+02	1.00E+01	9.00E+01
Total Flow kg/hr	9.88E-01	2.08E+03	4.57E+02	1.63E+03	9.98E-01	2.08E+03	4.57E+02	1.63E+03
Temperature C	4.00E+01	2.50E+01	2.50E+01	2.50E+01	4.00E+01	2.50E+01	2.50E+01	2.50E+01
Enthalpy kcal/mol	-1.09E+02	-6.81E+01	-6.63E+01	-6.82E+01	-1.09E+02	-6.81E+01	-6.63E+01	-6.82E+01
Entropy cal/mol-K	-1.06E+02	-4.30E+01	-8.22E+01	-3.89E+01	-1.06E+02	-4.30E+01	-8.21E+01	-3.89E+01

*EG: ethylene glycol

Sensitivity analysis of feed stages on mass fraction of ethanol in the distillate indicates that the concentration of ethanol is maximized when streams S1 and S2 are fed at stage 3 and 12, respectively. The value of the vapor flow rate is adjusted to achieve desirable separation of water from ethanol.

Table 2 compares the operating conditions and configurations of columns T101 and T102 in the conventional and thermally integrated designs. The molar reflux ratio in the conventional column T101 is varied until it has reached a similar purity of ethanol in the distillate as in the thermally integrated column T101. The reboiler and condenser duties of the columns are compared in Table 2. There is no reboiler in column T101 of thermally integrated design. Stream VAPOR supplies energy to operate column T101. The total exergy losses for the columns in thermally integrated design are slightly higher due to the mixing taking place at stages where the interacting streams are introduced. For column T101, most of the exergy loss is due to the mixing on feed stages 3 and 12 as shown in Figure 3. However, significant exergy loss on stage 16 of column T101 in thermally integrated design is mainly due the vapor feed mixing with the liquid. Figures 3a and 3b compare the composition and temperature profiles of column T101. Both the designs exhibit similar trends.

Table 2

Comparison of operating conditions and configurations of columns T101 and T102 in the conventional and thermally integrated designs

	Conventional Design		Thermally integrated Design	
Condition & Configurations	T101	T102	T101	T102
Number of stages	16	6	16	6
S1 feed stage	3	—	3	—
S2 feed stage	12	—	12	—
WATGLY feed stage	—	4	—	—
VAPOR feed stage	—	—	16	—
VAPOR side withdraw stage	—	—	—	4
LIQUID withdraw stage	—	—	16	—
LIQUID feed stage	—	—	—	4
Molar reflux ratio	2.20	2	—	2
Distillate rate, kmol/hr	10	90	10	90
Condenser duty, kW	-346.73	-3065.83	-318.15	-3066.03
Condenser temperature, ^{0}C	77.95	99.23	77.94	99.19
Condenser pressure, bar	1	1	1	1
Reboiler duty, kW	464.59	3110.28	0.00	3546.24
Boilup rate, kmol/hr	40.43	205.50	38.00	234.46
Bottoms rate, kmol/hr	100	10	135	10
Reboiler temperature, ^{0}C	102.20	195.95	102.40	195.94
Reboiler pressure, bar	1	1	1	1
Exergy loss, kW	36.52	486.38	44.80	545.98

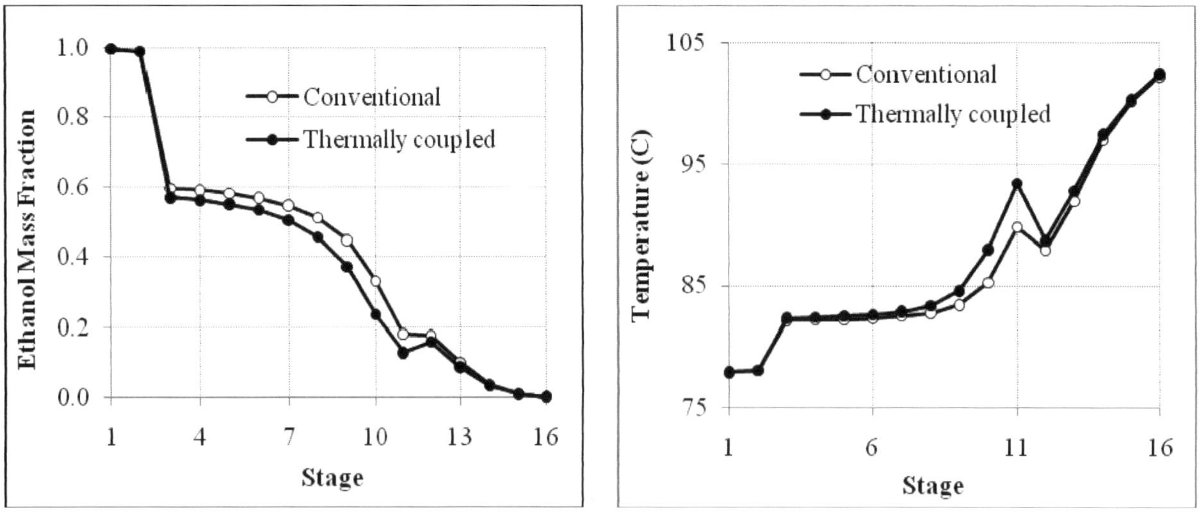

(a) (b)

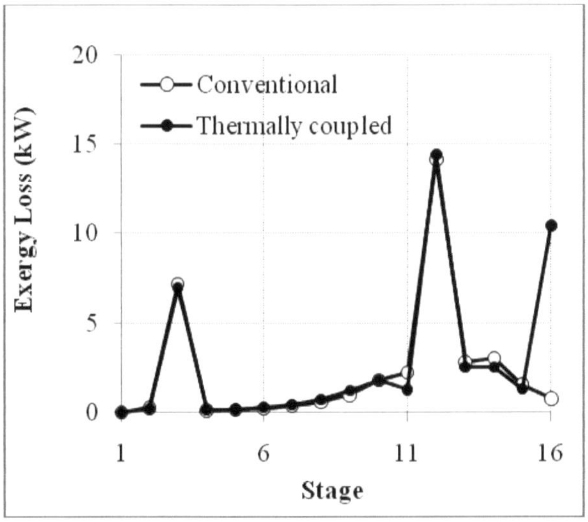

(c)

Figure 3. Comparison of operating conditions for column T101: (a) composition profiles; (b) temperature profiles; (c) exergy loss profiles.

Reboiler and condenser duties of Column T102 depend strongly on the number of stages. Preliminary analysis has indicated that the reboiler duty exceeds 3700 kW when the column operates with 5 stages, therefore, a column with 6 stages is used. Figure 4 compares the temperature and exergy loss profiles for column T102. The temperature profiles overlap and the exergy loss profile is slightly higher for the column in thermally integrated configuration due to the higher flow rate and mixing.

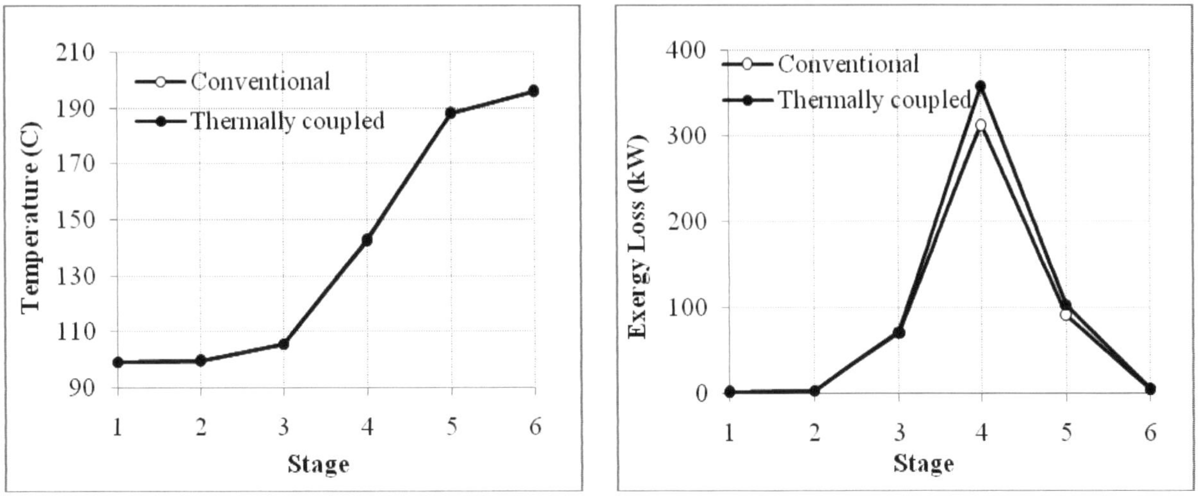

(a) (b)

Figure 4. Comparison of operating conditions for column T102: (a) temperature profiles; (b) exergy loss profiles.

4. Conclusions

Thermally integrated column sequences are constrained for azeotropic distillation due to the requirement of feed locations that entrainer feed to the top and water-ethanol mixture feed to the bottom of the column. Using thermally integrated side-stripper configuration is a better. However, exergy analysis of the columns indicates that thermally integrated design is not thermodynamically more efficient than the conventional sequence. Thermally integrated design eliminates one of the reboilers to save equipment cost but results in higher energy requirement. The total exergy loss in the conventional design is 67.88 kW less than the thermally integrated design.

References

1. Gil, I. D., Uyazán, A. M., Aguilar, J. L. (2008). Separation of ethanol and water by extractive distillation with salt and solvent as entrainer: process simulation. Brazilian J. Chem. Eng., 25, 207-215.

2. Gil, I. D., Uyazán, A.M., Aguilar, J. L. (2008). Simulation of ethanol extractive distillation with a glycols mixture as entrainer. 2nd Mercosur Congress on Chemical Engineering & 4th Mercosur Congress on Process Systems Engineering.

3. Hernández, S. (2008). Analysis of energy-efficient complex distillation options to purity bioethanol. Chem. Eng. Technol., 31, 597-603.

4. Agrawal, R., Fidkowski, Z. T. (1998). Are thermally coupled distillation column always thermodynamically more efficient for ternary distillation? Ind. Eng. Chem. Res., 37, 3444-3454.

5. Malinen, I., Tanskanen, J. (2009). Thermally coupled side-column configurations enabling distillation boundary crossing. 1. An overview and a solving procedure. Ind. Eng. Chem. Res., 48, 6387-6404.

6. Malinen, I., Tanskanen, J. (2009). Thermally coupled side-column configurations enabling distillation boundary crossing. 2. Effects of Intermediate heat exchangers. Ind. Eng. Chem. Res., 48, 6372-6386.

7. Brüggemann, S., Marquardt, W. (2004). Shortcut methods for nonideal multicomponent distillation: 3. Extractive distillation columns. AIChE J., 50, 1129-1149.

8. Lee, F. M., Pahl, R. H. (1985). Solvent screening study and conceptual extractive distillation process to produce anhydrous ethanol from fermentation broth. Ind. Eng. Chem. Process Des. Dev., 24, 168-172.

(315b) Integrated Biorefineries: Isolation and Characterization of Lignin From Ionic Liquid Pretreated Biomass for the Production of Value Added Lignin by-Products

Thehazhnan (Thihal) K. Ponnaiyan and Sasidhar Varanasi, Chemical & Environmental Engineering, University of Toledo, Toledo, OH

Plant (lignocellulosic) biomass is an abundant, inexpensive and sustainable source of organic carbon, for the production of renewable fuels and chemicals. However, many technical challenges exist in developing commercially viable processes for producing sugars from biomass that serve as precursors for fuel and chemical production. Pretreatment and enzymatic hydrolysis are critical steps, and most of the current pretreatment methods do not effectively disrupt the crystallinity of cellulose, which is a major barrier for efficient enzymatic hydrolysis. Ionic liquids (ILs) being non volatile, non-flammable, recyclable and designer friendly, are gaining wide recognition as green solvents, and their unique solvating properties make them ideal for pretreating lignocellulosic substrates like corn stover and poplar.

From an economic standpoint, it is of interest to facilitate the production of value added products from all components of the biomass, including lignin, given its interesting functionalities and properties owing to its complex aromatic nature. In integrated biorefineries, currently, residual lignin is primarily burnt for recovering heat. Suitable isolation of lignin from the ionic liquid pretreated biomass can help in producing value added lignin by-products like flame retardant materials, adhesives, high-strength fillers, surfactants, etc. In view of this, our study addresses: (1) the isolation of lignin from IL pretreated biomass subjected to enzymatic hydrolysis, using different enzyme mixtures, and (2) the characterization of recovered lignin compounds using thermal, chromatographic and spectroscopic methods, in order to assess its usability for further processing to produce value added products.

See more of this Session: Biorefinery: Separation & Transport in Lignocellulosic Conversion Processes
See more of this Group/Topical: Forest and Plant Bioproducts Division

(315c) Ionic Liquid Recovery and Reuse in Ligno-Cellulosic Biomass Pretreatment Processes

Xi Du and G. Glenn Lipscomb, Chemical & Environmental Engineering, University of Toledo, Toledo, OH

Ethanol production from ligno-cellulosic biomass is complicated by the recalcitrance of biomass to enzymatic hydrolysis. A promising pretreatment process that can increase hydrolysis rates dramatically uses an ionic liquid to convert crystalline cellulose into an amorphous state.

Process economics require recovery and reuse of the ionic liquid. Of particular concern is reducing the water content of the spent ionic liquid to an acceptable level. A number of processes for water removal are studied including evaporation, pervaporation, and stripping.

Process performance is dependent on water vapor pressure, fluid viscosity, and water diffusivity. The effects of these material properties on performance and associated energy and capital costs are examined.

See more of this Session: Biorefinery: Separation & Transport in Lignocellulosic Conversion Processes
See more of this Group/Topical: Forest and Plant Bioproducts Division

Flocculation Dynamics of Lignocellulosic Hydrolyzates and Floc Characteristics

Rakesh Yasarla[1], **Bandaru Ramarao**[2] and Thomas Amidon[2], (1) Paper & Bioprocess Engineering,

State University of New York, Syracuse, NY, (2) State University of New York, Syracuse, NY.

Abstract

Lignocellulosic materials such as wood present as renewable and sustainable resources for the production of fuels and plastics. In one pathway to the production of liquid fuels and plastics fermentable sugars are extracted by treatment with hot water after which the hydrolyzates are purified and detoxified. Purification of such liquors constitutes an important step in biorefinery processes. Raw lignocellulosic hydrolyzates produced by acid or water treatment contain significant quantities of colloidal material. The particulates are composed mostly of lignin and its derivatives and are anionic in nature. These colloidal particulates not only inhibit the fermentation activities of microorganisms but also foul any filtration membranes used for separation and purification of extracts. Flocculation of the colloidal particles can make separations easier and reduce the fouling tendencies of membranes. In this paper we investigated the dynamics of the wood extracts with particular emphasis on the action of three flocculating agents – a multivalent electrolyte, alum ($Al_2(SO_4)_3$) and two cationic polyelectrolytes: Polyethylene Imine (PEI) and poly Diallyl Dimethyl Ammonium Chloride (pDADMAC).

Extracts were susceptible to flocculation by either the electrolyte (alum) or the cationic polyelectrolytes (PEI and pDADMAC). Higher Concentrations of alum of the order of 0.1M was necessary for flocculation but the polymers effected flocculation at relatively low dosages in the 50ppm range. Flocculation was observed to occur rapidly when the zeta potentials of the extracts were close to zero showing that charge neutralization is the significant destabilizing mechanism. At higher dosages of the cationic polyelectrolytes, charge reversal occurred and the extracts were restabilized showing significant adsorption of the polyelectrolytes. Hence patching of the polyelectrolytes should also contribute to the flocculation process. The rate of sedimentation of aggregated particles was measured by turbidity of supernatants of the dispersed solution as a function of time. The pH Sensitivity of flocculating agents was tested by varying the pH of the wood extracts between 3.5 and 8.0. Flocculation particularly by PEI was sensitive to pH at higher values; higher dosages of PEI were required.

1. Introduction

Lignocellulosic materials such as wood are renewable and sustainable alternative resources for the production of fuels and plastics [1]. Accelerated industrial development around the globe has resulted in strong demand for petroleum as a fuel and also as a source for plastics and chemicals. Lignocellulosic feedstocks present an alternative to alleviate some of this pressure on petroleum resources in a particularly sustainable way. They present a significantly carbon neutral solution since biomass sequesters atmospheric carbon during its growth phase which is released during its combustion. Moreover, lignocellulosics present a source of fuels such as ethanol relieving the stress on corn, grain and such agricultural food sources. Societal awareness of such positive environmental benefits in addition to its obvious economic advantages has made the development and implementation of biobased energy from lignocellulosics imperative. Biomass processing is expected to occur in large biorefineries manufacturing a spectrum of fuel, chemical and material products in a scale efficient manner.

One class of biomass processes begin by hydrolyzing wood or the lignocellulosic raw material under different temperature and pressure conditions using dilute acid, hot water or mild alkaline solutions. The lignocellulosic hydrolyzates produced by this process consist of dissolved and colloidal oligomers of hemicelluloses, lignin and small quantities of extractives [Duarte et al., Mittal et al.]. The hemicelluloses in the hydrolyzates are transformed into Biofuels or biobased plastic products by fermentation or other routes. Hydrolyzates must be significantly purified and detoxified in order to conduct and maximize yields of the downstream fermentation processes. Today, one of the major obstacles to the large scale industrial fementation of hydrolyzates is indeed the lack of efficient and cost effective separation and purification methods. The major constituents of woody biomass (cellulose, hemicellulose and lignin) cannot be isolated simultaneously as polymers and several processes must be employed involving the degradation of at least one polymer. One approach is initially to treat the biomass to degrade hemicellulose by extraction or autohydrolysis in absence of mineral acids or caustics leaving both cellulose and lignin as essentially undegraded polymers [4, 5]. Autohydrolysis (hot water extraction) is of interest because water and biomass are the only reagents. The raw extracts from woody biomass consists of water soluble and insoluble substances mostly as monomers and oligomers of sugars, acetic acid, methanol, aromatic compounds, other low molecular weight extractable substances and fractions of lignin and residual particles[3,5].

Raw extracts produced by water treatment contains significant quantities of colloidal material. These particulates are composed mostly of lignin and its derivatives. These colloidal particles not only inhibit the fermentation activities of microorganisms but also foul any filtration membranes used for separation and purification of extracts. Separation and purification these components from hot water extracts is most important step in separation processes of biorefinery industry. The particulate phase of the wood extracts contains colloidal particles which can foul and plug membranes used in the separation and purification of wood extracts [6, 7]. Suspended colloidal particles are constantly and randomly bombarded from all sides by molecules of the liquid, making them move in a zigzag path. This type of movement is known as Brownian motion and increases in significance as particle size decreases. Since the mass of a colloidal particle is also small, its settling rate under the influence of gravity is slow. When the effect of Brownian motion dominates, it becomes very difficult and slow to separate the particles from the liquid by gravity sedimentation [6, 7].

Colloidal particles, which cannot be removed from a liquid by sedimentation within a short period of time (less than few hours), are typically converted into aggregates by coagulation or flocculation. The larger aggregates have more mass and the influence of gravity dominates over Brownian motion so that sedimentation occurs in a relatively short time. Flocculation and sedimentation of colloidal suspensions play an important role in separation of solids particles from liquid media. The particulate phase in wood extracts can be separated by treating with polyelectrolyte flocculating agents. Polymer induced flocculation enhances the separation of colloidal particles in wood extracts [6, 7, 8]. The effect of polymers on the colloidal stability depends on the peculiarities of the colloidal particle adsorption to the surface of the polymer. The polymers can destabilize the colloidal particles through charge neutralization, electrostatic patch and bridging flocculation. The time dependence and efficiency of the flocculation process is a function of many variables includes the structure of the molecule, its molecular mass, charge density, concentration of the polymer solution, content of the electrolytes and the mode of addition of polymer solution to suspension [8]. The kinetics of flocculation depends mostly on charge neutralization, and rate of adsorption (initial attachment) of polymer chains to the surface. The charge density and molecular weights of cationic polymers plays important role in the coagulation of negative particles. For oppositely charged polymers and particles, two main mechanisms can be involved in the particle flocculation i.e., charge neutralization and bridging flocculation. [9]. Low molecular weight and high charge density polymers such as poly-ethyleneimine (PEI), poly–diallyldimethyl ammoniumchloride (pDADMAC) are cationic polymers which are widely used for separation of colloidal particles. These polymers involve in aggregating the particles by charge neutralization and patch flocculation mechanisms. Cationic polyelectrolytes are subject to change in charge and size in solution upon alteration of pH and ionic strength. Furthermore, the absorbability of the polyelectrolytes on an oppositely charged surface may change with these solution properties. Since these polymers are polybase, addition of protons (reduction in pH) will result in protonation and subsequent expansion of polyion due to mutual charge repulsion. [8, 10]

In the present work, we report the flocculation dynamics of colloidal particles in sugar maple wood extracts by three flocculating agents: aluminum sulfate (alum), PEI and pDADMAC. Flocculation was followed by measuring the changes in turbidity, particle size and the settling behavior of the extracts. Final sediment volumes were also determined. The effect of changing extract pH, dosage of flocculants and the influence of electrolytes were studied.

2. Materials and Methods

The hot water extractions were carried out in a MK digester using 500 g oven dried sugar maple wood chips and 4:1 liquor (water) to wood ratio at 160°C for 2 hours. Particle size and zeta potential of the wood hydrolyzate were measured using a Brookhaven Particle Size and Zeta Potential Analyzer (90 Plus® and ZetaPlus®, Holtsville, NY). A Micro100 turbidimeter (HF Scientific Inc., Fort Myers, FL) was used to measure turbidity of the samples (Nephelometric turbidity units, NTUs). It was necessary to dilute the samples at least 10 fold to measure the turbidity, particle size and zeta potential. All the dilutions required were performed with filtrated (100 nm filter) reverse osmosis water.

The polyelectrolytes used for the study are alum, PEI and pDADMAC. Different concentrations of these polymers were added to the wood extracts for the study of flocculation kinetics. The concentrations of polymers were alum (0.01 M, 0.1 M, 0.25 M), PEI (0.5% and 1% v/v), pDADMAC (23.6 ppm and 47 ppm; weight of polymer per weight of extract) and a combination of alum and PEI (0.15 M + 0.5% v/v) was also used for the study. A 100ml of neat wood extract was taken in a glass beaker and flocculating agent was added and the mixture was agitated with magnetic stirrer. A 5 ml of sample mixture was collected to measure the particle size and zeta potential during process of mixing. Next the agitated mixture was processed for sedimentation in a 100ml graduated glass cylinder in a fixed position without any disturbance and turbidity of supernatant was measured for about 2 hours time period. Turbidity, particle size and zeta potential of the neat extract were measured initially for the reference. Besides the concentration of polymers, effect of wood extract pH on flocculation was also studied. The pH of wood extract was varied from 3.5 to 6.1 and 8.0 with dilute NaOH solution. The study was performed for various concentrations of alum, PEI and pDADMAC. The average particle size of colloidal particles in neat sugar maple wood hydrolyzate were around 260 – 290 nm and pH of the solution was 3.5. The zeta potentials of dispersed particles in extracts are between -18.6 to -21.0 mV which showed the particles are strongly anionic. The presence of negatively charged particles indicates that separation of these particles could be possible by flocculation with cationic polymers followed by sedimentation. The flocculation kinetics depends on several factors such as mixing conditions, adsorption on particles and concentration of polymers. The charge density and molecular weights of cationic polymers play an important role in coagulation of negatively charged particles.

3.1 Charge Neutralization Mechanism

The negatively charged particles in wood extract were neutralized by all the three flocculating agents. The effect was varied from polymer to polymer and also dosage of individual polymer. For oppositely charged polymers and particles charge neutralization mechanism is observed initially. The mechanism of charge neutralization was observed with all three polymers. A combination of polymers also showed the possibility of charge neutralization at optimal dosage.

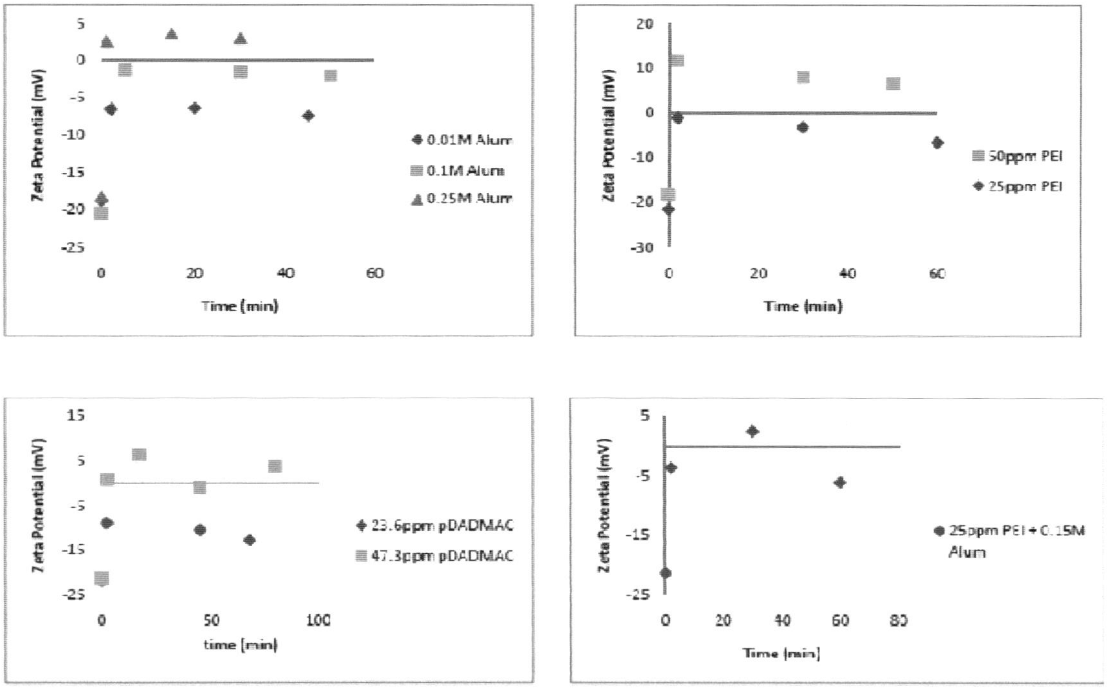

Fig 1. Zeta Potential of the particles vs time for different concentration of polymers (a) various concentrations of Alum (b)Various concentrations of PEI (c)Various concentrations of pDADMAC (d) Combination of PEI and alum

Fig.1. shows the effects of cationic polymer concentration on the zeta potential of colloidal particles. For lower concentration (0.01M) of alum charge was increased but not reached near the point of zero potential. For higher concentrations (0.1M and 0.25M) of alum the zeta potentials of the particles were increased significantly which were near to zero zeta potential which attributes to charge neutralization. For the polymers PEI the zeta potential of colloidal particles increases significantly as the polymer is added and the charge reversal was observed for optimal concentrations (0.5% and 1% PEI). For pDADMAC the zeta potentials were increased for higher concentration (47.3ppm) reached near zero zeta potential and charge reversal was observed but for lower concentration(23.6ppm) charge was increased but not significant. The combination of PEI and alum polymers showed that charge neutralization at optimal dosage (0.5% PEI and 0.15 M alum). The zeta potential of these two polymers combination was near zero zeta potential. The pH of the solution was changed from 3.5 to 3.2 with addition of alum and no was observed with PEI and pDADMAC.

3.2 Rate of sedimentation

The dynamics of flocculation was studied by rate of sedimentation of particles in wood extracts. The rate of flocculation represented by rate of increase of particle size and is dependent on polymer concentration. The average particle size in presence of polymer was above 3000nm with all the three polymers, but for 0.01M alum concentration the particle size 282nm and zeta potential were around -11.2 to -16.3 mV. The mechanism of patch flocculation was observed in process of sedimentation. Patching occurs when segments of polymer chain adsorb on different particles and help the particles to agglomerate into larger particles. The rate of sedimentation was measured by turbidity of supernatants of the solution and height of the sediment volume as a function of time.

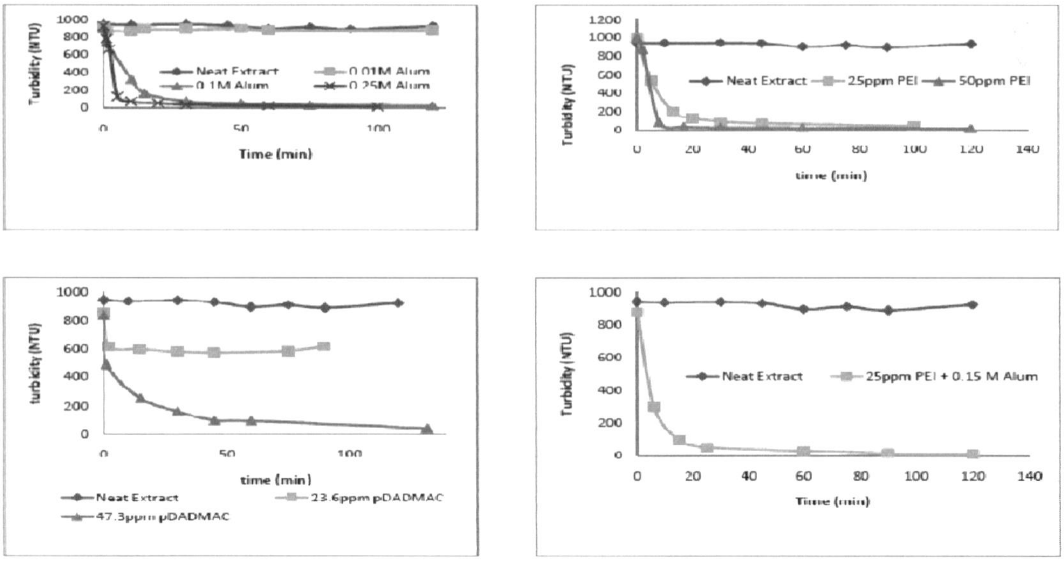

Fig 2. Turbidity of supernatant vs time for different concentration of polymers (a) various concentrations of Alum (b) Various concentrations of PEI (c) various concentrations of pDADMAC (d) Combination of PEI and alum.

Fig 2. shows the turbidity of supernatants for different concentrations of polymers as a function time and Fig 3. shows the height of the sediment volume in solution vs time for different concentrations of polymer.

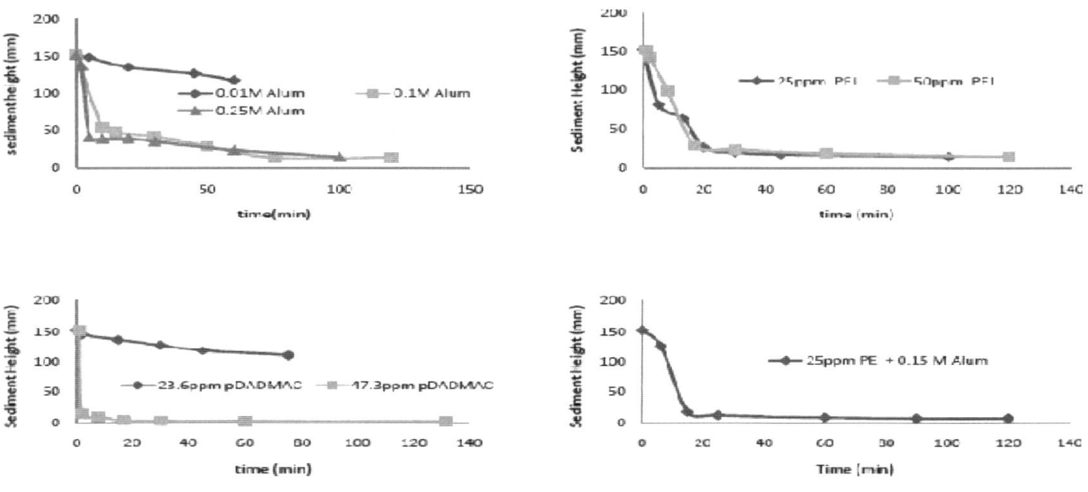

Fig 3. Height of the sediment vs time for different concentration of polymers (a) various concentrations of Alum (b) Various concentrations of PEI (c) various concentrations of pDADMAC (d) Combination of PEI and alum.

Sedimentation was rapid for higher concentrations of polymers compared to lower concentrations of polymers. The settling velocity of the particles was dependent on both charge neutralization and patching mechanism. The kinetics of flocculation varied from polymer to polymer, flocculation with PEI showed rapid sedimentation when compared with alum and pDADMAC. Sedimentation was not occurred for lower concentrations of alum (0.01M) and pDADMAC (23.6ppm). The height of the sediment volume was depended on rate of aggregation of particles (rate of adsorption of colloidal particles to polymer). Faster the aggregation and larger the particle size the sedimentation was rapid. For positively charged zeta potentials and near zero zeta potentials the settling velocity was rapid.

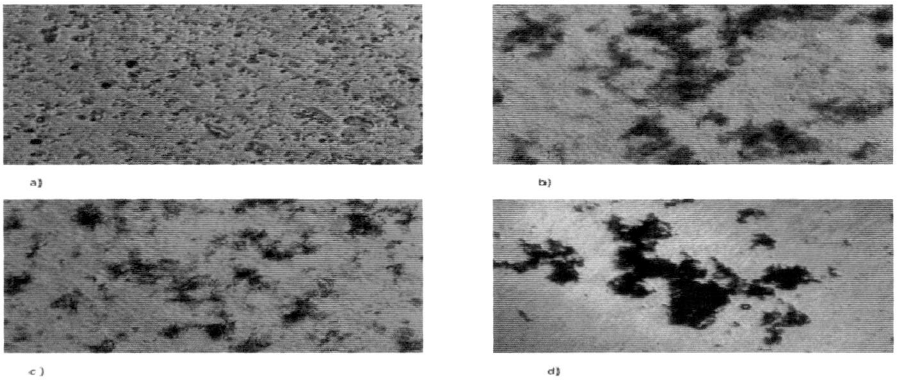

Fig 4. Microscopic images neat wood extract and flocculated particles by polymers a) neat wood extracts b) alum c)pDADMAC d) PEI

Fig 4. the microscopic images shows the size of particles were larger in presence of polymer than particles of neat wood extract. The flocculated particles with PEI were closely aggregated and larger than flocs of pDADMAC and alum. The sedimentation of the colloidal particles left the supernatants of the solution clearer which were represented by turbidity.

3.3 Effect of pH on polymer flocculation

The cationic polymers were sensitive to pH of the solution; they lose charge with variation in pH. We studied the effect of pH on three flocculating agents by varying the pH of wood extract. The normal pH of sugar maple wood extract was 3.5. The effect of pH on charge neutralization, it was observed that for pH 6.1 the zeta potential was increased to negative charge with low concentration and reduced negative charge for higher concentrations of PEI. The values of PEI at 0.5% and 1% concentrations were taken as controls for this study (fig 5).

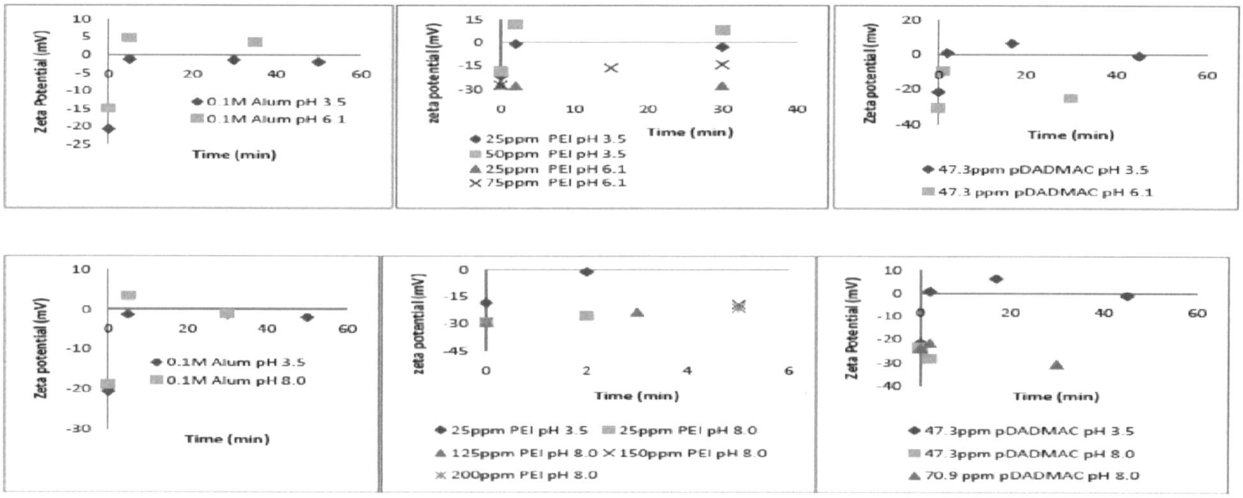

a) b) c)

Fig 5. Effect of pH on polymer: Zeta potential Vs time for different concentration of polymers at pH 6.1 and 8.0 (a) various concentrations of Alum (b) Various concentrations of PEI (c) various concentrations of pDADMAC.

The zeta potentials varied a lot for solution of extracts at pH 8.0. The average particle size of flocculated particles was larger than 3000nm for all concentrations of PEI at both pH conditions.

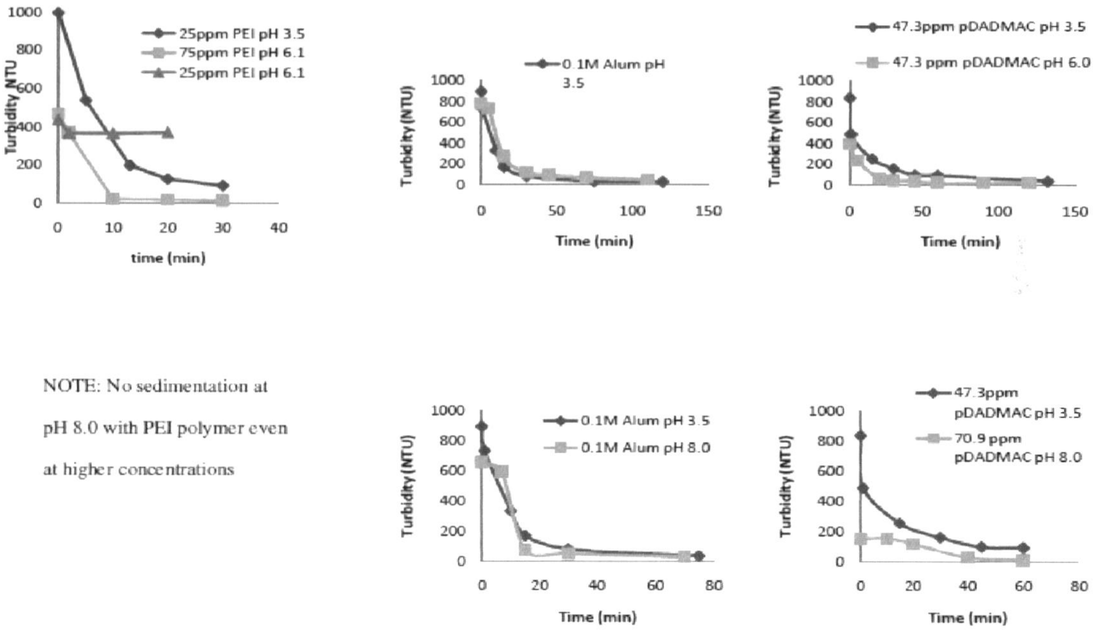

Fig6: Effect of pH on polymer: Turbidity of supernatant vs time for different concentrations of polymers at pH 6.1 and 8.0 (a) various concentrations of PEI (b) Various concentrations of Alum (c) various concentrations of pDADMAC.

The rate of sedimentation was rapid with higher concentration of PEI at pH 6.1 (fig 6). There was no sedimentation of particles at pH 8.0 even the size of particles are larger than 3000nm. The charge of pDADMAC (47.3ppm) polymer was reduced when the pH of wood extract was changed to 6.1 and 8.0. Zeta potentials of wood extract at these pH conditions were not similar to the pH 3.5 of same concentration (pH 3.5, 47.3ppm pDADMAC taken as control) of wood extract. The zeta potentials of aggregated particles were anionic and not even reached nearer to zero zeta potential values. The sizes of the aggregated particles were larger and the rate of sedimentation was faster at pH values of 6.1 and 8.0.

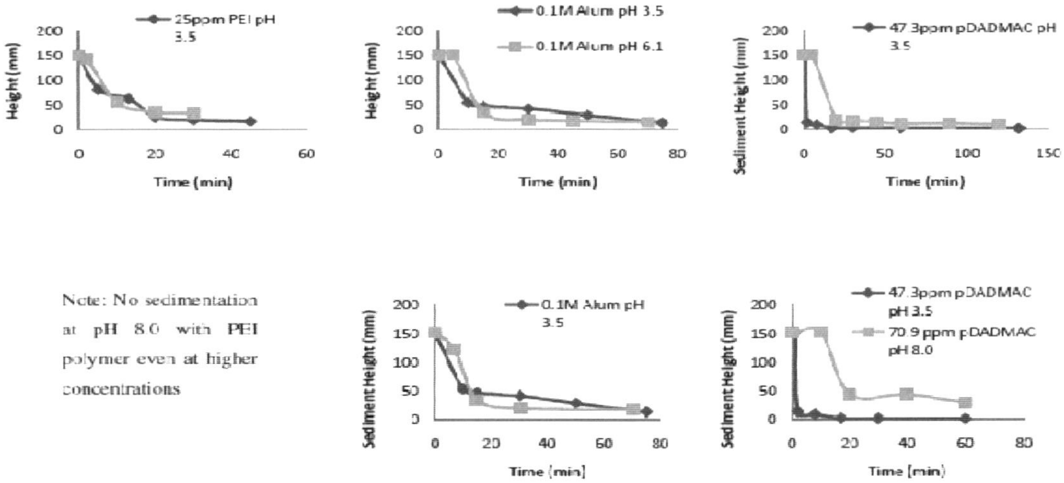

Fig 7: Effect of pH on polymer: Height of the sediment vs time for different concentration of polymers at pH 6.1 and 8.0(a) various concentrations of PEI (b) Various concentrations of Alum (c) various concentrations of pDADMAC

The turbidities of supernatants were reduced faster at these pH conditions than the original pH 3.5. But the heights of the sediment volumes were not lower at these pH conditions. This shows the initial rate of sediment was rapid at pH 6.1 and 8.0 and lowered with a period of time. The pH of the wood extract changed from 6.1 and 8.0 to 3.5 when treated with alum (0.1M). Zeta potential values at pH 6.1 and 8.0 were very nearer to zero zeta potential and approached values of zeta potential of control concentration (0.1M Alum, pH 3.5). This shows alum is not only flocculating agent but also a buffering agent. The rates of sedimentation were similar at these pH conditions also. The heights of sediment volumes and turbidity of supernatants were nearer with the control values. This study shows that polymers PEI and pDADMAC were sensitive to pH of the solution and alum was buffering agent.

4. Conclusion

This study shows the separation of colloidal particles (lignin and its derivatives) from hot water extracts of sugar maple wood extracts can be achieved by low molecular weight high charge density polymers. The dynamics of flocculation depends on both charge neutralization and patching mechanisms (adsorption of particles to the polymer). The optimal concentrations of polymers for separation of the colloidal particles from wood extracts of sugar maple were investigated by this study. The results showed the dynamics of flocculation was rapid with polymer PEI compared with alum and pDADMAC. Besides the dynamics of flocculation the sensitivity of pH on flocculating agents were investigated in this study. Low molecular weight and high charge density polymers PEI and pDADMAC were sensitive to pH of the solution and they lose charge with change in conditions of the pH. Alum acted as a buffering agent along with flocculation. Flocculation and separation of particulate phase from wood extracts can increase the efficiency of membranes of any filtration.

References

1. Kusch S., Morar M.V., 2009. Integration of lignocellulosic biomass into renewable energy generation concepts. ProEnvironment 2 (2009) 32-37.
2. Tunc, M. S., van Heiningen, A. R. P., 2008. Hemicellulose Extraction of Mixed Southern Hardwood with Water at 150°C: Effect of Time., Ind. Eng. Chem. Res., 47, 7031-7037.

3. Amidon.T.E and Liu.S. 2009. Water-Based Woody Biorefinery, Biotechnology Advances, 27(5), 542-550.
4. Jonsson A.S, Wallberg O. 2007 Cost estimation of kraft lignin recovery by ultrafiltration. Desalination 237 (2009) 254-267.
5. Grzenia L D, Schell .D. J, Wickramasinghe S.R. 2009. Detoxification of biomass hydrolystes by reactive membranes. J.Mem Sci ,348 , (2010) , 6-12.
6. Grzenia L D, Wickramasinghe S.R. 2007 . Adsorptive membranes and resins for acetic acid removal from biomass hydrolyzates , Desalination , 234, (2008), 144- 151
7. Huang, H. J., Ramaswamy, S., Tschirner, U. W., Ramarao, B. V. A review of separation technologies in current and future biorefineries. Sep. Pur. Tech. 62 (2008) 1-21.
8. Duarte G. V., Ramarao B.V, Amidon T.E,. Polymer induced flocculation and separation of particulates from extracts of lignocellulosic (sugar maple) materials. Bioresource Technology (under review).

(315e) Process Simulation of Pulp Mill-Based Integrated Biorefinery with Hemicellulose Pre-Extraction and Black Liquor Gasification

Hua-Jiang Huang and **Shri Ramaswamy**, Department of Bioproducts and Biosystems Engineering, University of Minnesota, Saint Paul, MN

Pulp and paper mill is a well-established major platform for effective and sustainable use of woody biomass. The main focus, however, has been around production of cellulose fiber (pulp and paper) and energy (steam and power) via combustion of black liquor. The current approach has some major disadvantages including partial dissolution and combustion of low heating value hemicelluloses and black liquor combustion and combined heat and power cycle is less efficient. Therefore, modification of the modern day pulp mills into integrated forest biorefineries (IFBR) by incorporating hemicellulose pre-extraction prior to pulping for biofuels and biochemicals production and black liquor gasification combined cycle (BLGCC) presents an excellent opportunity to produce, in addition to valuable cellulose fiber, co-products including fuel grade ethanol and additional energy. This is expected to contribute to increased revenue streams and profitability for the pulp and paper industry as well as potentially lowering the greenhouse gas emissions. Here we present a complex process model developed to simulate the pulp mill-based IFBR with hemicelluloses extraction prior to pulping for ethanol production as well as BLGCC. The simulation results can help better understand the economic and environmental benefits and impacts of modifying conventional pulp and paper mills into future integrated biorefineries and contribute to the design and development of future IFBR.

See more of this Session: Biorefinery: Separation & Transport in Lignocellulosic Conversion Processes
See more of this Group/Topical: Forest and Plant Bioproducts Division

(315f) Determining the Dilute Acid Hydrolysis of Natural and Hybrid Poplar Species Grown in Turkey

Bahattin Gürboy, Chemical of forest product, Istanbul University,Forest faculty, Istanbul, Turkey, Gülnur Mertoglu Elmas, Paper Techn.Program, Istanbul University Forest Faculty, Forestry Vacational High school, Istanbul, Turkey, öznur özden, Chemical of Forest Product, Istanbul Universitr Forest Faculty Engineering of Forest Industry, Istanbul, Turkey, Mahmut Bayramoglu, Chemical Engineering, Gebze Institute of Tech, Izmit, Turkey, Bahri Ovalý, Chemical engineering, Gebze Institute of Tech, Izmit, Turkey and Sacit Koçer, Poplar Research Institue, Izmit – Kocaeli, Turkey

The wood and annual plants that contain lignocellulosic raw material are formed by the natural polymers which contain the monomers that has mainly glucose, other hexose and pentose. It is known that ethyl alcohol is produced by hydrolysis of glucose as biomass from cellulosic material. In this respect bioethanol is the mostly considered source; however its cost still very high. In order to compete with other fossil fuels, the most economical production process should be developed for this biofuel. Another important point is the selection of proper raw material. It is thought that under the conditions of our country, wood, being a lignosellosic biomass, is more economical raw material than the other starch or sucrose based biomass. Chemical composition and morphologic structure of wood depends on biological variables of different species and growth conditions which effects hydrolysis rate and conversion to glucose . For high efficiency, biomass should be rich in cellulose. On the other hand for high hydrolysis rate, lignin ratio should be low and fiber structure should be suitable. The wood of broad-leaved tree has a lower lignin ratio compared to wood of coniferous tree. In this study, poplar tree, which is a broad-leaved tree growing in every region of Turkey and as fast grown species in the energy forests, has been chosen as lignocellulose biomass. At the first stage, samples of mostly widespread three species of poplar tree will be collected via field work. On these samples chemical composition and morphologic properties will be determined. At the second stage, wood chips will be hydrolysed under high temperature and pressure with dilute acid. For each species, the most suitable hydrolysis conditions will be determined. Afterwards hydrolysed samples will be measured viscosity,contained acid-ýnsoluble lignin in wood. Finally, the most suitable poplar species for bioethanol production will be selected. Hydrolysis experiments will be performed at a batch/continuous reactor will be used. Planning of experiments, analysis of results and determination of optimum conditions will be done by statistical methods. In this research, facilities of Istanbul University Faculty of Forestry and Gebze Institute of Technology will be utilized primarily. It is planned that in the research team there will be experts in the field of chemical engineering and forestry industries engineering.

See more of this Session: Biorefinery: Separation & Transport in Lignocellulosic Conversion Processes
See more of this Group/Topical: Forest and Plant Bioproducts Division

(387a) On Polydispersity of Plant Biomass Recalcitrance and Its Effects On Pretreatments

J.Y. Zhu, USDA Forest Service, Forest Products Laboratory, Madison, WI

abstract will be added later

See more of this Session: Biorefinery - Biochemical Conversion and Biomass Recalcitrance
See more of this Group/Topical: Forest and Plant Bioproducts Division

(387b) Application of An Integrated High Throughput Pretreatment and Enzymatic Hydrolysis (HTPH) Screening Tool to Investigate Radial Variation in Populus

Jaclyn D. DeMartini and Charles E. Wyman, Chemical and Environmental Engineering, University of California Riverside, Riverside, CA

A high throughput pretreatment and enzymatic hydrolysis (HTPH) method was developed to screen natural and genetically modified biomass types to identify those with lower recalcitrance to sugar release, define pretreatment conditions, and screen enzyme formulations. In support of this system, a scaled-down method was devised to accurately measure biomass compositions without prior calibrations, dramatically reducing the amount of material sacrificed for compositional analysis and allowing easier replication of samples. Application of these two methods facilitated the investigation of radial variation in composition, sugar release, and recalcitrance in two cross sections of Populus tremuloides. The data revealed that while substantial within tree radial variation existed for composition and sugar release (g sugar released / g biomass), sugar yields (g sugar released / g sugar available) were far less variable across the radial sections. Results from this study will be presented to provide insight on structural changes in wood resulting from juvenile to mature transitions that effect recalcitrance, results which may also have implications for determining the optimal harvest age of Populus trees for conversion to sugars.

See more of this Session: Biorefinery - Biochemical Conversion and Biomass Recalcitrance
See more of this Group/Topical: Forest and Plant Bioproducts Division

Biocatalysis and Biotransformation of Lignocellulose into Bioethanol: Process Intensification and Mechanism Research

Rongxin Su, Mingjia Zhang, Wei Qi and Zhimin He*

State Key Laboratory of Chemical Engineering, School of Chemical Engineering and Technology, Tianjin University, Tianjin 300072, PR China
**Corresponding Author's E-mail: surx@tju.edu.cn*

Keywords: Bioethanol; lignocellulose; Enzymatic hydrolysis; Simultaneous saccharification and fermentation; Machanism; Cellulase.

Extended Abstract

As one kind of non-renewable resources, fossil fuels will be exhausted in the near future. At the constant production and consumption, the presently known reserves of oil will last around 45.7 years, natural gas 62.8 years, and coal 119 years(1). In addition, global warming caused by greenhouse gas emission attribute mostly to the combustion of fossil fuels. Bioethanol is one of the most potential renewable energies. For large-scale usage, lignocellulosic material is a promising choice to produce a second generation bioethanol fuel(2). As the renewable source, lignocellulose not only has extensive sources, such as sugarcane bagasse, corn stover, corncob, wheat straw, sawdust, etc, but also significantly decrease net the carbon dioxide emission. However, there are a number of obstacles such as slow reaction rate, high production costs, non-cellulosic material block, and low ethanol concentration, which diminishing the enzyme performance and hinder the progress of biomass industrialization(3,4). To decrease reducing product inhibition effect on cellobiohydrolases and β-glucosidase, simultaneous saccharification and fermentation (SSF) is often used in bio-ethanol production by fermentable sugars to ethanol timely(5). At the same time, various strategies were performed for the purpose of cost-effective enzymatic conversion such as reducing enzymes production cost, producing cellulase with higher specific activity(6) and enzyme recycling including cellulase adsorption onto fresh substrate(7) and cellulase immobilization(8,9).

This paper summarizes our work on biocatalysis and biotransformation of lignocellulose into bioethanol aiming to enhance enzyamtic hydrolysis efficiency and increase final ethanol concentration, and the mechanism of cellulase on cellulose was also investigated.

(1) **Enzyme complex.** Hemicellulose and possibly pectin are thought to restrict the access of cellulases to their substrate in pretreated lignocellulose. Addition of xylanases and pectinases can degrade these non-cellulosic saccharides and thus increase cellulose conversion(10). The digestion of cellulose was increased by enzyme complexes of cellulase and accessory enzymes including β-glucosidase, xylanase and pectinase to hydrolyze cellobiose, hemicellulose and pectin, respectively (Fig. 1).

To enhance the conversion of the cellulose and hemicellulose, the corncob pretreated by aqueous ammonia soaking was hydrolyzed by enzyme complexes. Firstly, corncob were pretreated by aqueous ammonia soaking and the saturation limits for cellulase and β-glucosidase were determined, which is 15mg protein/g glucan (50FPU/g glucan) for cellulase (Spezyme CP). Second, the accessory enzymes (β-glucosidase, xylanase and pectinase) were supplemented to hydrolyze cellobiose (cellulase-inhibiting product), hemicellulose and pectin (the component covering the fiber surfaces), respectively. It was found that β-glucosidase (Novozyme 188) loading of 1.45mg protein/g glucan (30CBU/g glucan) was enough to eliminate the cellobiose inhibitor, and 2.9mg protein/g glucan (60CBU/g glucan) was the saturation limit. The supplementation of xylanase and pectinase can increase the conversion of cellulose and hemicellulose significantly. The yields of glucose and xylose enhanced with the increasing enzyme loading, but the increasing trend became low at high loading. Compared with xylanase, pectinase was more effective to promote the hydrolysis of cellulose and hemicellulose. The supplementation of pectinase with 0.12mg protein/g glucan could increase the yields of glucose and xylose by 7.5% and 29.3%, respectively.

(2) **Usage of Membrane bioreactor.** In order to reduce the high cost of enzymes, different strategies were used to decrease use cost by enzyme recycling and reuse. MBR is advantageous for cellulose enzymatic hydrolysis, because it can simultaneously remove the product inhibitor and reuse the cellulose enzyme(11).

Corn stover was chosen to be the substrate for both its abundance in China and low in price. The enzymatic hydrolysis of corn stover was performed in an integrated MBR. To remove the majority of lignin, pretreatment methods of soaking in aqueous ammonia (SAA) and in diluted sulfuric acid–sodium hydroxide (H_2SO4–NaOH) were used, and their effects were compared. Enzymatic hydrolysis of corn stover was performed in an integrated membrane bioreactor (MBR) incorporating a 10 KDa flat sheet polysulfone membrane to increase cellulose conversion and to reduce enzyme dosage. A reservoir (Mimi-reservoir RC800, Amicon, MI) was used to supply the buffer by pressurized gas to keep the reaction volume constant at 100 mL. (Fig. 2). Various pretreatment methods and semi-continuous MBR were examined to investigate their effect on the increasing productivity, product yield and enzyme utilization efficiency. By optimize the operation condition, cellulase, mainly cellobiohydrolase (CBH) can be recycled and the recovery efficiency was reached 85% by the process of cellulase adsorption onto fresh material; β-glucosidase can be reused four times by immobilization; and by the method of fed-enzyme membrane bioreactor (MBR), utilize efficiency was 3.31-fold of conventional batch reator (CBR), and 1.32-fold of fed-CBR for corn stover pretreated by diluted sulfuric acid–sodium hydroxide.

(3) **SSF or SSCF at high DM.** High dry matter is essential to increase production concentration(12). To obtain the cellulosic ethanol with high concentration, the combined pretreatments of corncob were performed as well as the simultaneous saccharification and fermentation (SSF)(13). The combination of acid and alkaline pretreatments (strong acid-strong-alkali or weak acid-weak alkali) was proved to be the effective pretreatment processes for production of ethanol with high concentration from corncob in SSF setup. On one hand, it can remove most of non-cellulosic material to make cellulose more accessible to the enzymes that convert the carbohydrates into fermentable sugars. On the other hand, the density of substrate increased after pretreatment, which did not increase viscosity of slurry so much and but make SSF at high DM concentration possible.

The ethanol concentration reached 69.2 g/L in the batch SSF with 19% DM using acid-alkali pretreated corncob, resulting in 81.2% overall ethanol yield. Moreover, a fed-batch process at high-solids concentrations was also carried out to alleviate the high viscosity problem caused by high DM concentration which is crucial in order to reach high ethanol concentration. Applying a prehydrolysis at 50°C, fresh substrate pretreated by dilute sulfuric acid-sodium hydroxide was added at different feeding strategies during the first 24h giving a final dry matter content of 25% (w/v). Under the SSF conditions (cellulose loading: 22.8 FPU/g glucan, dry yeast (Saccharomyces cerevisiae) loading: 5 g/L, substrate supplement every 4 hours), the highest ethanol concentration of 84.7g/L was obtained after 96h SSF, corresponded to 79% overall ethanol yield, which was higher than 4% (v/v), the low level of economically feasible distillation (Table 1, Table 2). To our knowledge, it is the highest ethanol concentration which has been reported and exceeds the technical and economic limit of the industrial-scale alcohol distillation.

Approximately 90% of the ethanol was produced during the first 48h of the fermentation. From a process-economic aspect, it may be desirable to shorten the residence time. It seemed that longer residence time was required at higher DM concentration, i.e. the hydrolysis yield was lower. This might be due to some inhibitory effects, e.g. an increase in end-product inhibition as the sugar uptake rate was lower, but could also depend on less efficient mass transport.

Table 1. Final ethanol concentrations and yields of fed-batch SSF at different interval adding time

Label	DM (%)	Interval adding time (h)	Ethanol conc. (g/L)	$Y_{Ethanol}$ (%)	$Y_{Cellulose}$ (%)
1	19+6	8	80.2	76.8	80.1
2	19+6	4	84.7	79.6	82.3
3	19+6	2	83.3	79.3	81.3

Table 2. Final ethanol concentrations and yields of fed-batch SSCF at different substrate additions

Lable	DM (%)	Substrate additions (%)	Ethanol conc. (g/L)	$Y_{Ethanol}$(%)	$Y_{Cellulose}$ (%)
1	10	1.0	52.74	87.03	89.27
2	10	2.5	55.45	85.26	87.60
3	10	5.0	57.04	80.42	83.59
4	10	7.5	59.22	81.46	76.16
5	10	10.0	60.52	81.39	70.11

Moreover, simultaneous saccharification and co-fermentation with a recombined Z.mobilis was carried on and a final ethanol concentration of 60.52 g/L was obtained for 10% DM supplementation giving a total DM of 25% (Table3).

Table 3. Final ethanol concentrations and ethanol yields of fed batch SSCF at different substrate additions

Substrate additions (%)	Ethanol conc. (g/L)	$Y_{Cellulose}$ (%)	$Y_{Ethanol}$(%)
1.0	52.74	89.27	87.03
2.5	55.45	87.60	85.26
5.0	57.04	83.59	80.42
7.5	59.22	76.16	81.46
10.0	60.52	70.11	81.39

Zymomonas mobilis was genetic improved for pentose utilization, and simultaneous saccharification and co-fermentation (SSCF) of lignocellulose was carried out at high dry matter with recombinant strain. A xylose-utilizing strain of Z. mobilis CP4 was constructed involving expression of XI (xylA), XK (xylB), TA (tal), and TK (tktA) genes from E.coli. It showed good performance on both soluble sugar mixture (glucose and xylose) and NaOH pretreated corncob. Using sugar mixture as substrate, nearly all the glucose and xylose were consumed, and an ethanol concentration of 22.5 g/L was achieved, corresponding to 86.3 % of theoretic ethanol yield based on glucose and xylose content. Then the recombinant strain was used in SSCF of NaOH pretreated corncob at high dry matter (DM) of 15%. The suitable operation condition was determined, which was Z.mobilis loading of

0.3 g/L at pH 5.5 and 30 °C, when the final ethanol concentration is 49.24 g/L. Higher final ethanol concentrations were obtained when SSCF operated at fed-batch mode. Several amounts of substrate were supplemented from 1% to 10%, and the highest final ethanol concentration, 60.52 g/L, was obtained for 10% DM supplementation giving a total DM of 25%.

In order to explore the reaction mechanism and direct the process development, multi-scale properties of substrate during enzymatic hydrolsis were investigated.

(1) **Reaction mode of CBH and EG.** Various researches have been carried out to understand the enzyme-substrate interactions, most of which were probed by the scission reaction kinetics derived from the measurements of soluble sugar release. By size exclusion chromatography (SEC), molecular weight (MW) and molecular weight distribution (MWD) of insoluble solid residues during hydrolysis can be determined(14). According to the changes in MWD and degree of polymerization (DP), the mechanism of cellobiohydrolase (CBH) and endoglucanases (EG) were extrapolated. As previously reported, EG randomly cleave internal glycosidic bonds of cellulose molecule resulting in a rapid decrease in the degree of polymerization and generating new chain ends upon which CBH acts, whereas in a progressive manner, CBH cleaves cellulose chains at the ends to release soluble cellobiose or glucose(15). In our work, enzymatic hydrolysis of cellulose by a complex cellulase was re-examined by using size exclusion chromatography coupled with multi-angle laser light scattering (SEC-MALLS) in order to gain insight into the course of hydrolysis. By the established method of SEC/MALLS coupled with differential refractive index, microcrystalline cellulose and filter paper cellulose were characterized to reflect different reaction pattern of cellobiohydrolase (CBH) and endoglucanase (EG) on substrate. As shown in Figure 3, no significant change in MCC chains was found during the whole reaction period, indicating that CBH digestion follows a layer-by-layer solubilization manner. This reaction mode might be the major reason for slow enzymatic hydrolysis of cellulose. On the other hand, the DP of FPC chains decrease rapidly in the initial reaction, indicating EG digestion follows a random scission manner, which could create new ends for CBH easily. Both the slopes of the conformation plots for MCC and FPC increased gradually, which indicated the stronger chain stiffness of cellulose during hydrolysis.

(2) **Key factors of substrate that limit cellulose hydrolysis.** The effects of several physicochemical properties on lignocellulose hydrolysis had been typically investigated in bulk papers, but which factors playing a decisive role on enzymatic degradation are still not well known. After different pretreatments, cellulase hydrolysis and measurements of physicochemical characteristics by column solute exclusion, particle size analysis, X-ray diffraction (XRD), Fourier transform infrared spectroscopy (FTIR) and solid state 13C nuclear magnetic resonance (NMR) were carried out in the present study. Partial least squares (PLS) was then applied to seek the key factors limiting the rate and extent of cellulose digestion. According to the PLS results, the most important factor for cellulose digestion was accessible interior surface area, followed by delignification and the destruction of the hydrogen bonds. The cellulose digestion at 2h and 24h were improved with the increased accessibility of interior surface area to the reporter molecules of 5.1nm diameter. Removal of lignin and breaking of hydrogen bonds were also found to significantly promote cellulose conversion. Other properties including the break-down of intramolecular hydrogen bonds, cellulose crystallinity and hemicellulose content had less effect on the efficiency of enzymatic hydrolysis. According to results of the partial least square analysis (PLA), the most important factor for cellulose digestion was accessible interior surface area, followed by delignification and the destruction of the hydrogen bonds (Fig.4).

Figures:

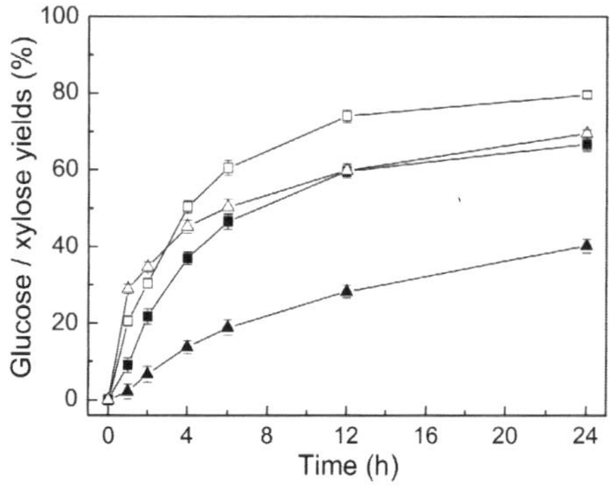

Figure 1. The yields of glucose (■) and xylose (▲) from enzymatic hydrolysis of SAA-treated corncobs by Spezyme CP and Novozyme 188, while those of glucose (□) and xylose (△) by such cellulases supplemented with Multifect Pectinase (0.12 mg protein/g glucan).

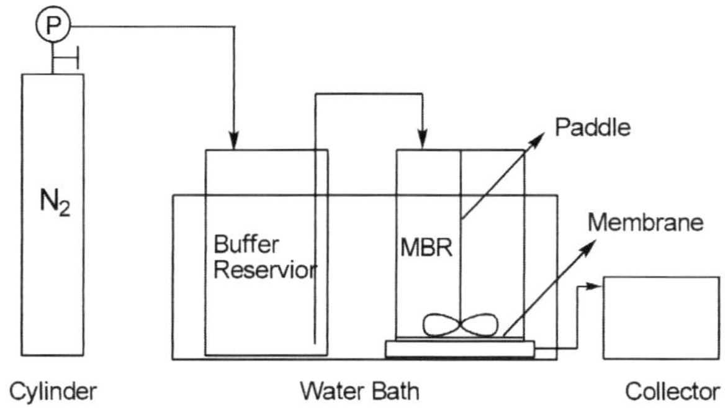

Figure 2. The schematic diagram of the membrane apparatus.

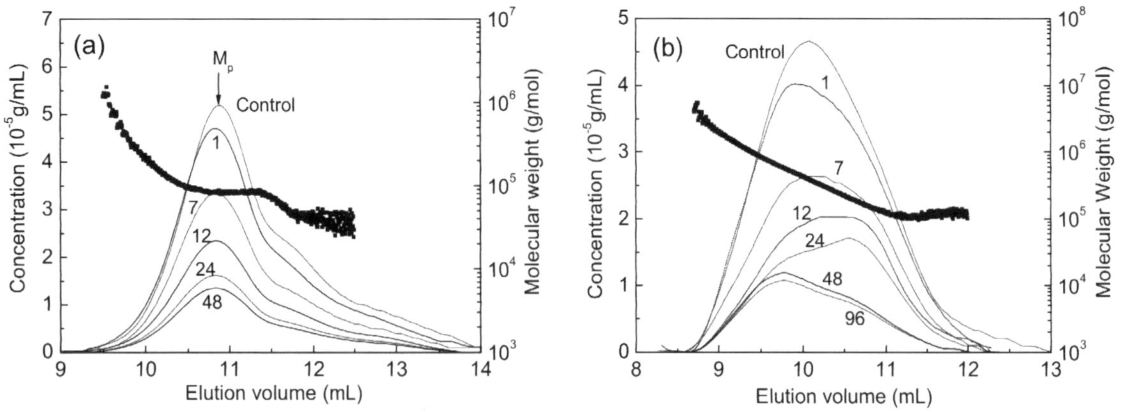

Figure 3. SEC/MALLS/RI chromatograms of (a) MCC and (b) FPC digested by cellulase for different time.

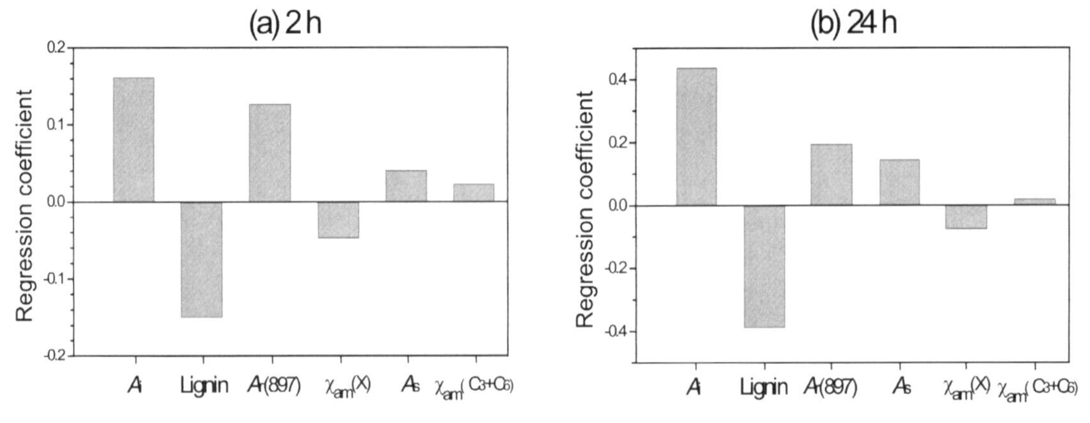

Figure 4. The result of regression coefficients of enzymatic digestibility with lignocellulosic characteristic parameters (a: 2h; b: 24h).

Acknowledgement: The authors acknowledge the financial supports received from Natural Science Foundation of China (Nos. 20976130 and 20806057), National Key Technology R&D program (2007BAD42B02), the International Science and Technology cooperation program (2006DA62400), and the Ministry of Education (Nos. IRT0641 and B06006).

References

1. BP. BP Statistical review of World Energy June 2010. available online (www.bp.com/statisticalrewiew). 2010.
2. Gray, K. A., Zhao, L. S. and Emptage, M. (2006), Bioethanol, *Current Opinion in Chemical Biology*, 10, 141-146.
3. Jorgensen, H., Kristensen, J. B. and Felby, C. (2007), Enzymatic conversion of lignocellulose into fermentable sugars: challenges and opportunities, *Biofuels Bioproducts & Biorefining-Biofpr*, 1, 119-134.
4. Zhang, Y. H. P. and Lynd, L. R. (2004), Toward an aggregated understanding of enzymatic hydrolysis of cellulose: Noncomplexed cellulase systems, *Biotechnology and Bioengineering*, 88, 797-824.
5. Tomas-Pejo, E., Oliva, J. M., Ballesteros, M. and Olsson, L. (2008), Comparison of SHF and SSF processes from steam-exploded wheat straw for ethanol production by xylose-fermenting and robust glucose-fermenting Saccharomyces cerevisiae strains, *Biotechnology and Bioengineering*, 100, 1122-1131.
6. Zhang, Y. H. P., Himmel, M. E. and Mielenz, J. R. (2006), Outlook for cellulase improvement: Screening and selection strategies, *Biotechnology Advances*, 24, 452-481.
7. Tu, M. B., Chandra, R. P. and Saddler, J. N. (2007), Recycling cellulases during the hydrolysis of steam exploded and ethanol pretreated lodgepole pine, *Biotechnology Progress*, 23, 1130-1137.
8. Tu, M. B., Zhang, X., Kurabi, A., Gilkes, N., Mabee, W. and Saddler, J. (2006), Immobilization of beta-glucosidase on Eupergit C for lignocellulose hydrolysis, *Biotechnology Letters*, 28, 151-156.
9. Li, C. Z., Yoshimoto, M., Fukunaga, K. and Nakao, K. (2007), Characterization and immobilization of liposome-bound cellulase for hydrolysis of insoluble cellulose, *Bioresource Technology*, 98, 1366-1372.
10. Berlin, A. and al., e. (2007), Optimization of enzyme complexes for lignocellulose hydrolysis, *Biotechnology and Bioengineering*, 97, 287-296.

11. Belafi-Bako, K., Koutinas, A., Nemestothy, N., Gubicza, L. and Webb, C. (2006), Continuous enzymatic cellulose hydrolysis in a tubular membrane bioreactor, *Enzyme and Microbial Technology*, 38, 155-161.
12. Jorgensen, H., Vibe-Pedersen, J., Larsen, J. and Felby, C. (2007), Liquefaction of lignocellulose at high-solids concentrations, *Biotechnology and Bioengineering*, 96, 862-870.
13. Ohgren, K., Rudolf, A., Galbe, M. and Zacchi, G. (2006), Fuel ethanol production from steam-pretreated corn stover using SSF at higher dry matter content, *Biomass & Bioenergy*, 30, 863-869.
14. Eremeeva, T. (2003), Size-exclusion chromatography of enzymatically treated cellulose and related polysaccharides: a review, *Journal of Biochemical and Biophysical Methods*, 56, 253-264.
15. KlemanLeyer, K. M., SiikaAho, M., Teeri, T. T. and Kirk, T. K. (1996), The cellulases endoglucanase I and cellobiohydrolase II of Trichoderma reesei act synergistically to solubilize native cotton cellulose but not to decrease its molecular size, *Applied and Environmental Microbiology*, 62, 2883-2887.

(387d) Ionic Liquid Solvent Properties as Predictors of Lignocellulose Pretreatment Efficacy

Thomas V. Doherty, Materials Science and Engineering, Rensselaer Polytechnic Institute, Troy, NY, J. Mauricio Mora-Pale, Dept of Chemical & Biological Engineering, Rensselaer Polytechnic Institute, Troy, NY, Sage E. Foley, Biology, Rensselaer Polytechnic Institute, Troy, NY and Jonathan S. Dordick, Chemical & Biological Engineering, Rensselaer Polytechnic Institute, Troy, NY

Effective pretreatment of lignocellulosic biomass is vital to its bioconversion to a usable liquid fuel. A growing body of work has focused on using room temperature ionic liquids (RTILs) to pretreat lignocellulose for subsequent fermentation. However, little is known about the physicochemical parameters of RTILs that promote effective pretreatment. In this work we examine the relationship between the Kamlet-Taft α, β, and π* solvent polarity parameters of different RTILs ([Emim] [OAc], [Bmim] [OAc], and [Bmim] [MeSO4]) and effective pretreatment of lignocellulosic biomass. We find the β parameter is an excellent predictor of pretreatment efficacy. Acetate containing RTILs ($\beta > 1.0$) remove > 32% of lignin from maple wood flour and significantly reduce cellulose crystallinity, resulting in > 65% glucose yields after 12 h cellulase hydrolysis. Pretreatment in [Bmim] [MeSO4] ($\beta = 0.60$) results in the removal of only 19% of the wood flour's lignin with no decrease in crystallinity, and no improvement in sugar yield over untreated wood flour. The addition of water and the dilution of the acetate anion with the methyl sulfate anion decrease the β value and subsequently have a negative impact on lignin extraction, cellulose crystallinity, and sugar yields.

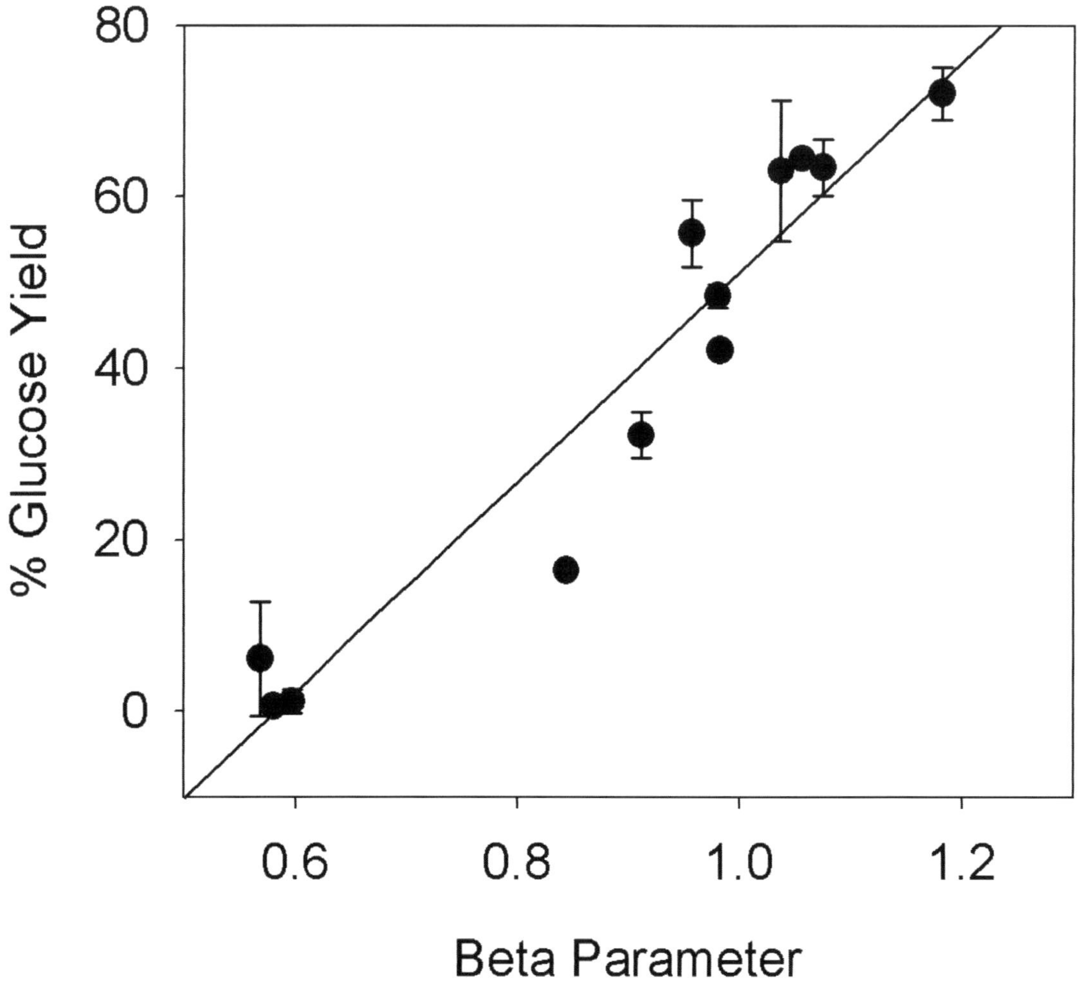

See more of this Session: Biorefinery - Biochemical Conversion and Biomass Recalcitrance
See more of this Group/Topical: Forest and Plant Bioproducts Division

ENZYMATIC BIOPROCESSING OF CELLULOSE IN NMMO

John Collier*, Subramanian Ramakrishnan, Samuel Grant, John Telotte
Rilwan Oyetunji, Brett Robbins, Daniel Morales, Elizabeth Yau, Gary Brodeur, & Kimberly Badal
Chemical and Biomedical Engineering
FAMU-FSU College of Engineering
Florida State University
Tallahassee, Florida 32310
jcollier2@fsu.edu

ABSTRACT

Enzymatic conversion of cellulose was accomplished by dissolving cellulose in the lyocell solvent, N-methylmorpholine N-oxide/water (NMMO/H_2O) near monohydrate, followed by gel formation caused by increased water content due to pH adjustment and enzyme addition. The initial dissolution and subsequent phase separation upon dilution is similar to the commercial Tencel® fiber production in which more than 99% of the solvent is recovered due to the essential lack of vapor pressure for NMMO/H_2O monohydrate. Previous rheological studies by the presenting author indicate that lignin is not soluble in NMMO/H_2O and can be separated by simple filtration, whereas cellulose and hemicelluloses are soluble in NMMO/H_2O as well as in the ionic liquids tested: 1-ethyl-3-methylimidazolium acetate ([Emim]Ac), 1-ethyl-3-methylimidazolium diethyl phosphate ([Emim]DEP), and 1-butyl-3-methylimidizoluim chloride ([Bmim]Cl). NMR, MRI and X-ray diffraction were used to characterize the formation of a gel phase for the cellulosic solutions and related diffusion characteristics after dilution with water. High cellulosic loading in the initial solvent (at least 15%w cellulose) and yields of reducing sugars (up to 90% conversion) were accomplished by dissolution in a modified twin screw extruder. Hydrolysis was similar in both a reactor after the extruder and in the extruder when equipped with recycle. The commercial enzyme mixtures (primarily Accelerase™1000 by Genencor) were more active in NMMO/H_2O than in the three ionic liquids tested or in aqueous systems with suspended cellulose. Initial reaction rates are much more rapid in the gel state than in suspension due to the accessibility of the cellulose by the enzymes. Two-way analysis of variance of the data (ANOVA) was performed using MATLAB. At the 95% confidence level, conversion of cellulose to sugars after dissolving in NMMO/H_2O were higher at higher cellulose concentrations, higher at lower pH, and dependent upon the general form of the cellulose (gel, suspended or regenerated). Also at the 95% confidence level, the conversion to sugars was higher in NMMO/H_2O than in the ionic liquids and higher than in cellulose suspended in H_2O, and the initial conversion rate in NMMO/H_2O was also higher than cellulose re-suspended after pretreatment with NMMO/H_2O. However, at the 95% confidence level, the overall conversion in NMMO/H_2O gels is not significantly different from cellulose re-suspended after pretreatment with NMMO/H_2O. Trend line analysis for R^2 correlation factors using Excel indicated most of the data indicated a dependence of conversion on the logarithm of time at the 95% confidence level except at 60 °C, high enzyme loading and high pH. The preliminary results at 60 °C and HPLC results imply subsequent reactions could follow hydrolysis in NMMO/H_2O.

INTRODUCTION

A chemical process is being developed to convert lignocellulosics to hydrocarbons or to H_2 and CO_2 using a continuous enzymatic saccharification (i.e. conversion to simple sugars) process.

The higher concentration needed for economic considerations are being formed in a twin screw extruder because these concentrations are highly viscous and viscoelastic, similar to the polymer melts for which the extruders are designed. The final goal is a continuous process using a second stage reactor fed by the extruder discharge to produce the hydrocarbons or H_2 without a difficult separation process. In this research, the lignocellulosic saccharification is being developed using dissolving pulp (degree of polymerization, DP, 1160) as an initial stage and the conversion to hydrocarbons or H_2 and CO_2 as a subsequent second stage. (DP is the number of glucose units in cellulose) The ultimate focus is on lignocellulosic sources from forest wastes or agricultural residues that have been accumulated at a processing facility for extraction of a different primary product, e.g. bagasse from sugar cane processing etc. There are currently two research grants supporting the saccharification phase of this research, a SunGrant through the US Department of Transportation [1] and a matching industrial grant from Bush Brothers & Co. [2] with in-kind support from various other industries.

Materials and Equipment

The primary solvent used is N-methylmorpholine oxide, NMMO, in its near monohydrate form. The structure of NMMO and the ionic liquids studied are shown in Figure 1.

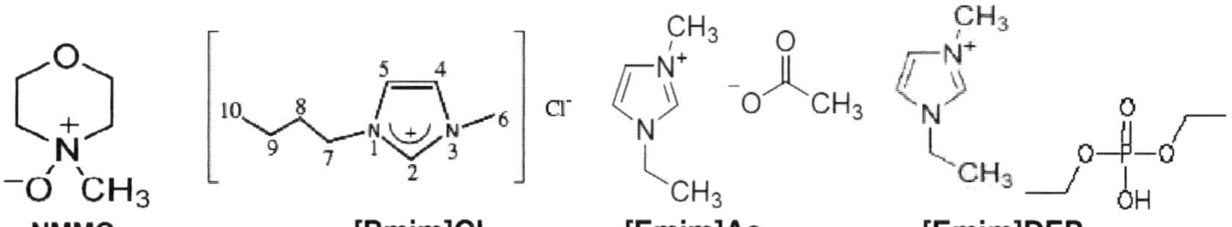

NMMO [Bmim]Cl [Emim]Ac [Emim]DEP

Figure 1: NMMO and Ionic Liquids

When interacting with cellulose, NMMO apparently is inserted between cellulose molecules that were previous bonded through hydrogen bonds with each other. Another solvent considered was the ionic liquid 1-butyl-3-methylimidizoluim chloride, [Bmim]Cl (Figure 1); however, due to strong ionic bonds, this solvent has been reported to deactivate enzymes [3], which was confirmed in this study. Two other ionic liquids considered were 1-ethyl-3-methylimidazolium acetate, [Emim]Ac, and 1-ethyl-3-methylimidazolium diethyl phosphate, [Emim]DEP, as shown in Figure 1. These solvents were obtained from Aldrich. The cellulosic source used was V-67 dissolving pulp from Buckeye Technology with an 1160 average DP and 98.3% alpha cellulose.

NMR and MRI measurement were made on an 11.75-Tesla (500-MHz) Bruker system. The HPLC were made on a 1200 series unit from Agilent technologies with a Bio-Rad HPX-87P column for plant sugars.

Phase Equilibrium Considerations

When dissolved in NMMO/H_2O or ionic liquids, cellulose is precipitated by adding water. However, NMMO/H_2O monohydrate and ionic liquids essentially have no vapor pressure. Therefore, the solvents easily and nearly completely can be recovered by evaporation of water. The processing conditions employed for dissolving cellulose in NMMO are dependent on the phase equilibrium of the quasiternary system of cellulose, NMMO and H_2O. Chanzy and coworkers [4] have characterized this behavior with the phase diagram reproduced in Figure 2, and a broader range binary diagram of NMMO and H_2O was characterized by Biganska and Navard [5]. NMMO is reported [5] to form three compounds in the solid state, the monohydrate (one H_2O molecules per NMMO molecule), either the disesquihydrate (2.5 H_2O molecules per NMMO molecule) or the dihydrate (two H_2O molecules per NMMO molecule), and the octahydrate (eight H_2O molecules per NMMO molecule) [4,5]. When the water content is intermediate between the mono and disesquihydrate, below the liquidus curve, and above 40 °C, a two phase region exists as shown in Figure 2. Because thermodynamic equilibrium is difficult to attain, data at 27.8% H_2O may indicate the compound disesquihydrate form with a eutectic that could be at the dihydrate composition of 23.5 °C. However, Biganska and Navard [5] reported the dihydrate rather than the disesquihydrate form. In the primary operative region for dissolving cellulose in NMMO/H_2O, the water content should be between 4 and 17% H_2O and above 80 °C. At these conditions and in the two phase region between the monohydrate (13.3% H_2O) and the dihydrate (23.5% H_2O) forms [6-8], cellulose prefers the near monohydrate composition.

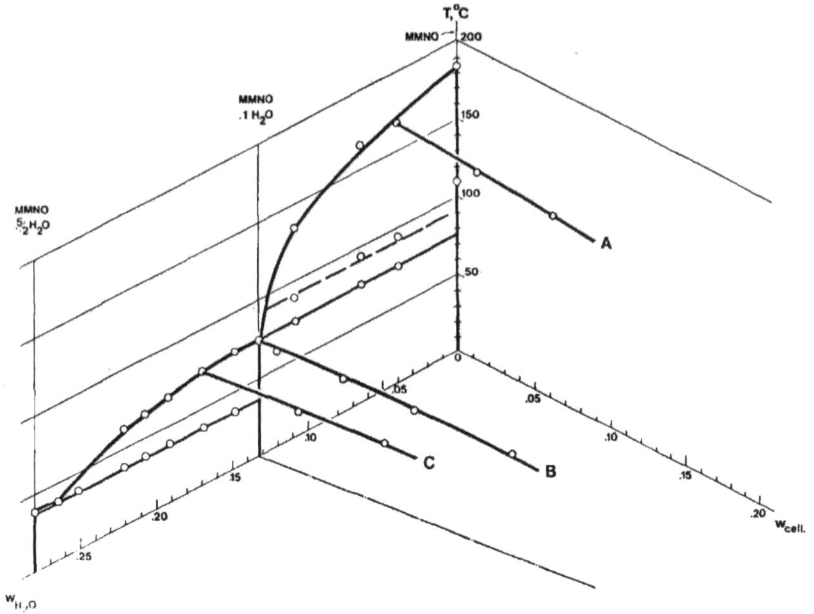

Figure 2: Phase Equilibrium of Cellulose in NMMO/H_2O from Chanzy et al.
(Note that Chanzy [4] uses MMNO rather than NMMO for N-methylmorpholine N-oxide.)

Conversely, cellulases favor a more aqueous liquid phase due to their general hydrophilic nature. For H_2O concentrations greater than 17%, cellulose is not soluble, and H_2O concentration less than 4% requires dissolution temperatures close to the decomposition point of NMMO [4]. The PI and his associates, as well as commercial applications, generally use the near monohydrate concentration (13.3%) and dissolve the cellulose above 80 °C. For removal of H_2O in compositions starting at greater than 13.3% and temperatures at and above 80 °C, the evaporation rates of H_2O slows significantly as the 13.3% is approached. To approach the anhydrous NMMO form requires extraction with acetone or other substance. The composition dependence of evaporation rates indicates that there is a strong association between H_2O and NMMO near the monohydrate compound composition even at temperatures above the compound melting point. For water contents between 13.3 and 27.8% H_2O, at and below 39 °C, a different two phase mixture of the monohydrate and the disesquihydrate exists. However, the phase equilibrium is sensitive to the presence of cellulose dissolved in NMMO/H_2O. For instance, the liquidus curve for 83%/17% NMMO/H_2O occurs at 75 °C, but with the same ratio of NMMO to H_2O, the liquidus curve occurs at 68 °C with 5% of 120 DP cellulose and at 65 °C with 10% of that cellulose. Both NMMO and the three ionic liquid solvents prefer water over cellulose. Therefore, cellulose is rejected from these solutions when water is added that is in excess of 17% H_2O for the NMMO solutions and any water for the ionic liquid solutions. The research team for this proposal has experience in dissolving cellulose, a hemicellulose (xylose) and organisol lignin in NMMO/H_2O and [Bmim]Cl and characterizing the rheology of both type solutions [9-11].

Enzyme Activity in Cellulose Solution

The commercial enzyme mixture from Genencor Spezyme® CP has been shown to enhance the saccharification of corn stovers in NMMO to 44.8% conversion to simple saccharides compared to the same enzyme in water at 32.1%, even though the NMMO solvent contained 25% water at 80 °C [12]. This water content is not in the range for dissolving cellulose; therefore, a higher yield and more rapid conversion should occur in the gel state. The enzyme from Clostridium acetobutylicum was reported to convert cellulose to simple sugars with 90% glucose yield [13]. The enzyme Trichoderma reesei in NMMO/H_2O has been reported to be effective in removing lyocell cellulosic sausage casings [14] and on saccharification of cellulose pretreated with either NMMO•H_2O [15] or [Bmim]Cl [16-18].

Genencor has released other cellulases, Accellerase™1000, Accellerase® 1500, and Accellerase® DUET for lignocellulosic biomass hydrolysis. Spezyme® CP and Accelerase™1000 have been the enzyme complexes of choice in this project. The pH and temperature ranges for these enzymes are critical, and for Accelerase™1000 in aqueous systems at 50 °C, the pH should be 4.8 to 5.2. The pH of the dissolving pulp cellulose dissolved in NMMO is near 11, thereby requiring buffering of the solutions after dissolving the cellulose. Various buffering compounds are used for enzymes including sodium acetate, and dibasic potassium phosphate mentioned in the technical bulletins from Genencor [19], and other compounds such as citric and acetic acids, and pectin. In this study, deionized water and acetic acid were used for buffering. Lignocellulosics contain pectin, lignin, and hemicelluloses that should have a lower pH in solution than dissolving pulp that is very pure cellulose. The lead author and colleagues have demonstrated that various lignocellulosics except their lignin components are soluble in NMMO monohydrate [20-24].

Gel Phase Formation in Cellulosic Solutions

A gel phase forms when H_2O is added to cellulose solution in NMMO/H_2O near monohydrate, and in the ionic liquids [Emim]Ac, [Emim]DEP, and [Bmim]Cl. Figure 3 is an MRI image of water added to a 2.3% cellulose in NMMO/H_2O near monohydrate, and Figure 4 is a photograph of vials of cellulose solutions in NMMO, [Emim]DEP. [Emim]Ac after water addition. Rheological data by the corresponding author and his colleagues [11] on [Bmim]Cl also indicate gel formation even without H_2O addition that is favored at lower temperatures, higher cellulose concentrations and higher cellulose DP, i.e. higher average molecular weight. NMR was used to determine diffusion coefficient changes (Figure 5) that indicate the progression of gel structural changes and also the effect of NMMO diffusion into the bulk H_2O phase at the top of the image. The diffusion values are appropriate for gels [25]. Repeated MRI and NMR measurements were recorded from the same sample longitudinally following the addition of H_2O to the top of the lyocell solution. Figure 3 displays regions of interest in the initial bulk H2O phase (ISA 1,2 & 3), bulk lyocell phase (ISA 5, 6 & 7) and the interface (ISA 4), which were quantified with respect to diffusion in Figure 5. Figure 6 contains MRI T_2-weighted time series images of the phase separation as H_2O was added to the cellulose/NMMO/ H_2O near monohydrate. In these images, the gel phase is the intermediate region between the upper H_2O rich phase and the lower cellulose and NMMO rich phase. After three hours (time stamp for each image in upper right), the gel phase has not quite reached the bottom of the test vial. The last image is shown in the same size as others for comparison and enlarged in color for more detail. Although there is apparent convective mixing in the H_2O rich phase, the viscoelastic gel and solution phases are not as affected.

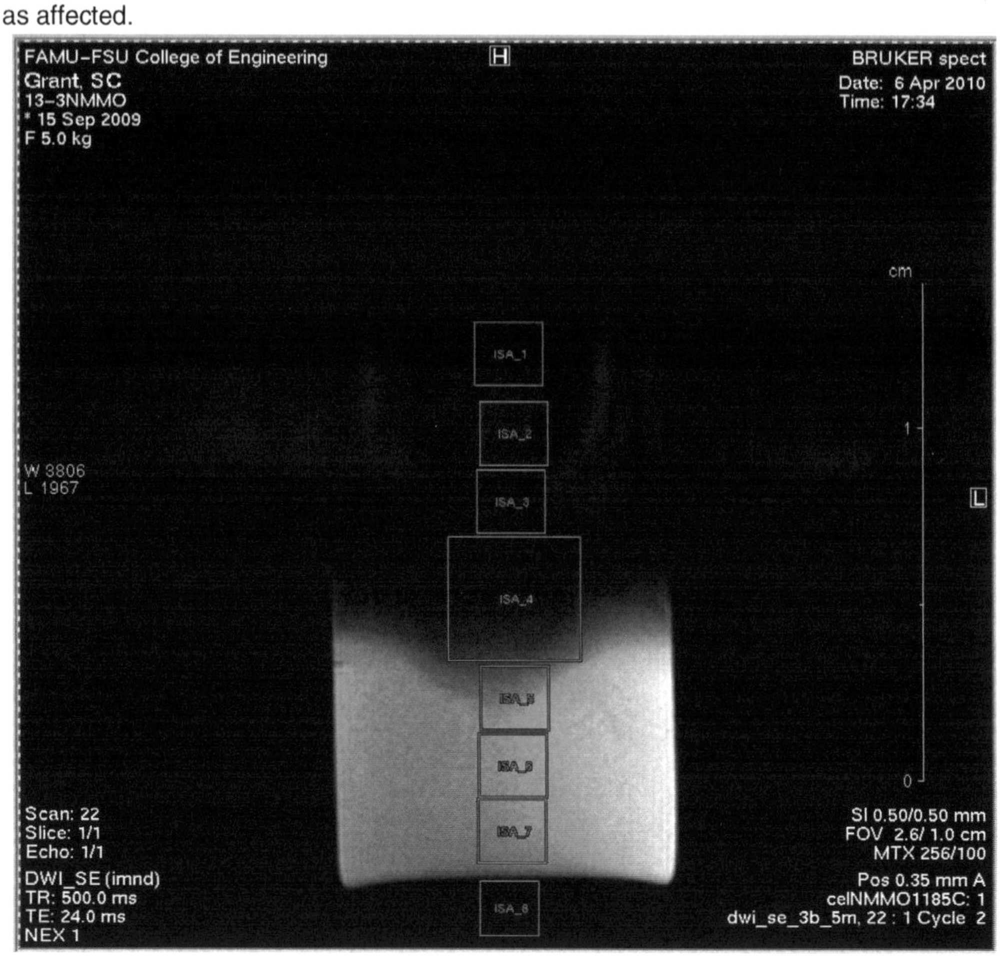

Figure 3: MRI of NMMO/H_2O/Cellulose, top H_2O, middle gel, bottom cellulose solution

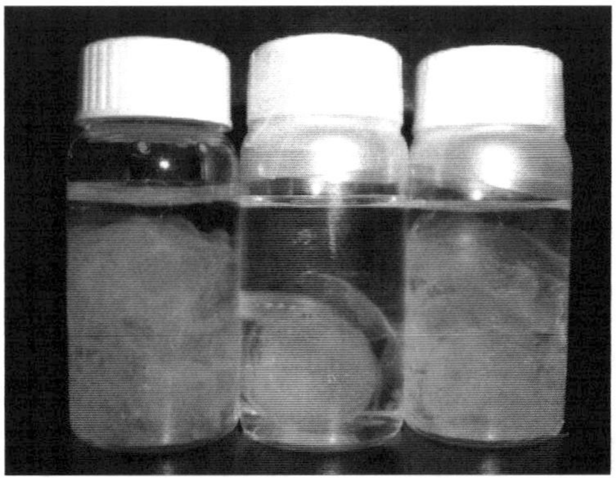

Figure 4: Gel Phase with H2O addition in NMMO, [Emim]DEP. [Emim]Ac

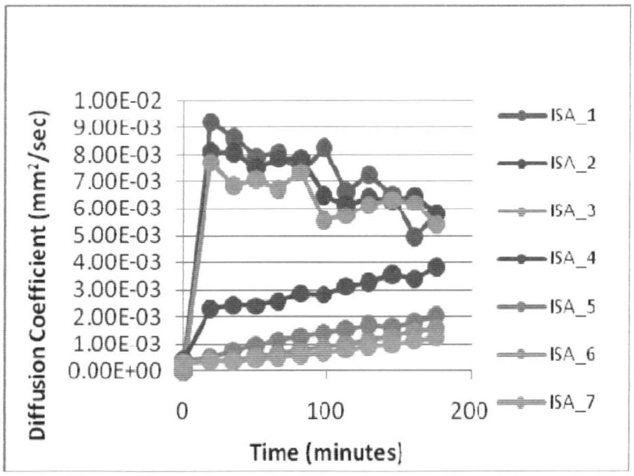

Figure 5: Diffusion Coefficients as function of time for ISA regions in Figure 3

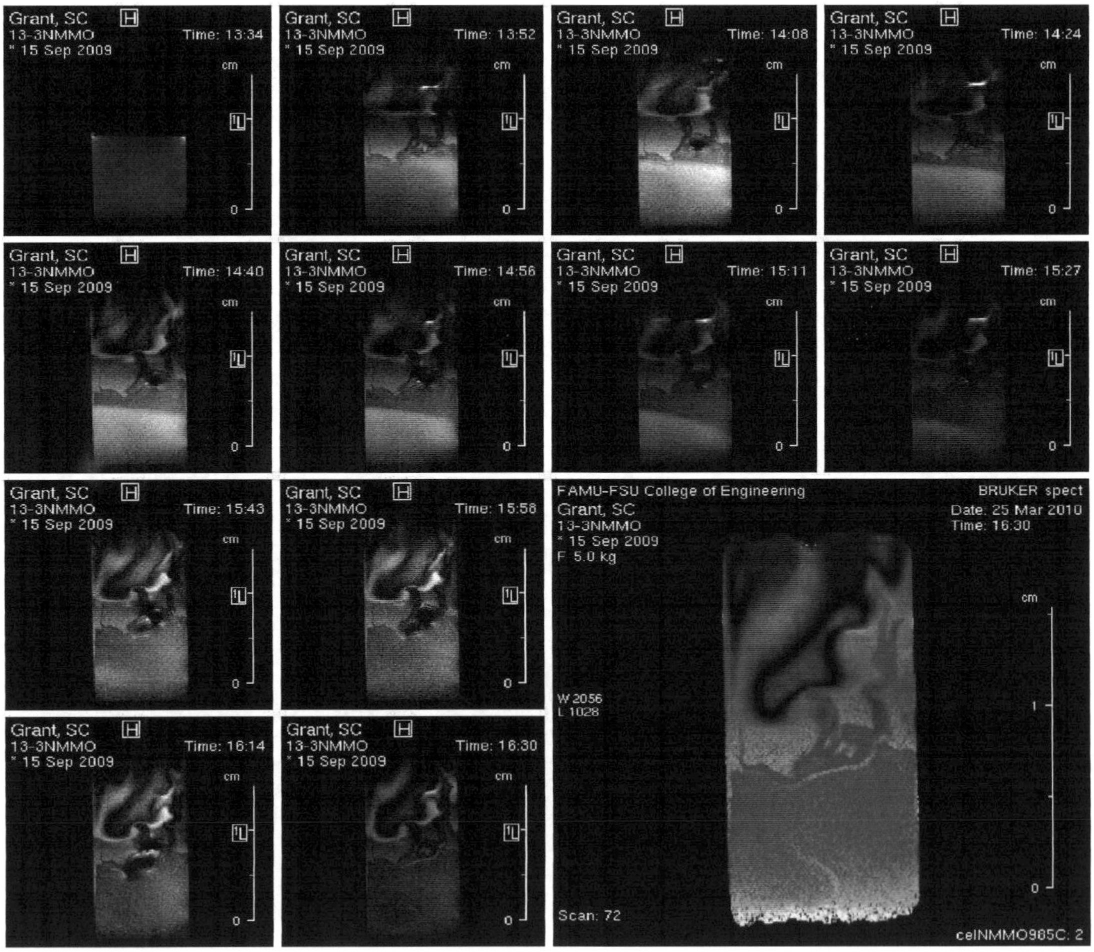

Figure 6: T_2 map: Regeneration of cellulose by adding H_2O to $NMMO/H_2O/Cellulose$

Hydrolysis Analysis

In the enzymatic results presented in Figures 7 through 11, the solutions were stirred when the aqueous enzyme and buffers were added in order to enhance the access of the cellulose by the enzymes. Room temperature X-ray diffraction measurements on the [Emim]Ac and [Emim]DEP gels indicate that these gels are non-crystalline. Both [Emim]Ac and [Emim]DEP solvents are still liquid at room temperature. However, as indicated in Figure 2, NMMO/H_2O near monohydrate solidifies before cooling to room temperature. As noted later in this paper the consolidation of the gel phase apparently affects the access of the cellulose by the enzymes.

The initial analysis procedure was a DNS assay in which the aromatic compound 3,5-dinitrosalicylic acid (DNS or DNSA) reacts with reducing sugars and other reducing molecules to form 3-amino-5-nitrosalicylic acid, which absorbs light strongly at 540 nm. A reducing sugar is any sugar that, in basic solution, forms aldehyde or ketone groups. Reducing sugars include glucose, fructose, mannose, lactose and maltose, but not sucrose. For the DNS assay, it is assumed that all of the glucosidic groups in cellulose are converted to a monosaccharide, which has one aldehyde or ketone group per molecule. If disaccharides such as lactose and maltose that form only one aldehyde or ketone were present, then a lower apparent conversion would be indicated by the DNS assay. Also if non-reducing sugars or other subsequent reaction of the monosaccharide molecules occur, the apparent conversion indicated by the DNS assay would be even lower. The agreement of DNS assay and high performance liquid chromatography (HPLC) measurements is shown in Figure 7 for 1% cellulose in NMMO/H_2O monohydrate gel with an enzyme concentration of 122 FPU/g, pH of 5.7 and at 50 °C. Enzyme activity is characterized in terms of Filter Paper Units (FPU), where a FPU is defined as the activity that releases 1 μmol of glucose equivalents per minute from a Whatman No. 1 filter paper.

Enzymatic Conversion of Cellulose

As shown in Figure 8, the conversion of cellulose to reducing sugars is much more rapid in the NMMO/H_2O gel than for cellulose suspended in H_2O with the same cellulose and enzyme concentration, temperature and pH. The conversion is also more rapid in the first five hours when the cellulose is in the gel state with NMMO/H_2O than when the cellulose was pretreated with NMMO/H_2O, the solvent removed, the sample dried, and then suspended in H_2O. Figure 9 indicates that the same enzymes in NMMO/H_2O under the same conditions are more active both initially and at longer times in NMMO/H_2O than in any of the three ionic liquids studied. The ionic liquid [Emim]AC has similar initial reactivity as NMMO probably because both NMMO/H_2O and [Emim]AC form a relatively stable open gel phase upon dilution by buffering agents and aqueous enzyme suspensions. However, [Emim]AC does not have as high of conversion even during the first measurement at half an hour and continues to lag significantly behind NMMO for the full 24 hours of the experiments. The ionic liquid [Emim]DEP has significantly lower initial conversion than NMMO and [Emim]AC probably because a more consolidated gel phase forms upon dilution by buffering agent and aqueous suspensions as shown in Figure 4, making the cellulose less accessible to the enzymes. After about six hours, the [Emim]DEP begins to outperform [Emim]AC, probably after the enzymes have significantly penetrated the gel. The third ionic liquid, [Bmim]Cl, is not shown on the graph because it deactivated the enzymes immediately, and no significant conversion occurred. Figure 10 indicates that the activity of

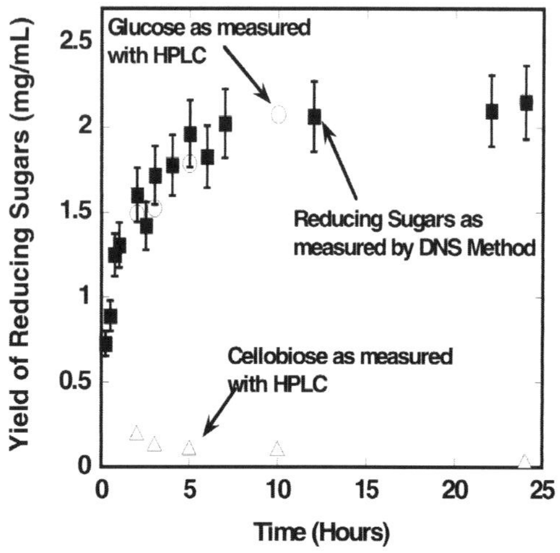

Figure 7: Comparison of DNS to HPLC

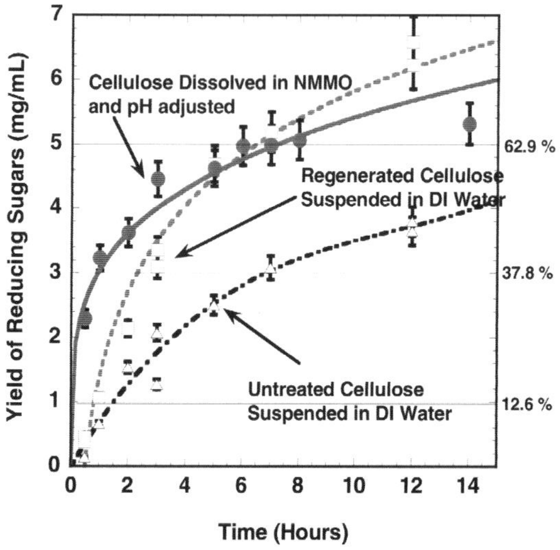

Figure 8: Effect of regeneration of cellulose, pH=5.7, T=50° and C122 FPU/g

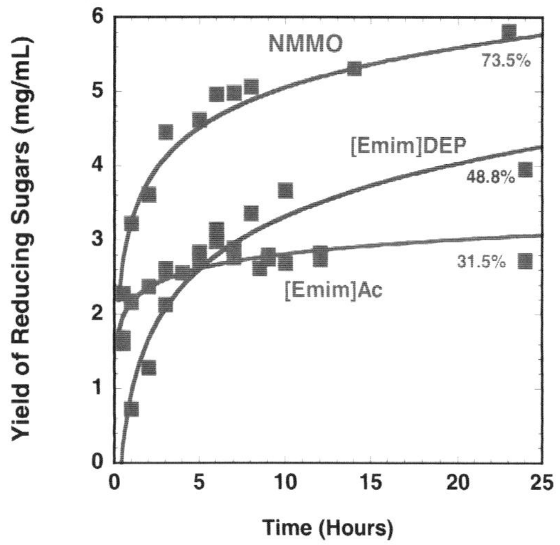

Figure 9: NMMO & Ionic Liquids.
7.22 mg/mL cellulose, T = 50 °C, 122 FPU/g

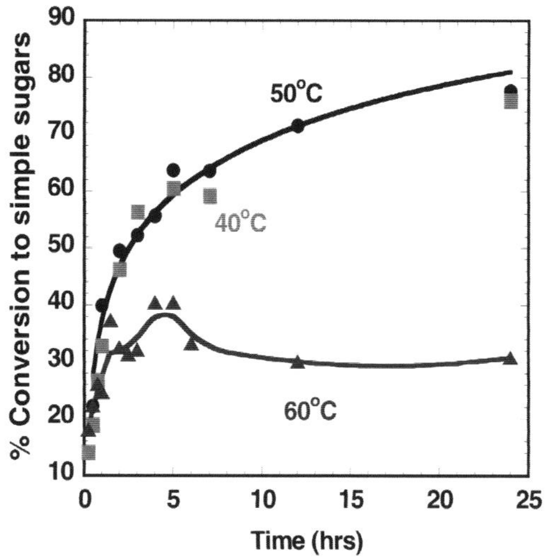

Figure 10: Temperature Effect.
1% Cellulose, 122 FPU/g, pH 5.7

enzymes NMMO/H_2O is not significantly different at 40 and 50 °C but is significantly lower at 60 °C with an apparent maximum around five hours. As noted earlier, the DNS assay only measures the concentration of ketone and aldehyde groups, and assumes conversion to monosaccharides. If subsequent reactions occur or other saccharides are present, then the DNS assay could exhibit a maximum, and higher temperature would favor subsequent reactions. Some subsequent reactions also would be favored if the monosaccharides had a significant fraction in the chain rather than cyclic form. The cyclic form is strongly favored in near neutral pH aqueous systems. However, as the HPLC preliminary measurements depicted in Figure 11 for NMMO at 50 °C may suggest, the chain structure is more prevalent in the more basic NMMO solutions. Also the pH increases with conversion for higher cellulose concentrations. Cellulose consists almost entirely of glucosidic units, and therefore, the presence of mannose in Figure 11 indicates that at least epimerization reactions are occurring in the NMMO system. Epimerization reactions of the chain form of glucose to the chain form of fructose occur in pH basic systems, and as a result, the stereo-site characterizing the difference between glucose and mannose loses its stereo-specific character. The fructose then can transform back to glucose or to mannose, and an equilibrium of glucose, mannose and fructose could results.

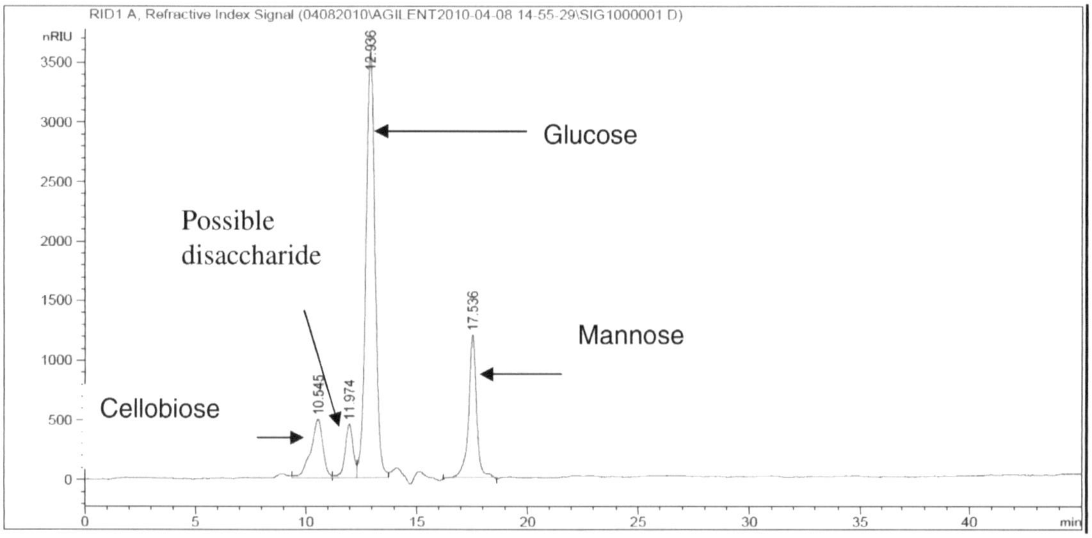

Figure 11: HPLC of NMMO hydrolysis results

Conclusions

The enzymatic conversion of cellulose to reducing sugars is more effective in NMMO/H_2O systems because cellulose form a gel phase upon dilution with buffering agents and aqueous enzyme suspensions. Although rapidly separating from the solvent upon dilution, the cellulose does not solidify immediately. As a result, the cellulose appears to be more accessible and reactive than suspended partially crystalline cellulose in aqueous systems with the same enzymes. Except for the initial effect of gel consolidation on reactivity that forms in [Emim]DEP, the four solvent systems tested affect the enzyme activity as anticipated based upon their bonding characteristics [26]. The activity of the three ionic liquids is apparently dependent on the character of their ionic bonds. [Bmim]Cl with a small Cl anion is the most detrimental to the enzymes, exhibiting no reactivity [3]. [Emim]AC with a larger acetate anion that has a less localized charge than a Cl anion is not as detrimental on the enzymes immediately, but has limited long time conversions. [Emim]DEP, with the largest and least localized charge on the anion, is the more effective than [Emim]Ac at longer times but still displays much less activity than NMMO that does not have ionic bonding.

There may be subsequent reactions of the reducing sugars in the NMMO/H_2O system indicating the potential for carrying out further reaction in this solvent system. These subsequent reactions are the focus of additional work that takes advantage of the solubility of reducing sugars in NMMO/H_2O, thus avoiding a difficult separation step. If such reactions can be performed creating alcohols, H_2 or hydrocarbons that have significant vapor pressures, these products could be removed in the solvent regeneration scheme, which would behave as a steam distillation separation. As such, the recovery of low vapor pressure NMMO/H_2O by water evaporation (yielding more than 99% solvent recovery) could be coupled with the direct recovery of highly desired biofuels.

Two-way analysis of variance of the data (ANOVA) was performed using MATLAB. At the 95% confidence level, conversion of cellulose to sugars after dissolving in NMMO/H_2O were higher at higher cellulose concentrations, higher at lower pH, and dependent upon the general form of the cellulose (gel, suspended or regenerated). Also at the 95% confidence level, the conversion to sugars was higher in NMMO/H_2O than in the ionic liquids and higher than in cellulose suspended in H_2O, and the initial conversion rate in NMMO/H_2O was also higher than cellulose re-suspended after pretreatment with NMMO/H_2O. However at the 95% confidence level, the overall conversion in NMMO/H_2O gels is not significantly different from cellulose re-suspended after pretreatment with NMMO/H_2O. Trend line analysis for R^2 correlation factors using Excel indicated a dependence of conversion on the logarithm of time at the 95% confidence level except at 60 °C, high enzyme loading and high pH. The preliminary results at 60 °C and HPLC results imply subsequent reactions could follow hydrolysis in NMMO/H_2O. Confirmation of these preliminary results is underway.

.

References

1. US DOT Sun Grant, Enzymatic and Multiphase Solution Processing of Lignocellulosic Biomass, Subramanian Ramakrishnan, John Collier, Sam Morton, $250,000, 10/1/07-9/30/10, FAMU/FSU through FAMU.
2. Bush Brothers & Company, Industrial Support for Biomass Research, John Collier, $50,000, 10/1/07, FAMU/FSU through FAMU, match to Sun Grant.
3. Turner, Megan B., Scott K. Spear, Jonathan G. Huddleston, John D. Holbrey and Robin D. Rogers, Ionic liquid salt-induced inactivation and unfolding of cellulase from Trichoderma reesei, Green Chemistry, 5, 443–447, 2003.
4. Chanzy, H.; Nawrot, S.; Peguy, A; Smith, P. "Phase Behavior of the Quasiternary System N-Methylmorpholine-N-Oxide, Water, and Cellulose." Journal of Polymer Science, 20, 1909-1924, 1982.
5. Biganska, Olga, and Patrick Navard, Phase diagram of a cellulose solvent: N-methylmorpholine-N-oxide-water mixtures, Polymer 44, 1035-1039, 2003.
6. Rosenau, T; Hofinger, A.; Potthast, A; Kosma, P. "On the Conformation of the Cellulose Solvent N-Methylmorpholine-N-oxide (NMMO) in solution." Polymer. 44, 6153-6158, 2003.
7. Kim, S.O.; Shin, W.J.; Cho, H. Kim, B.C.; Chung, I.J. "Rheological Investigation on the Anisotropic Phase of Cellulose-MMNO/H20 Solution System." Polymer, **40**, 6443-6405, 1999.
8. Laity, P.R.; Glover, P.M.; Hay, J.N.; "Composition Phase Changes Observed by Magnetic Resonance Imaging During Non-solvent Induced Coagulation of Cellulose." Polymer, **43**, 5827-5837, 2002.
9. Sammons, Rhea J., John R. Collier, Timothy G. Rials, Simioan Petrovan, Rheology of 1-

Butyl-3-Methylimidazolium Chloride Cellulose Solutions. I. Shear Rheology, Journal of Applied Polymer Science **110**, 1175–1181, 2008.
10. Collier, John R., Joy L. Watson, Billie J. Collier, and Simioan Petrovan, Rheology of 1-Butyl-3-Methylimidazolium Chloride Cellulose Solutions. II. Solution Character and Preparation, Journal of Applied Polymer Science, **111**, 1019–1027, 2009.
11. Sammons, Rhea J., John R. Collier, Timothy G. Rials, Simioan Petrovan, Rheology of 1-Butyl-3-Methylimidazolium Chloride Cellulose Solutions. III. Elongational Rheology, Journal of Applied Polymer Science, **110**, 3203-3208, 2008.
12. Burdette, Jill, Brian Vande Berg, Brian Carr, Nicholas B. Duck, Michael G. Koziel, Nadine Carozzi, and Paresma R. Patel, US Patent 20040231060, Method to enhance the activity of lignocellulos-degrading enzymes, November 25, 2004.
13. Lee, Song, F., Cecil W. Forsberg, and L. N. Gibbens, Cellulolytic Activity of Clostridium acetobutylicum, Applied and Environmetal Microbiology, **50**, (2), 220-228, 1985.
14. Lacoste-Bourgeacq, Jean-Francois, Skeeter Meranda Bassue-Hyligar, Roger Aceto, Method for removing cellulosic casing from sausages, US Patent 5914141, March 28, 2000.
15. Carrillo ,F. X. Colom, M. Lo´pez-Mesas, M.J. Lis, F. Gonza´lez, J. Valldeperas, Cellulase processing of lyocell and viscose type fibres: kinetics parameters, Process Biochemistry **39,** 257-261, 2003.
16. Dadi, P., Sasidhar Varanasi, Constance A. Schall, Enhancement of Cellulose Saccharification Kinetics Using an Ionic Liquid Pretreatment Step, Biotechnology and Bioengineering,. **95**, 904-910, 2006.
17. Liu, Liying, Hongzhang Chen, Enzymatic hydrolysis of cellulose materials treated with ionic liquid [BMIM]Cl, Chinese Science Bulletin., **51,** 2432—2436, 2006.
18. Liying, Liu, & Chen Hongzhang, Enzymatic hydrolysis of cellulose materials treated with ionic liquid [BMIM]Cl, Chinese Science Bulletin, **51**, 2432-2436, 2006.
19. Accellerase™1000, Genencor® Product Information, October 2007.
20. Dever, Maureen, Billie J. Collier, Simioan Petrovan, and John R. Collier, Lyocell Solutions from Undervalued Cellulose Sources, Clothing and Textiles Research Journal, **21** (4), 167-173, 2003.
21. Collier, B.J. , M. Dever, S. Petrovan, J.R. Collier, Z. Li, and X. Wei, Rheology of Lyocell Solutions from Different Cellulose Sources, Journal of Polymers and the Environment, **8**, (3), 151-154, 2000 (©2002).
22. Petrovan, Simioan , Ioan I. Negulescu and John R. Collier, Elongational and Shear Rheology of Cellulosic and Lignocellulosic Solutions in N-Methylmorpholine Oxide Monohydrate, Cellulose Chemistry and Technology, **35** (1-2), 89 -102, 2001.
23. Petrovan, S., J.R. Collier, and I.I. Negulescu, Rheology of Cellulosic N-Methylmorpholine Oxide Monohydrate Solutions of Different Degrees of Polymerization, Journal of Applied Polymer Science, **79**, 396-405, 2001.
24. Petrovan, S., J.R. Collier, and G.H. Morton, Rheology of Cellulosic N-Methylmorpholine Oxide Monohydrate Solutions, Journal of Applied Polymer Science, **77**, 1369-1377, 2000.
25. Simpson, N. E., S. C. Grant, S. J. Blackband and I. Constantinidis, NMR Properties of Alginate Microbeads, Biomaterials. **24** (27), 4941-4948, 2003.
26. Lungwitz, Ralf, Veronika Strehmel and Stefan Spange, The dipolarity/polarisability of 1-alkyl-3-methylimidazolium ionic liquids as function of anion structure and the alkyl chain length, New J. Chem., 34, 1135–1140, 2010.

(387f) Comparative Molecular Dynamics Study of Cellulose I-Beta and III(I) Allomorphs

Giovanni Bellesia, T6 and CNLS - Los Alamos National Laboratory, Los Alamos, NM, Shishir Chundawat, Chemical Engineering and Materials Science, Michigan State University, Lansing, MI, Paul Langan, Biosciences Division - Los Alamos National Laboratory, Los Alamos, NM and Sandrasegaram Gnanakaran, T6 - Los Alamos National Laboratory, Los Alamos, NM

Recent experiments show that when cellulose I-beta is converted into cellulose III(I) (via liquid NH3 pretreatment) the enzymatic hydrolysis rate increases 2-5 times. Interestingly the same experimental studies show a reduced non--specific enzyme binding for cellulose III(I) respect to cellulose I-beta.

Motivated by these experimental findings, we present the first comparative, long--time (200 ns) molecular dynamics study on the cellulose I-beta and III(I) allomorphs. The main purpose of our extensive computational study is to gain a detailed understanding of the relative differences between the two cellulose allomorphs in terms of structural, thermal and solvation properties and to relate those differences to enzyme binding and digestibility.

Our simulations reveal that the thermal fluctuations along both the intersheet and the intrasheet directions within the cellulose fibrils are larger in cellulose III(I) respect to cellulose I-beta and that the extent of the water--cellulose hydrogen bond network (hydration) was larger in cellulose III(I) than in cellulose I-beta. We will show how these results can provide a detailed physico-chemical description of the differences between cellulose I-beta and cellulose III(I) in terms of both enzyme binding and activity.

See more of this Session: Biorefinery - Biochemical Conversion and Biomass Recalcitrance
See more of this Group/Topical: Forest and Plant Bioproducts Division

(452a) Fast Pyrolysis Bio-Oils From High Protein Containing Feedstocks

Charles A. Mullen and Akwasi A. Boateng, Eastern Regional Research Center, Agricultural Research Service, USDA, Wyndmoor, PA

Typically biomass considered for conversion to biofuels is comprised mostly of carbohydrates and lignin, with small amounts of other materials such as protein, lipids, and ash. Fast pyrolysis of such lignocellulosic materials generally produces bio-oils that are acidic, thermally unstable and have approximately half the energy content of petroleum fuels. However, the properties of bio-oils produced from biomass that contains a higher percentage of protein are different. Fast pyrolysis was performed on several high protein feedstocks, including distiller's grains with solubles (DDGS) from grain fermentation to ethanol, extracted oil seeds (pennycress, camelina), and animal rendering by-product. We found that during the pyrolytic breakdown of protein, basic organo-nitrogen compounds are formed and therefore the bio-oils have a higher pH and lower acid number than typical bio-oils. They also tend to have higher energy contents than wood pyrolysis oils. In this presentation, the results of these experiments, including mass and energy balances, carbon conversion efficiencies, and detailed product characterizations will be presented. Bio-oil chemical characterizations include elemental analysis, GC/MS, GPC, NMR and stability assessments. The fuel properties of the bio-oils including heating values and viscosities will also be presented. Finally, the bio-char co-products of high protein feedstocks will be compared with bio-chars from wood-like feedstocks.

See more of this Session: Biorefinery - Thermochemical Conversion of Biomass I
See more of this Group/Topical: Forest and Plant Bioproducts Division

Fast pyrolysis of lignocellulosic biomass to produce biooil and biochar : Parameter optimization and physicochemical analysis of them

Kwang Ho Kim, Seoul National University, Seoul, South Korea
In Yong Eom, Seoul National University, Seoul, South Korea
Sun Joo Moon, Seoul National University, Seoul, South Korea
Tai Seung Kim, Seoul National University, Seoul, South Korea
Soomin Lee, Korea Forest Research Institute, Seoul, South Korea
Hwangmyeong Yeo, Seoul National University, Seoul, South Korea
In Gyu Choi, Seoul National University, Seoul, South Korea
Joon Weon Choi, Seoul National University, Seoul, South Korea

Introduction

Fast pyrolysis, which is characterized by its simple process and short reaction time, has the capability of transforming inhomogeneous and recalcitrant lignocellulosics into a liquid form of fuel, called biooil [1]. This process is one of the most promising thermal technologies in the world. Recently, liquid biooil has been highly regarded both as an alternative energy [2] and green chemical resources. The fast pyrolysis process also has another advantage, as compared to other biofuel production processes, all different types of biomass can be considered as potential raw materials for fast pyrolysis [3]. A wide variety of biomass has been studied for pyrolysis, from agricultural residues to lignocellulosic biomass [4-11].

Fast pyrolysis is operated in the absence of oxygen producing gas, liquid and char in different proportions, according to experimental conditions. During fast pyrolysis, temperature, residence time of volatiles, heating rates, and particle size are well known as crucial parameters affecting the yield and fuel properties of biooil, as well as its chemical composition. Uzun et al. [12] studied the effect of temperature and nitrogen gas flow rates on the yield of biooil using soybean residues after the extraction of biodiesel. Zheng [13] produced biooil from rice husks in different temperature ranges and analyzed the fuel properties of the biooil.

In this study, yellow poplar wood, a representative fast-growing species of woody biomass, was pyrolyzed at various temperatures and residence times, using a fluidized bed-type reactor. The goal of this study was to elucidate the effect of temperature and residence time during fast pyrolysis on the conversion rate of woody biomass to biooil and to analyze the physicochemical properties of the biooil, as well as the biochar, which is a co-product of fast pyrolysis of woody biomass

Materials and methods

Feedstock

Yellow poplar wood *(Liriodendron tulipifera)* was used in this experiment as the raw lignocellulosic biomass. Bark-free stems were chipped and filtered through a 0.5mm aperture screen. The wood meal was air dried to about 8% moisture content.

Pyrolysis

Fast pryolysis of yellow poplar wood meal was performed using a fluidized bed reactor. A schematic diagram of the fluidized bed reactor is shown in figure 1. Before the experiment, the reactor was purged for 30 min with nitrogen to create an oxygen-free atmosphere. A sample weighing 150g was fed into the reactor for 1hr. After the experiment, the produced biooils were kept in the refrigerator at 4°C for further analysis.

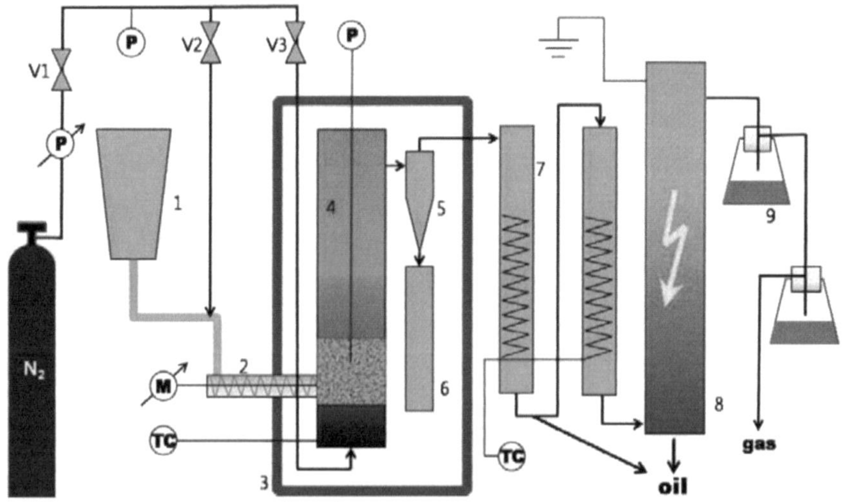

Figure 1. Schematic diagram of fluidized bed reactor (1: feeder, 2: screw feeder, 3: heater, 4: reactor, 5: cyclone, 6: char collector, 7: cooler, 8: electrostatic precipitator, 9: filter)

Analytical methods

Several physical and chemical analyses were performed to characterize biooil and biochar. The list of analytical methods for biooil is shown in Table 1. Elemental analysis and calorific value of biochars were measured with same equipment applied to biooil analyses. The surface morphology of the biochars was investigated by a field-emission scanning electron microscope (FE-SEM, SUPRA 55VP of Carl Zeiss).

Table 1. Analytical methods

Property	Method	Unit
Water content	Karl-Fischer titration	%(w/w, wet basis)
pH	pH meter	pH unit
Viscosity (@ 40°C)	Capillary type viscometer	cSt
Calorific value	Bomb type calorimeter (Parr 6400, Parr instrument company)	MJ/kg
Elemental analysis	Elemental analyzer (CHNS-932 Leco Corp.)	%C, %H, %N (w/w, wet basis)
Gas chromatography	Column type : DB 5 Dimensions : 30 m × 0.25 mm × 0.25 μm Injector : 250 ℃, split ratio 1:20, FID detector : 280 ℃ Oven : 45°C(4min. hold), 3 °C /min. to 280 °C (20min. hold)	

Results and discussion

Influence of pyrolysis temperature

Fast pyrolysis of yellow poplar wood meal was carried out in the temperature range of 400°C to 550°C at intervals of 50°C, and each yield of pyrolytic products (gas, biochar and biooil) is shown in Figure 2. This experiment revealed that the quantitative composition of the pyrolysis products was clearly affected by temperature. The yield of biooil, which was the target product of the fast pyrolysis of woody biomass, gradually increased up to 500°C, then slightly decreased at 550°C. The yield of biooil was maximized at approximately 68.5% at 500°C, based on the weight of yellow poplar. On the other hand, when the temperature of fast pyrolysis increased, the yield of non-condensable gases also continuously increased from 20.9% at 400°C to 28.2% at 550°C, while the formation of char decreased from 28.9% at 400°C to 7.9% at 550°C.

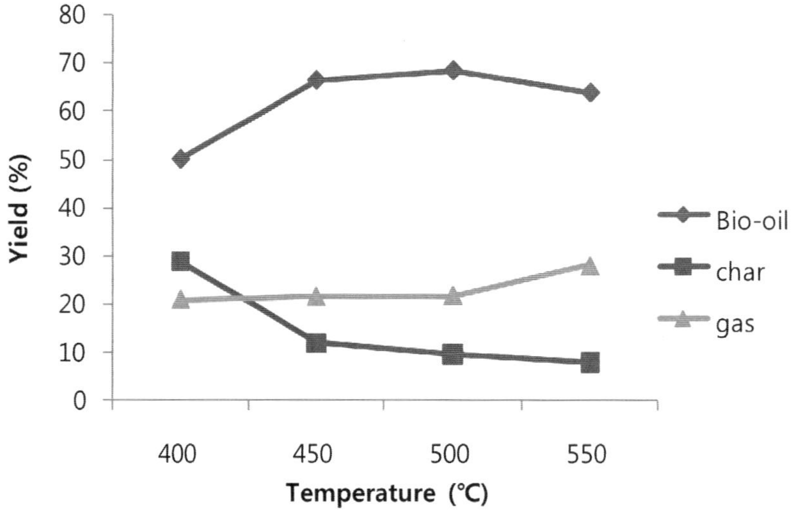

Figure 2. The distribution of product yields as a function of pyrolysis temperature (wet basis, residence time: 1.2sec)

According to previous studies [14-16], the highest yields of biooil from woody biomass feedstock were obtained at temperatures between 500°C and 520°C, which means that the energy required for biomass degradation to low molecular weight compounds was not sufficient below 500°C, but char formation was possible. In other words, a lower pyrolysis temperature prevents full decomposition of the biomass [7]. However, the components of biomass could decompose too much at temperature ranges over 500°C to be condensed into biooil. Higher pyrolysis temperatures cause secondary cracking of primary pyrolytic products, which leads to a decrease in biooil yield and an increase in gas yields [14, 15].

Influence of residence time

Residence time of pyrolytic products in the reactor is another critical parameter in the fast pyrolysis process. Pyrolysis was performed at different residence times, between 1.2s and 7.7s, which was adjusted by the nitrogen gas flow rate. The residence time is defined as the void volume of the reactor divided by the inlet gas flow rate at reactor conditions [8]. Four different nitrogen flow rates,

2.5ℓ/min, 5.0ℓ/min, 10.0ℓ/min, and 15.0ℓ/min, were chosen to examine the effect on pyrolytic product distribution. Each rate indicates 7.7s, 3.8s, 1.9s and 1.2s of residence time, respectively. In this case, the temperature was fixed at 500°C, at which the yield of biooil was the highest. The distribution of product yields, as a function of residence time at 500°C, is shown in Figure 3.

The yield of biooil was influenced by the residence time of the pyrolytic products. There was little difference in the yield of biooil at the residence times of 1.2s and 1.9s. However, when the residence time was prolonged to 3.8s and 7.7s, biooil production was clearly reduced to 61.8% and 51.6%, respectively. In the case of char and gas formation, however, their yields increased from 8.9% to 17.9%, and from 21.8% to 30.5%, respectively. This result could be explained by the residence time prolonging the primary degradation products to further exposure of secondary cracking reactions, thus converting the biomass to non-condensable gases and not biooil [7].

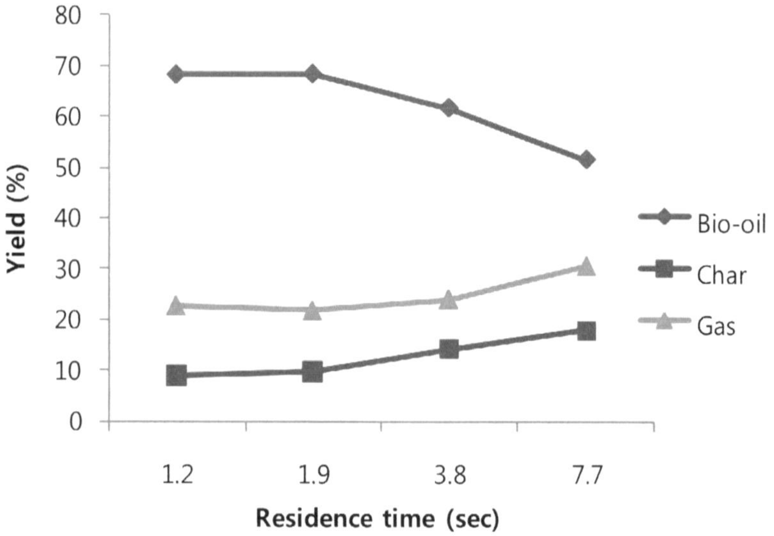

Figure 3. The distribution of product yields as a function of residence time (wet basis, pyrolysis temperature: 500°C)

Physicochemical properties of biooil

The water content in the biooil varied from 21.6 wt% to 42.6 wt%, depending on pyrolysis conditions. Considering the water content (approximately 8%) of the yellow poplar wood meal, the amount could be explained by the considerable amount of water formed during the fast pyrolysis of biomass, especially by a dehydration reaction [17]. Several kinds of acids, such as acetic acid and benzoic acid were formed from the wood components during fast pyrolysis, and pH values of the biooil were between 1.9 and 2.9 [17]. The heating values of the biooils, produced under different temperatures and residence times, were found to be between 10.4 and 17.2 MJ/kg. The relatively low heating values of the biooils were attributed to the high water content and oxygen-containing degradation products.

Characterization of biochar

Elemental composition and calorific value of the pyrolysis biochar was analyzed, and is presented in Table 2. As can be seen in table, the biochar was composed of more than 80% carbon in all experimental runs having a calorific value of 27.1~29.9MJ/kg, which is comparable to coal (approximately 30MJ/kg). The amount of carbon in biochar increased with decreasing pyrolysis temperature and increasing residence time. Like routine carbonization processes, the formation of biochar with a high carbon content and calorific value was favored at lower temperatures and longer residence times.

Table 2. Elemental analysis of biochar coproduced from fast pyrolysis of yellow poplar

Elemental analysis (wt%)	Residence time fixed at 1.9 sec				Temperature fixed at 500°C			
	400°C	450°C	500°C	550°C	7.7s	3.8s	1.9s	1.2s
Carbon	83.4	83.1	82.7	82.0	84.3	83.3	82.7	80.0
Hydrogen	2.0	1.4	1.1	0.8	1.6	1.3	1.1	1.0
Nitrogen	trace	trace	trace	trace	trace	trace	trace	trace
Higher heating value (MJ/kg)	29.9	29.0	28.4	27.8	29.8	28.9	28.4	27.1

A field-emission scanning electron microscope (FE-SEM) was applied to investigate the effect of thermal treatment on the biomass morphology. Figure 4 shows the FE-SEM images of feedstock and biochar obtained from pyrolysis. Significant structural differences can be seen when comparing both images. The biochar looked distorted, and its surface was rough, due to the thermal treatment.

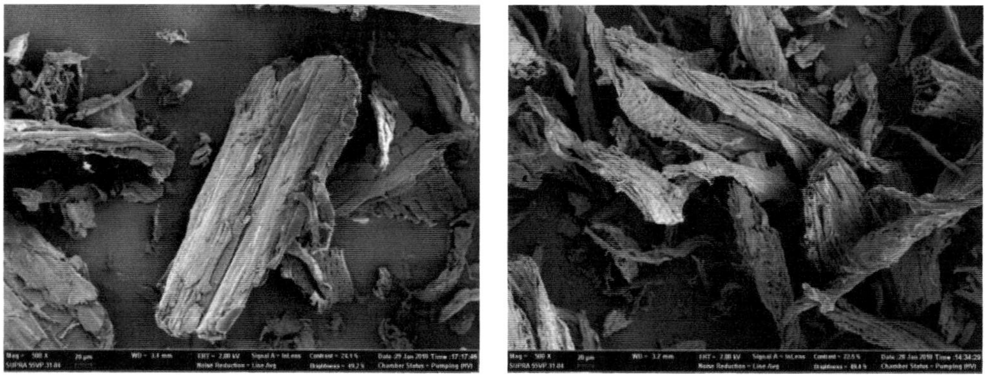

Figure 4. SEM pictures (×500) of yellow poplar wood meal (left) and biochar (right)

Conclusion

The yellow poplar wood meal was pyrolyzed to convert it into biooil and biochar under different experimental conditions using a fluidized bed reactor, and their physicochemical properties were characterized. The mass balance of pyrolytic products, biooil, biochar and gases, was significantly influenced by pyrolysis temperature and residence time. The maximum yield of biooil was 68.5%, based on a biomass weight at 500°C and with a residence time of 1.9s. Pyrolysis of yellow poplar wood for a prolonged residence time, as well as below 500°C, resulted in lower yields of biooil but activated the formation of co-products.

The yield of biochar was 8.0~28.9% on a mass basis. Biochar distribution was highly dependent on the pyrolysis conditions. The analytical results showed that the elemental composition of biochar was affected by experimental conditions. The carbon content of biochar increased with decreasing pyrolysis temperature and increasing residence time, which is similar to the carbonization mechanism.

References

1. Westerhof R.J.M., Kuipers N.J.M., Kersten S.R.A., Swaaij W.P.M. (2007) "Controlling the water content of biomass fast pyrolysis oil." *Industrial & Engineering Chemistry Research*, 46(26): p. 9238-9247.
2. Scott D.S., Majerski P., Piskorz J., Radlein D. (1999) "A second look at fast pyrolysis of biomass - the RTI process." *Journal of Analytical and Applied Pyrolysis*, 51(1-2): p. 23-37.
3. Bridgwater, A.V. (2003) "Renewable fuels and chemicals by thermal processing of biomass." *Chemical Engineering Journal*, 91(2-3): p. 87-102.
4. Putun, A.E. (2002) "Biomass to bio-oil via fast pyrolysis of cotton straw and stalk." *Energy Sources*, 24(3): p. 275-285.
5. Mullen C.A., Boateng A.A., Goldberg N.M., Lima I.M., Laird D.A., Hicks K.B. (2009) "Bio-oil and bio-char production from corn cobs and stover by fast pyrolysis." *Biomass and Bioenergy*, 34(1): p. 67-74.
6. Tsai W.T., Lee M.K., Chang Y.M. (2007) "Fast pyrolysis of rice husk: Product yields and compositions." *Bioresource and Technology*, 98(1): p. 22-8.
7. Xu R., Ferrante L., Briens C., Berruti F. (2009) "Flash pyrolysis of grape residues into biofuel in a bubbling fluid bed." *Journal of Analytical and Applied Pyrolysis*, 86(1): p. 58-65.
8. Piskorz J., Majerski P., Radlein D., Scott D.S., Bridgwater A.V. (1998) "Fast pyrolysis of sweet sorghum and sweet sorghum bagasse." *Journal of Analytical and Applied Pyrolysis*, 46(1): p. 15-29.
9. Park H.J., Park Y.K., Dong J.I., Kim J.S., Jeon J.K., Kim S.S. (2009) "Pyrolysis characteristics of Oriental white oak: Kinetic study and fast pyrolysis in a fluidized bed with an improved reaction system." *Fuel Processing Technology*, 90(2): p. 186-195.
10. Garcia-Perez M., Wang X.S., Shen J., Rhodes M.J., Tian F.J., Lee W.J. (2008) "Fast pyrolysis of oil mallee woody biomass: Effect of temperature on the yield and quality of pyrolysis products." *Industrial & Engineering Chemistry Research*, 47(6): p. 1846-1854.
11. Schnitzer M.I., Monreal C.M., Facey G.A., Fransham P.B. (2007) "The conversion of chicken manure to biooil by fast pyrolysis I. Analyses of chicken manure, biooils and char by 13C and 1H NMR and FTIR spectrophotometry." *Journal of Environmental Science and Health B*, 42(1): p. 71-7.
12. Uzun B.B., Putun A.E., Putun E. (2006) "Fast pyrolysis of soybean cake: product yields and compositions." *Bioresource and Technology*, 97(4): p. 569-76.
13. Zheng, J.L. (2007) "Bio-oil from fast pyrolysis of rice husk: Yields and related properties and improvement of the pyrolysis system." *Journal of Analytical and Applied Pyrolysis*, 80(1): p. 30-35.
14. Bridgwater, A.V. (1999) "An overview of fast pyrolysis of biomass." *Organic Geochemistry*, 30(12): p. 1479-1493.
15. Venderbosch, R.H, Prins, W. (2010) "Fast pyrolysis technology development." *Biofuels Bioproducts & Biorefining-Biofpr*, 4(2): p. 178-208.
16. Mohan D., Pittman C.U., Steele P.H. (2006) "Pyrolysis of wood/biomass for bio-oil: A critical review." *Energy & Fuels*, 20(3): p. 848-889.
17. Lu Q., Li W.Z., Zhu X.F. (2009) "Overview of fuel properties of biomass fast pyrolysis oils." *Energy Conversion and Management*, 50(5): p. 1376-1383.

(452c) Types of Carbon in Biochars From Slow Pyrolysis, Fast Pyrolysis and Gasification

Catherine E. Brewer, Center for Sustainable Environmental Technologies, Iowa State University, Ames, IA, Robert C. Brown, Center for Sustainable and Environmental Technologies, Iowa State University, Ames, IA and Klaus Schmidt-Rohr, Iowa State University, Ames, IA

Biochars, sustainably produced charcoals used for soil amendment and carbon sequestration applications, contain both organic (carbon) and inorganic (ash) constituents. The composition of the inorganic fraction is mostly dependent on the mineral matter present in the feedstock. The types and quantities of carbon in the organic fraction of the biochars, however, are highly dependent on processing conditions such as temperature, heating rate, reaction atmosphere (inert or oxic), and residence time. This study compares the types of carbons in biochars from the slow pyrolysis, fast pyrolysis, and gasification of switchgrass, corn stover and various hardwoods. Among the analytical techniques used to characterize the chars are CHNOS elemental analysis, proximate analysis, Fourier-transform infrared spectroscopy (FTIR), and quantitative ^{13}C solid-state nuclear magnetic resonance spectroscopy (NMR) using direct polarization magic angle spinning (DP/MAS) techniques. Three biochars from the fast pyrolysis of corn stover were also used to conduct a soil incubation study to measure carbon bioavailability based on CO_2 evolution rates. Overall findings suggest that biochars can be tailored to specific applications and that not all chars created from the thermochemical conversion of biomass will be suitable for use as biochars.

See more of this Session: Biorefinery - Thermochemical Conversion of Biomass I
See more of this Group/Topical: Forest and Plant Bioproducts Division

Product and Economic Analysis of Direct Liquefaction of Swine Manure

Mitchell James Minarick, University of Illinois, Urbana, IL, USA. Additional authors: Yuanhui Zhang, Zhichao Wang, Guo Yu, and Ted Funk.

Abstract

Hydrothermal processing of biomass is a technology that has shown promising results in treating waste and producing oil. A batch hydrothermal processing system that was previously developed at the University of Illinois at Urbana-Champaign was used to treat swine manure and it successfully converted up to 70% of swine manure volatile solids into oil and reduced manure chemical oxygen demand by up to 75%. A continuous reactor was developed and it resulted in similar conversion rates to the batch process and it indicated the potential for the process to be scaled-up. These results were also documented previously.

Following these results, a slurry pump was designed and built to enable longer test durations and lower repair costs, and it was used to conduct 24-hour continuous tests and produce quantities of oil suitable for more detailed analysis.

The hydrothermal process utilizes manure containing around 25% solids content which is not available in typical swine confinement facilities so a preliminary investigation of dewatering methods was conducted to determine the best dewatering method in terms of the resultant oil yields and the effects on fertilizer values. The best dewatering method appears to be a coarse filtering process or the use of a low-cost additive. A coarse filtering process followed by the hydrothermal process results in an oil product as well as a fertilizer product that retains the majority of its nitrogen value with a reduced level of phosphorus as compared to the unprocessed swine manure.

Energy and economic analyses of the dewatering and hydrothermal processes were conducted for swine manure as well as general analyses for other domestic biomass feedstocks. The feasibility of the conversion of swine manure was judged to be high, dependent mainly on the validity of the assumptions for the value of the oil product.

Introduction

Using waste materials to produce biofuel sidesteps the controversial food versus fuel debate and gives the production process a competitive advantage over methods using traditional feedstocks since they can be acquired much more cheaply, if not for free. It makes sense to research methods of biofuel production from waste streams alongside research of biofuels produced from more valuable feedstocks.

Waste Conversion technologies

The conversion of biomass to fuel generally follows one of two routes: biochemical or thermochemical. A summary of the primary biomass conversion technologies is provided by He (2000) in Figure 1 below.

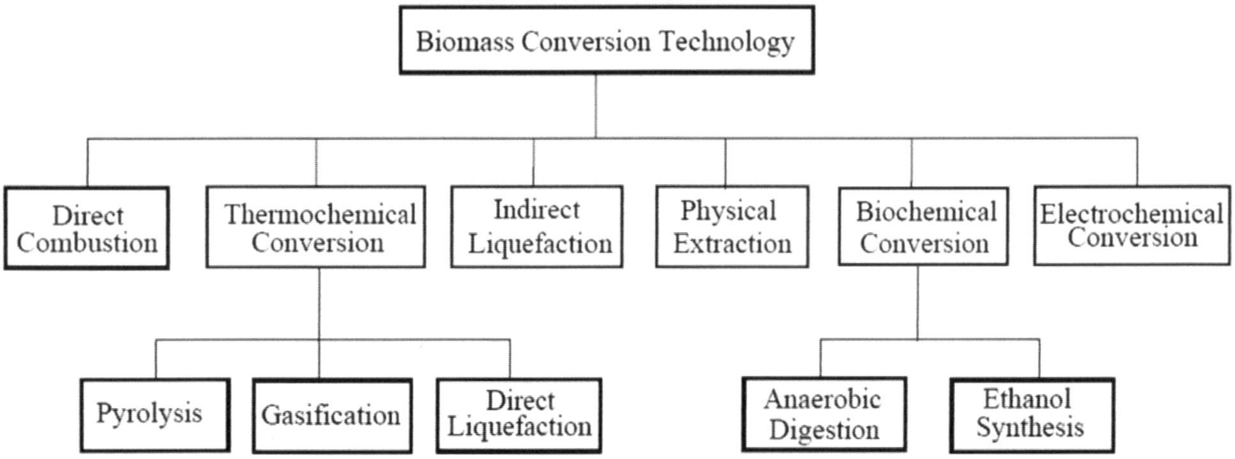

Figure 1. Biomass conversion technologies.

Direct liquefaction is a type of thermochemical conversion that has been studied extensively during the past 10 years at the University of Illinois. Similar to pyrolysis, direct liquefaction is well-suited for the conversion of livestock waste because it does not require a dry feedstock. Previous research teams headed by Dr. Yuanhui Zhang have concluded that direct liquefaction could be a viable solution for the disposal of swine manure, and they have attempted to determine the parameters that optimize the oil yield.

Ocfemia (2005) applied these parameters to a continuous reaction process to determine the feasibility of scaling the system up beyond the batch scale. His system had a capacity to process up to 48 kg of manure slurry per day, and successfully produced oil at equal or greater efficiencies than the batch process used by Brian He.

Water-separation considerations

The desired solids content of the feedstock for the hydrothermal process is around 25% according to He (2000). This is approximately the solids content contained in fresh (that is, taken directly from the confinement floor) swine manure. However, according to Fulhage (2001) most swine manure is handled as a liquid, utilizing shallow storage pits beneath slotted floors that allow the producer to apply manure to fields when it is most beneficial. This slurry has a higher water content than what is optimal for the hydrothermal conversion process (He, 2000) and since additional water translates into additional heating and processing costs, it is necessary to remove as much water as possible without affecting the oil yield.

For this technology to be useful on a farm, it is necessary to develop Ocfemia's continuous process (2005) into a working prototype that utilizes swine manure taken directly from the production facility. In a typical production facility, the manure as well as urine and spilt food and water falls through the slotted floor to the shallow-pit, where it is stored until it is pumped into an external tank for disposal every few weeks. Initial tests (Wang, 2007) showed that conversion into oil of the manure slurry as taken from the pit was impaired when compared to the conversion of manure taken from behind the pig, so it is necessary to determine the effect of different water separation methods on oil yields in order to dovetail thermochemical conversion technology with manure collection systems that are currently in place.

Methods

Water Separation Methods

As previously discussed, swine manure is commonly stored in a shallow pit, where it becomes diluted to less than 10% solids. The thermochemical process requires a feedstock with a solids content of around 20%, so a practical means for attaining higher solids content needed to be determined.

Evaporation is the simplest method of water separation, but also the most time consuming. The swine manure was simply stored in open containers with a fan circulating air over the tops of the containers. The solids content was measured periodically until it reached the desired level.

Coarse filtration was accomplished using a standard aluminum window screen as the filter medium. The screen had square openings measuring 1.13 x 1.30 mm, and it was used as a liner inside a round laundry basket that served as the filter holder. The manure from the pit was placed in the filter assembly and the laundry basket was rolled by hand until the flow of aqueous product out of the filter ceased.

The difficulties associated with this water separation method are discussed in section 3.3.4, but in summary fine filtration was defined as any filtration with a smaller pore size than the coarse filter (1.13x1.3 mm). The portion of the swine manure that was subjected to the fine filtration was the aqueous portion of the swine manure that was not already removed by the coarse filter.

For the low-speed centrifugation, a vertical-shaft solid-basket centrifuge manufactured by Lavin Centrifuge (model 12-413V) was used to subject the aqueous product from the coarse filtration to 2,250 Relative Centrifugal Force (2,250 RCF=2,250 times the force of gravity) for 5 minutes. The solids were collected and the aqueous portion was saved for further separation.

The aqueous solution remaining after the low-speed centrifugation was placed in capped 500mL Nalgene polycarbonate centrifuge bottles (Part #3140-0500, 69.5 mm diameter, 158.8 mm length, 13,700 maximum allowed RCF) and carefully weighed to make sure they would be balanced in the centrifuge. The centrifuge was a Dupont Sorvall centrifuge equipped with a GS-3 rotor and was used to subject the samples to 10,816 RCF for 15 minutes. The solids formed a dense layer on the bottom of the centrifuge bottles, so the aqueous solution was easily separated and both portions were saved.

Thermochemical Conversion Process

The thermochemical conversion was conducted using similar conditions as those used in the experiments by He (2000) and Ocfemia (2005). The temperature was the main control variable and was held at 305°C for 30 minutes of residence time in a Parr 2-L reactor. The oxygen was evacuated by pressurizing the reactor with 100 psi of nitrogen gas and then releasing the pressure. This was repeated three times before filling the reactor with 100 psi of nitrogen gas which was used as a process gas throughout the experiment.

These operating conditions resulted in a peak reactor pressure of approximately 1650 psi when the reactor was at its peak operating temperature. No catalyst was used during these tests.

Oil Analysis

The thermochemical process results in a raw oil product that is very crude and unrefined. The water content of the bio-oil was determined using ASTM standard D95-99 and subtracted from the oil yields. The toluene solubility was determined using ASTM standard D4072-98 and used to define a 'refined oil yield' which makes up roughly 35-45% of the raw oil. The refined oil yield gives an idea of both the quantity and quality of the oil and is a more robust measurement of the oil production.

Assumptions

Energy

The following assumptions were used when calculating energy balances:

The bio-oil has a heating value of 30,000 kJ/kg (Ocfemia, 2005).

The energy contained in the feedstocks before undergoing TCC was negligible or unavailable. In other words, it is assumed that the only energy input is heat via electricity, and the chemical energy in the feedstock is not considered. In most cases the feedstock had too high of a water content to be burned so this was an accurate assumption for the feedstocks considered.

The reactor is perfectly insulated. Although perfect insulation is not truly possible, the amount of heat loss through the reactor walls could be negligible with good insulation.

The internal energy values of the feedstocks are equal to those of water (i.e. the energy needed to heat/pressurize the feedstock is equal to that needed for water). This is a reasonable and conservative assumption because the feedstocks contained 75-95% water, and the specific heat of water is higher than most materials. For instance, the specific heat of water is over 4 times that of sawdust. Thus, more energy is needed to heat pure water to a given temperature than would be needed to heat the same amount of water mixed with sawdust. These figures for the change in internal energy are therefore higher than the actual values, and the actual energy needed to heat the different solutions should be less than (but fairly close to) what would be calculated using the following values for water:

Reaction conditions are approximately 100 bar (1450 psi) and 303°C,
Internal energy: Uf (303°C) = 1350.5 kJ/kg
Uf (22°C) = 92.32 kJ/kg
ΔU for reaction = 1258 kJ/kg
Energy used to pre-process the feedstock is negligible.

No heat is recaptured with a heat exchanger unless otherwise stated. When a heat exchanger is used in calculations, the effectiveness is assumed to be 75%.

Economic

For the following calculations (unless otherwise specified), the cost of electricity was estimated to be $0.05/kW-hr. If the reactor was only operated during certain times of the day, this is a reasonable estimate for industrial power usage, although in practice this value will vary.

Additionally, the value of crude oil varies greatly. It is reasonable to assume that the value of the bio-oil will be related to the value of crude oil since they are competitive products. Consequently, the value of the bio-oil can be expected to vary significantly based on the fluctuations in the price of crude oil, which can be volatile. For example, in just a 12-month span from winter of 2008 to winter of 2009 the price of oil varied from \$40-\$140. This variability certainly has an effect on the value of the bio-oil and on any alternative fuel for that matter.

For most of the calculations that follow, the price of oil is assumed to \$50/barrel unless otherwise specified. If this price were assumed to be higher, it would result in more favorable results for the thermochemical process.

Estimating the actual value of the bio-oil is difficult. It is hard to find the true market value of an oil product which does not have an established market yet. For the purpose of calculation, the value of the refined bio-oil (i.e. the toluene soluble portion) was estimated to be 60% of the value of petroleum light crude oil. This value was chosen in an attempt to be conservative so as not to overestimate the value of the bio-oil. This value might well be higher since the crude oil is unrefined and the toluene-soluble portion of the bio-oil has already been subjected to one refinement step (hence the choice of the term 'refined oil yield' to describe it). This assumption is admittedly difficult to defend. Without an existing market, the value of the bio-oil will not be precisely known. However, taking 40% of the value off of the unrefined petroleum crude hopefully allows adequate room for this uncertainty.

Most petroleum chemists divide crude oil into four fractions: saturates, aromatics, resins and asphaltenes. Asphaltenes are, broadly speaking, the fraction of oil that are insoluble in heptane or pentane and soluble in benzene/toluene. Appleford (2004) analyzed the bio-oil from the thermochemical conversion processing of swine manure and found that the oil was primarily resins and asphaltenes as seen in Table 1 below where the first two columns of results represent crude oil from swine manure:

Table 1. Properties of bio-crude oil from swine manure vs. natural crude oil.

Fraction (%)	C-305-1500-60	N-305-1500-60	Pooled Sample	Natural Crude Oil*	Natural Bitumen*
Saturates	2.93	2.48	1.32	>30	15
Aromatics	1.84	1.67	1.60	>30	34
Resins	44.65	45.87	35.87	<20	34
Asphaltenes	44.73	43.39	54.81	<10	17
Water + Loss	5.85	6.59	6.40	varies	varies

As seen in Table 1, the bio-crude from the swine manure contains around 50% asphaltenes and around 35% resins. Therefore an alternative valuation for the bio-oil can be obtained by assuming the products will be used as asphaltanes rather than as a refined oil product. Asphalt prices, like oil prices, fluctuate dramatically. Table 2 shows the price of asphalt used for paving for 2007 and 2008 (California Department of Transportation, 2009):

Table 2. Monthly price of asphalt for 2007-2008.

Month-Year	$/ton	Month-Year	$/ton
Jan-07	$287.10	Jan-08	$497.80
Feb-07	$267.70	Feb-08	$434.40
Mar-07	$295.30	Mar-08	$510.00
Apr-07	$317.40	Apr-08	$504.10
May-07	$314.90	May-08	$567.70
Jun-07	$321.70	Jun-08	$645.30
Jul-07	$355.70	Jul-08	$716.90
Aug-07	$384.30	Aug-08	$630.40
Sep-07	$370.70	Sep-08	$547.40
Oct-07	$393.40	Oct-08	$486.10
Nov-07	$464.50	Nov-08	$288.10
Dec-07	$439.30	Dec-08	$199.40

Importantly, the use of the bio-oil as an asphalt would eliminate many of the expenses, technological hurdles, and uncertainties associated with refinement of the oil and finding a market for the finished product.

The net present value (NPV) of an investment is a standard method for using the time value of money to appraise long-term projects. Essentially it involves calculating the sum of the present values of future cash flows based on an assumed discount rate (or the expected interest rate of investment alternatives), and if the NPV is a positive number the investment is considered favorable to the investment alternatives. The NPV for a given set of cash flows can be calculated with a few assumptions. Labor costs could be limited to one employee at $35,000 per year. Maintenance costs can be estimated at $20,000 per year based on an assumed capital cost of $200,000. The annual discount rate (i.e. the rate of return of investment alternatives) is assumed to be 15% and the life expectancy of the equipment is 10 years.

Results

Figure 2 below shows the summarized results of the TCC tests using swine manure and a variety of different dewatering methods in terms of their refined oil yields. However, refined oil yield is not the only variable determining a successful test. Certainly energy input costs have a significant effect. For a feedstock with low solids content a significant amount of energy is spent heating and pressurizing water in addition to the solids. Manure slurry taken straight from the pit and sent to the reactor without any dewatering suffers from its high water content. Figure 3 below summarizes the relative processing costs associated with oil production based on the different dewatering methods.

Figure 2. Refined oil yields.

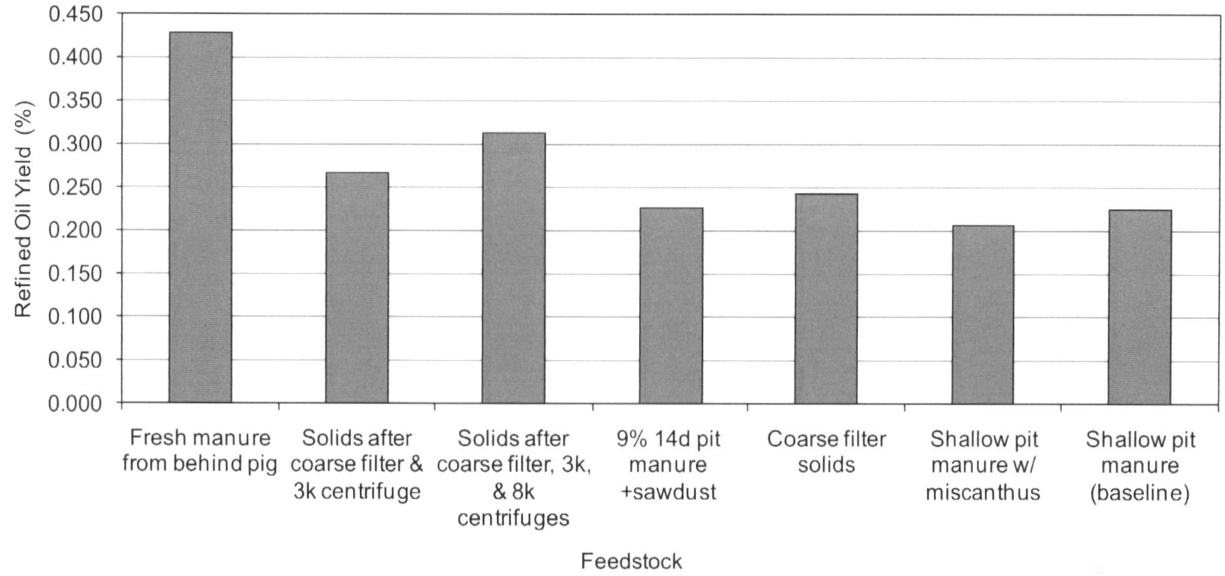

Figure 3. Relative processing costs per oil production.

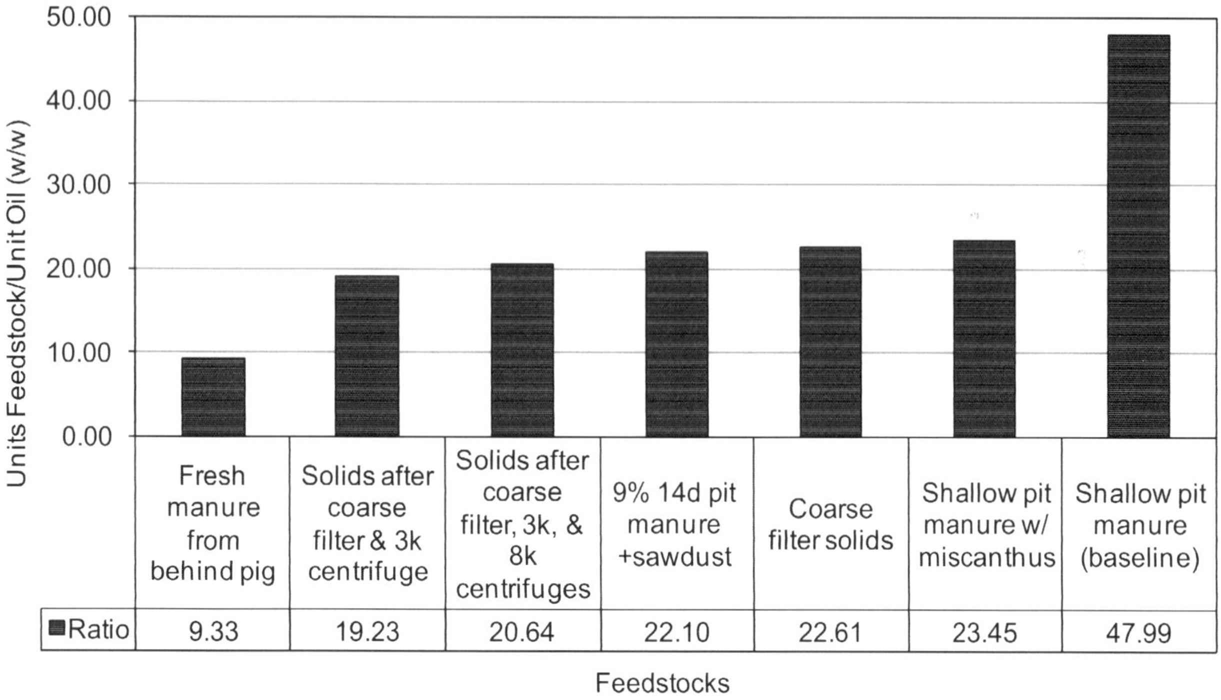

In summary, the best dewatering method appears to be mixing the slurry with additives such as sawdust or miscanthus if the additive is cheap or free. If the additive is not cheap, rough filtration is probably better.

Figure 4 below demonstrates the energy balances of the oil production process in terms of processing energy in versus oil-based energy out, both with and without a heat exchanger. The balances are favorable for all cases except the raw manure straight from the pit. It is important to remember that although most of the ratios seem to show a negligible energy gain, the main value added by the process is not due to an addition of net energy but rather a conversion of the form of the energy, from heat via electricity or combustion to a liquid fuel product.

Figure 4. Energy out versus energy in.

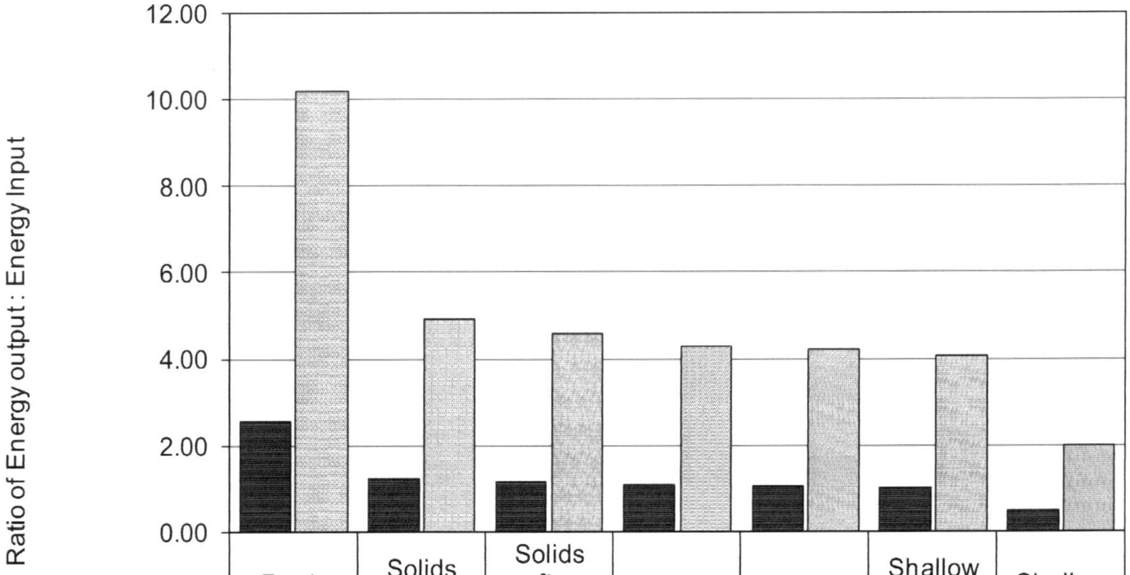

	Fresh manure from behind pig	Solids after coarse filter & 3k centrifuge	Solids after coarse filter, 3k, & 8k centrifuges	9% 14d pit manure +sawdust	Coarse filter solids	Shallow pit manure w/ miscanthus	Shallow pit manure (baseline)
■Energy output : energy input	2.55	1.24	1.16	1.08	1.05	1.02	0.50
□Energy out:in w/ heat exch.	10.22	4.96	4.62	4.32	4.22	4.07	1.99

Feedstocks

Figure 5 below shows that the fertilizer values of the aqueous products after the TCC test are not generally destroyed by the conversion process, with the exception of phosphorus content. The bars on the left side of each set of data represent the fertilizer values contained in the manure before processing. Nitrogen is fairly well preserved in most forms, except in Case 4 with the miscanthus as an additive. Potassium was retained well in all cases. This is an important piece of information because it shows that a farmer can process manure thermochemically to obtain bio-oil, and still retain over half of the original fertilizer value. Additionally, less phosphorus is applied to the field after the TCC process, lowering the risk of phosphorus runoff.

Figure 5. Fertilizer values remaining in aqueous solution.

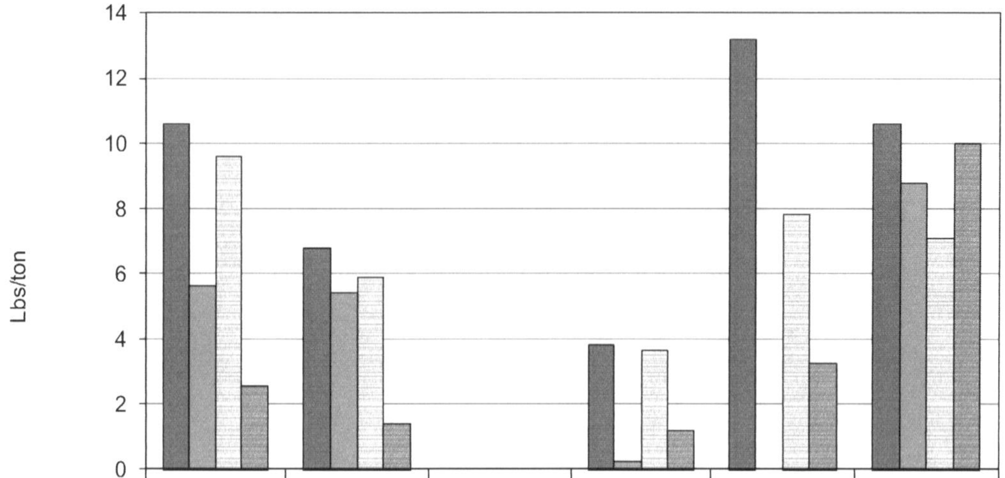

	Total Nitrogen	Ammonium Nitrogen	Nitrate Nitrogen	Organic Nitrogen	Phosphorus	Potassium
■ Raw Manure from pit, Case 1	10.60	6.80	0.00	3.80	13.20	10.60
■ Raw Manure after TCC, Case 2	5.64	5.41	0.00	0.23	0.00	8.80
□ Raw Manure after coarse filter & TCC, Case 3	9.61	5.90	0.00	3.66	7.84	7.09
■ Raw Manure w/ Miscanthus after TCC, Case 4	2.56	1.40	0.00	1.17	3.26	10.02

Components

Economics of Swine Manure Conversion

For an operation with 10,000 hogs, a feedstock supply that does not cost anything to acquire, and an empirically-determined (Wang, 2007) refined oil yield of 42.9% for manure from the floor behind the pig, 6,552 barrels of refined oil per year could be produced.

Assuming the value of refined oil is equal to 60% of the value of crude oil and a barrel of crude oil is worth $50, the annual gross income for the operation is $196,560. The cost to process this amount of biomass assuming a fresh manure solids content of 25%, an electrical energy cost of $0.05/kW-hr, and no heat recapture is $178,101 annually. Thus an annual income of $196,560-$178,101=$18,459 would be expected in this case study. This is probably not even enough income to pay for the associated labor costs, and a system like this is not economically feasible.

Factors that would decrease this amount include: oil prices lower than $50/bbl, refined oil worth less than 60% of crude oil, oil conversion rate of less than 42.9%, energy input costs more than $0.05/kW-hr, labor and capital costs, heat losses to the atmosphere, etc. Factors that would increase this amount include: oil prices higher than $50/bbl, refined oil worth more than 60% of crude oil, energy input costs less than $0.05/kW-hr, and heat recapture via a heat exchanger. Of these factors, the most significant changes would probably come in the estimates for energy input costs. If the operator was careful to run the reactor at hours of low electrical demand, it is conceivable that the cost of electricity would average less than $0.05/kW-hr. There are other energy input options as well. Natural gas was estimated to cost Midwest customers $1.11/therm or $0.0378/kW-hr during 2006 (Energy Information

Administration, 2007), although the efficiency of the boiler would not be 100% so that value would be increased slightly in a practical application.

The simplest way to reduce energy input costs, however, would be to install a heat exchanger to take some of the energy from the hot exit stream and use it to preheat the cool incoming feedstock.

When heat recovery assumptions are applied to the economic calculations, the numbers improve dramatically. The oil yield and gross income of $196,560 would remain the same, but the energy input costs are quartered due to the assumed energy recovery of 75%. The energy input costs become $44,525 annually. This drives the annual profit up from $18,459 to $152,035 before labor, capital, and maintenance costs and the NPV is then $486,996 which means the project is profitable (i.e. returns more than 15%) if the capital costs are less than the NPV. These results are very promising and would be even more compelling for larger hog farms because with larger volumes they would generate greater annual profits yet the labor and capital costs would remain virtually the same. This result is dependent on multiple assumptions but they were kept conservative and overall the feasibility is judged to be very high.

If the refined oil product was able to be used as an asphalt substitute, it could be conservatively priced at $200/ton. The gross income of the operation then becomes $241,098. This drives the annual profit up to $196,573 and the NPV becomes $710,522. To recap then, the NPVs for the manure from the floor are as follows:

- Manure from floor used to make oil: NPV=negative
- Manure from floor used to make oil, heat exchanger used: NPV=$486,996
- Manure from floor used to make asphalt, heat exch. used: NPV=$710,522

The benefits of using a heat exchanger and exploring the opportunities for using the bio-oil as an asphalt are apparent, and these modifications will be used in the remaining economic analyses.

NPVs using manure from shallow pit

The gross income using manure from a shallow pit and adding sawdust would be $323,479, however this is assuming that the additive does not cost anything. In reality, sawdust can cost up to $30 per ton in some areas, reducing the gross income by $137,394 to $186,085 per year. The operating costs would then be $141,815 and the gross profit is only $44,270, which is not even enough to pay for labor and maintenance expenses.

If the excess water was removed with rough filtration instead of using sawdust as an additive, then the gross income would be $42,201 with processing costs of $17,179 annually. The ratio of the income to the expenses is 2.46. This is a very acceptable ratio, but the gross profit of the TCC process ($42,201-$17,179=$25,022) is not even large enough to support a full-time employee. The high profitability ratio, however, indicates good potential for scaling up the system. If the thermochemical conversion system was mounted on a portable trailer and used to process the manure from 5 farms, the net profit would jump to $125,110 and the NPV would be $351,866.

To recap then, the NPVs for the manure from the shallow pit are as follows:

- Manure from pit, heat exchanger, asphalt product and $30/ton sawdust: NPV=negative

- Manure from pit, heat exchanger, asphalt product, rough filtration: NPV=negative after labor/maintenance costs are added, but the profitability ratio is high, meaning more volume is needed for profitability.
- Manure from pit, heat exchanger, asphalt product, rough filtration throughout 5 farms: NPV=$351,866

These results are very promising because they represent conservative calculations for the net present values of the types of system that would be used with existing production facilities without modification. The numbers would be even more compelling for larger hog farms, again because larger volumes would generate greater annual profits yet the labor and capital costs would remain virtually the same.

The primary downside risk to these NPV figures is the difficulty in determining an accurate value of the bio-oil. As stated previously, the refined bio-oil was sometimes valued at 60% of petroleum crude oil in these calculations, but that number is simply an estimate and not backed by firm data (since there really isn't any firm data extant). Other times the refined bio-oil was considered as a substitute for petroleum asphalt selling for $200/ton. This is probably more accurate than the other approximation since the oil product more closely resembles asphalt, but firmer data regarding the suitability of the bio-oil as a substitute for asphalt would solidify these estimates considerably.

References

Appleford, J. M. 2004. Analysis and optimization of the thermochemical conversion process to produce oil from biomass. Masters thesis. Urbana, IL: University of Illinois at Urbana-Champaign, Department of Agricultural Engineering.

ASAE Standards. 2005. D384.2: Manure production and characteristics. St. Joseph, Mich.: ASAE.

ASTM. 2004a. ASTM D95-99. Standard test method for water in petroleum products and bituminous materials by distillation. In: Annual Book of ASTM Standards. West Conshohocken, PA: Am. Soc. for Testing Materials.

ASTM. 2004b. ASTM D473-02. Standard test method for sediment in crude oils and fuel oils by the extraction. In: Annual Book of ASTM Standards. West Conshohocken, PA: Am. Soc. for Testing Materials.

ASTM. 2004c. ASTM D4072-98. Standard test method for toluene-insoluble (TI) content of tar and pitch. In: Annual Book of ASTM Standards. West Conshohocken, PA: Am. Soc. for Testing Materials.

California Department of Transportation. 2009. California statewide paving asphalt price index. Sacramento,CA. Available at: http://www.dot.ca.gov/hq/esc/oe/asphalt_index/PREVIOUS_YEARS/07-08_AC_INDEX.pdf.

Dong, R. 2007. Personal communication. Department of Agricultural and biological engineering. Urbana, IL.: University of Illinois at Urbana-Champaign.

Elliott, D. C., et al. 1994. Chemical processing in high-pressure aqueous environments: continuous-flow reactor process development experiments for organics destruction. *Industrial and Engineering Chemistry Research*. 33: 566-574.

Energy Initiative Administration. 2007. Natural gas annual 2006. Washington, D.C.: United States Department of Energy, Energy Initiative Administration.

Fulhage, C., and D. Pfost. 2001. Swine manure management systems in Missouri. Columbia, MO.: University of Missouri-Columbia Extension.
He, B. 2000. Thermochemical conversion of swine manure to produce oil and reduce waste. PhD dissertation. Urbana, IL: University of Illinois at Urbana-Champaign, Department of Agricultural Engineering.
Hrubant, G. R., R. A. Rhodes, and G. H. Sloneker. 1978. Specific composition of representative feedlot wastes: a chemical and microbial profile. Washington, D.C.: Science and Education Administration, U.S. Department of Agriculture.
Kreith, F., and D. Y. Goswami. 2004. The CRC handbook of mechanical engineering. 2nd edition. CRC Press. Boca Raton, FL.
Moran, M. J., and H. N. Shapiro. 2008. *Fundamentals of engineering thermodynamics*. 6th edition. John Wiley and Sons. Hoboken, NJ.
Ocfemia, K. 2005. Hydrothermal process of swine manure to oil using a continuous reactor system. PhD dissertation. Urbana, IL: University of Illinois at Urbana-Champaign, Department of Agricultural Engineering.
Ott, L. S., B. L. Smith, and T. J. Bruno. 2008. Advanced distillation curve measurement: application to a bio-derived crude oil prepared from swine manure. *Fuel* 87: 3379-3387.
Reece, M. 2004. Special issue paper 9: Densities of oil products. Energy Statistics Working Group Meeting, International Energy Agency. Paris, France.
Scurlock, J. 2009. Bioenergy feedstock characteristics. Oak Ridge, TN. Oak Ridge National Laboratory, Bioenergy Feedstock Development Program.
Speight, J. G. 2001. *Handbook of Petroleum Analysis*. New York: Wiley Interscience.
Sweeten, J. M. 1992. Livestock and poultry waste management: A national overview. In J. Blake et al., eds., National Livestock, Poultry and Aquaculture Waste Management, pp. 4–15. ASAE Publication 03–92. American Society of Agricultural Engineers, St. Joseph, MI.
USDA-ERS. 2009. Livestock, dairy, & poultry outlook. Washington, D.C.: USDA Economic Research Service.
USDA-NASS. 2008b. Quarterly hog and pigs report. Washington, D.C.: USDA National Agricultural Statistics Service.
USDA-NASS. 2009. Crop production 2008 summary, p. 13. Washington, D.C.: USDA National Agricultural Statistics Service.
USDOE and USDA, 2005. Biomass as feedstock for a bioenergy and bioproducts industry: the technical feasibility of a billion-ton annual supply, p. 7. Washington, D.C.: USDOE and USDA.
U.S. Environmental Protection Agency Office of Solid Waste. 1999. Biosolids Generation, Use, and Disposal in the United States. EPA530-R-99-009, p. 2.
U.S. Environmental Protection Agency Office of Solid Waste. 2008. Municipal Solid Waste in the United States: 2007 Facts and Figures. EPA530-R-08-010, p. 1.
Vogel, K. P., and R. A. Masters. 1998. Developing switchgrass into a biomass fuel crop for the midwestern USA. Lincoln, NE. University of Nebraska. Agricultural Research Service, United States Department of Agriculture.
Wang, Z. 2007. Unpublished data. Department of Agricultural and Biological Engineering. Urbana, IL.: University of Illinois at Urbana-Champaign.
Wang, Z. 2007. Personal communication. Department of Agricultural and Biological Engineering. Urbana, IL.: University of Illinois at Urbana-Champaign.
Yu, G. 2008. Unpublished data. Department of Agricultural and Biological Engineering. Urbana, IL.: University of Illinois at Urbana-Champaign.

(452e) Effect of Torrefaction on the Fast Pyrolysis Using Forest Resources

Jiajia Meng, Sunkyu Park, David Tilotta and Junyeong Park, Forest Biomaterials, North Carolina State University, Raleigh, NC

Bio-oil has a great potential for multiple uses including heat/electricity generation and chemical production. This thermochemical technology appears to be a practical, effective, and environmentally sustainable means of producing renewable fuels while simultaneously reducing emissions of greenhouse gases. In addition, pyrolysis has a key advantage over biochemical methods in that it is less sensitive to variation in feedstocks, which makes supply-chain development less challenging. However, the quality of pyrolysis bio-oil is not sufficient to directly produce biofuels and chemicals due to high oxygen content, high acidity, and poor long-term stability. In this study, we investigated the potential of integrating fast pyrolysis with torrefaction to produce high quality bio-oil from forest resources. The benefits of this integration along with the preliminary experimental results will be discussed.

See more of this Session: Biorefinery - Thermochemical Conversion of Biomass I
See more of this Group/Topical: Forest and Plant Bioproducts Division

LIQUID PRODUCTS OF LIQUID PHASE-PYROLYSIS

Nikolaus Schwaiger[1], Verena Mertlitz[1]; Kerstin Zahel[2], Peter Pucher[2], Edgar Ahn[2], Matthäus Siebenhofer[1],
[1]Graz University of Technology; Institut für Thermische Verfahrenstechnik und Umwelttechnik; Inffeldgasse 25/C/II 8010 Graz/Austria;
[2]BDI-BioEnergy International AG; Grambach/Austria

Ligno-cellulosic feed is expected to contribute significantly to production of liquified combustibles in future, because of the quantity and the variety of feed material. Therefore the pyrolytic degradation properties of spruce wood and its building blocks glucose and cellulose were investigated. The interaction between heat carrier, biomass and biomass products in the condensed liquid and vapour phase was investigated.

The process was carried out in a semi-batch reaction vessel under isothermal conditions at a temperature of T = 623 K and at ambient pressure. For optimum heat transfer the pyrolysis process was carried out in liquid heat carrier phase. Gaseous products as well as condensibles were removed on top of the reaction vessel. Condensation temperature was T = 20°C. After recording flux and composition the gaseous phase was incinerated in a flare.

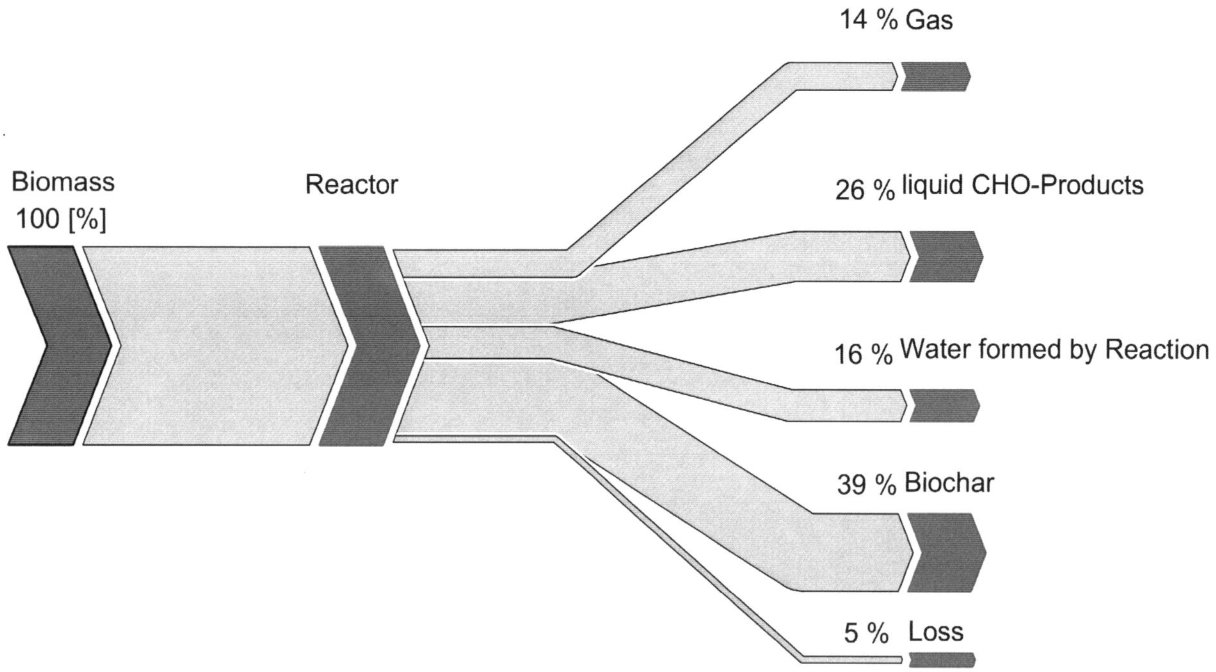

Figure 1: Biomass mass balance

The condensed products were analyzed by GC-MS and NMR spectroscopy as well as elemental analysis (CHNO).

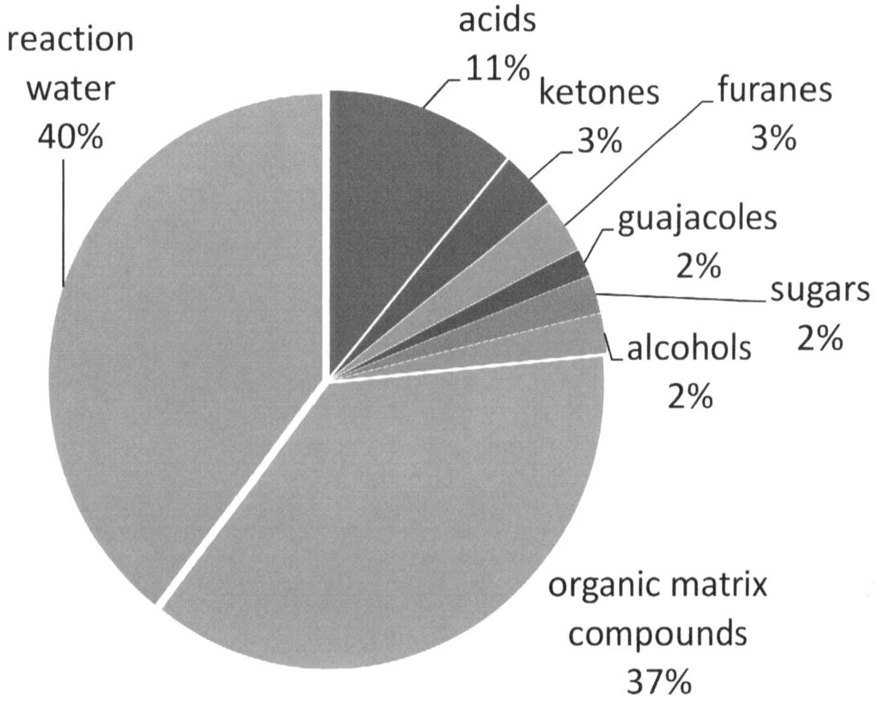

Figure 2: Composition of liquid products measured by GC-MS and Karl Fischer Titration

The solid product residues were pretreated by solid-liquid extraction and afterwards analyzed by elementary analysis, FTIR-ATR, Raman spectroscopy, scanning electron microscope and energy dispersive X-ray spectroscopy.

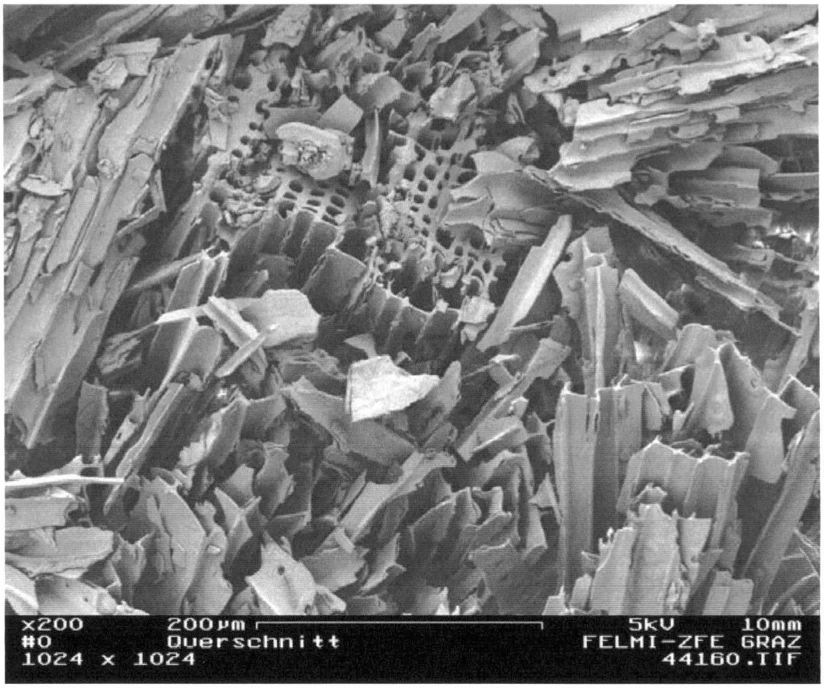

Figure 3: Scanning electron microscope photo of a Biochar cross section

Several operation parameters were investigated. Representatively the effect of particle size is briefly mentioned.

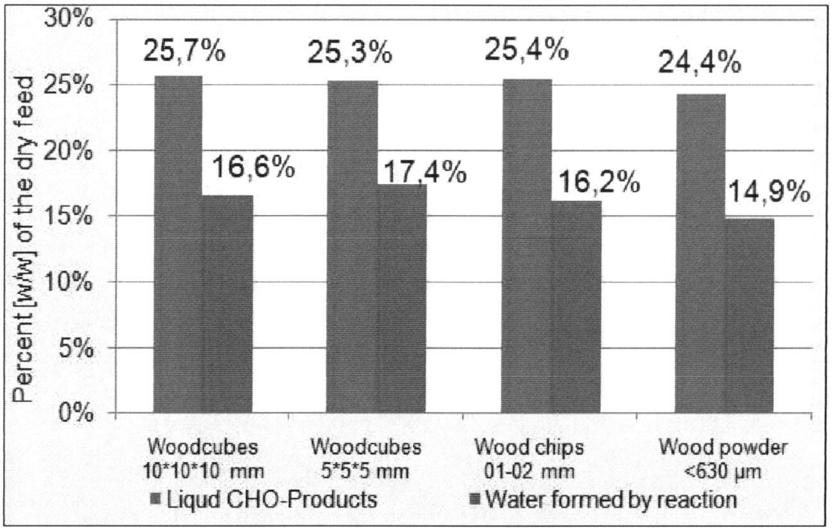

Diagram 1: Effect of feed particle size on the amount of CHO-products and the amount of water formed by thermal dehydration of softwood

Following the results shown in Diagram 1 variation of particle size does not have an effect on the amount of liquid CHO-products, when pyrolysing softwood. Water formation decreases for feed particles of smaller particles.

As shown in Diagram 2 the total amount of liquid CHO-products seems to be constant and independent of the feed particle size. The chemical composition of liquid phase pyrolysis products differs.

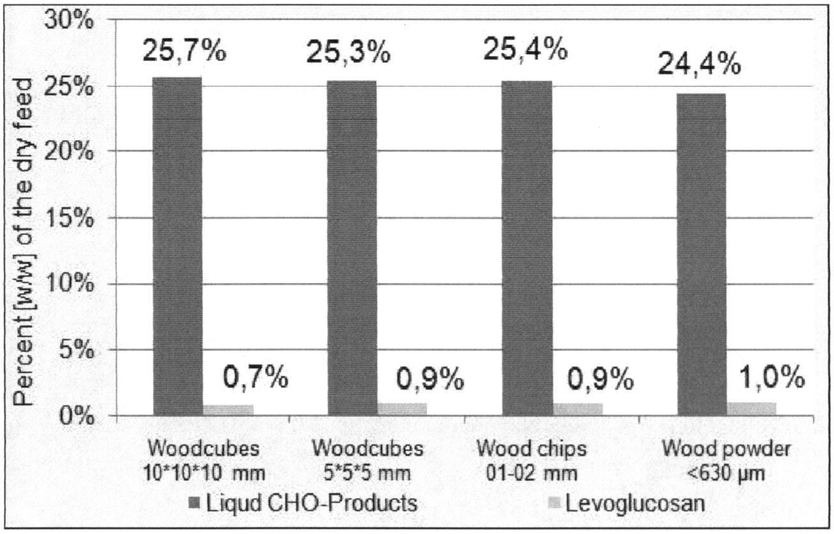

Diagram 2: Effect of feed particle size on non aqueous condesibles and products

Based on the detailed mass balance the effect of operation parameters on pyrolytic degradation of softwood, product formation and energy balance was determined.

(467a) Ethanol Production by Bioconversion of Softwood Prehydrolysates Supplemented with Pulp Mill Sludges

Li Kang[1], Sunghoon Yoon[1], Gopal A. Krishnagopalan[2] and Y. Y. Lee[3], (1)Chemical Engineering, Auburn University, Auburn, AL, (2)Chemical Engineering, Auburn University, Auburn University, AL, (3)Department of Chemical Engineering, Auburn University, Auburn University, AL

Eighty percent of the hemicellulose fraction of pulp mill feedstock is released into black liquor during kraft pulping process. In pulping process, the black liquor is burned to recover chemicals and generate steam and electricity. It is technically feasible to remove part of the hemicellulose before pulping keeping the quality of pulp unaffected. The hemicellulose portion of the pulp mill feedstock can be selectively converted to soluble sugars (termed as prehydrolysate) by water or mild acid hydrolysis. The prehydrolysate thus prepared contain pentose, hexose, and their oligomers. In this study, bioconversion of the prehydrolysate to ethanol was investigated using Saccharomyces cerevisiae. The prehydrolysate produced from wood also contain toxins, primarily organic acid, lignin and sugar degradation products. These toxins strongly inhibit microbial action. The prehydrolysate was therefore detoxified by over-liming and by addition of peroxide. The detoxified prehydroysate was further treated with pectinase to hydrolyze the mannose oligomers to monomer. After these treatments, the hysrolyzate was efficiently converted to ethanol. Use of mixture of prehydrolysate and pulp mill sludge as the fermentation feed has given significant increase of the ethanol concentration in the bioreactor over that of prehydrolysate fermentation.

See more of this Session: Innovations in Bio-Refinery Processes
See more of this Group/Topical: Process Development Division

(467b) Molecular Design of Solvents for 2nd Generation Biofuels Separation Processes

Ana Silveira[1], José Scilipoti[1], Martin Cismondi[2] and **Esteban A. Brignole**[3], (1)Chemical Engineering, PLAPIQUI - Universidad Nacional del Sur - CONICET, Bahía Blanca, Argentina, (2)Chemistry, Facultad de Ciencias Exactas y Naturales, Universidad Nacional de Córdoba, Córdoba, Argentina, (3) Plapiqui, PLAPIQUI - Universidad Nacional del Sur - CONICET, Bahía Blanca, Argentina

The second generation biofuels are based on the use of biomass , mainly from lignocellulosic sources, as raw materials for the production of biofuels and for the chemical and paper industry. Many challenging separations problems are faced for the conversion of cellulosic materials in fuels and chemicals. Even though there are different wood processing, in the present work we focus mainly on the process of detoxification of hydroxilates; that will be used for the production of ethanol and other chemicals by fermentation processes.

In this work, the ECOFAC molecular design of solvents program (Cismondi and Brignole [1]) is upgraded to deal with the separation of organic chemicals found in wood hydroxilates, that are toxic for the fermentation process. For this purpose, the last UNIFAC parameter data bank revision are updated with new parameters reported in the literature for the components of interest. Examples of selection of solvents for different toxic components are presented and a strategy, for solvent selection with multiple objetive optimization goals, is presented. The ECOFAC program is also updated for the prediction of environmental properties, among others: octanol – water partition coefficients and water solubilities of organic compounds.

The program also incorporates a more general molecular design procedure to generate families of lignocellulosic derivatives that are toxic for the fermentation process. In this way potential solvents can be screened in their capacity to separate these products. As a difference with a typical extraction process the raffinate is in this separation problem the product to be recovered and for this reason the primary solvent properties are more restrictive than in conventional extraction processes.

Another option under development for this particular type of application is the use of A-UNIFAC, an group contribution approach taking into account association.

References [1] M.Cismondi and E.A.Brignole, Molecular Design of Solvents: An efficient search algorithm for branched molecules. Ind. Eng. Chem. Res. 43(2004) 784-790

See more of this Session: Innovations in Bio-Refinery Processes
See more of this Group/Topical: Process Development Division

REACTIVE EXTRACTION AS A METHOD TO PRODUCE BIODIESEL FROM *JATROPHA CURCAS* SEEDS

Farizul H. Kasim, Adam P. Harvey, Newcastle University, Newcastle, UK

Abstract

Jatropha curcas (jatropha) produces oil that can be converted to biodiesel by transesterification. However, because of its high free fatty acid value (>2 %), acid catalysts are usually used in the process, rather than alkalis. Because of this, longer reaction times are required. Furthermore, jatropha oil contains phorbol ester, a carcinogen that can be exposed when the seeds are extracted for the oil. The application of reactive extraction as a method to produce biodiesel from jatropha, enables alkali based catalyst to be utilized, and shortens the reaction time without compromising on conversion and quality of biodiesel. Since seeds are used as raw material instead of oil, the phorbol ester exposure is greatly reduced. This study evaluates the parameters involved in reactive extraction of jatropha seeds. The molar ratio of alcohol to oil was found to be the most important parameter in this reaction. The yield in 1 hour was low at 100 molar ratio (2%) but then increased sharply when the ratio was increased to 200 (55%), 300 (75%) and 400 (82%). Only small increments were observed when the ratio was increased further to 500 (85%) and 600 (90%). Methyl ester yield was a strong function of seed particle size. Of 5 different sizes of particle size, the smallest (0.5 mm), produced the highest yield (86%) and the biggest (2.0-4.0 mm) produced the lowest (35%). On agitation speed parameter, low speed agitation (100 rpm) was unable to produce a high yield compared to high speed agitation (300, 400 rpm). The study on temperature meanwhile revealed that there was little change in yield when the reaction was performed at 30°C (90%), 40°C (89%), 50°C (91) and 60°C (98). Interestingly, it was shown that the process was complete relatively quickly, typically after 30 minutes. The methyl ester content of the samples demonstrate that high purity of methyl ester (<90%) can be obtained from this method. An implication of this is that reactive extraction could be a method of producing biodiesel from seeds with high free fatty acid contents.

Introduction

Current biodiesel production technology converts refined oils to biodiesel. The oils undergo a number of processes: mechanical crushing and solvent extraction (using hexane) to produce the raw oil, followed by degumming and esterification steps to clean up the oil prior to transesterification to biodiesel. If, instead of oil, oilseeds themselves can be converted directly to biodiesel, all of these steps can be missed out. This should greatly simplify the process, which should make it suitable for small-scale distributed production, as the capital cost and service requirements will be significantly reduced. This should allow oilseed farmers to realise more of the full value of their crop, and simultaneously reduce local biodiesel prices. The process is "greener" as hexane is no longer required.

The aim of this work is to develop a process for directly converting jatropha seeds to biodiesel, via "reactive extraction". The specific objectives are: to demonstrate that this can be done and to map out the parameter space affecting the process. Jatropha is of particular interest as it can grow on marginal or degraded land, so does not compete with food crops for land. Due to the lower land quality requirement, it does not require land clearance, as is so often the case with palm and soya. Another benefit in terms of the food vs. fuel debate is that it is inedible, so is not in the same market as the edible oils.

Materials and methods

Preparation of J. curcas Seeds

Jatropha seeds were provided by Indian Institute of Petroleum (IIP), Dehradun, India. The seeds were stored in dark air-tight containers. The oil content and free fatty acids (FFA) value of the seeds were determined prior to experiments and whole jatropha seeds (shell and kernel) were utilised. The seeds were ground using a blender (Kenwood BL440, UK) to reduce the particle size. A vibrator sieve-shaker (Retsch, Germany) was then employed to separate the ground seeds into different sizes. These seeds then were placed in a drying pan and were dried in a drying oven (Memmert, Germany), at 105°C for an hour.

Chemicals and Reagents

Methanol (99.8% purity) was purchased from Fisher Scientific, UK. Sodium hydroxide (NaOH), in pellet form was procured from Merck, Germany. Hexane and acetic acid (glacial) which were used in post-reaction steps were acquired from Fisher Scientific, UK as well. Heptane, which was used as a solvent in gas chromatography sample preparation and the internal standard, methyl heptadecanoate, were purchased from Sigma Aldrich, UK.

Reactive Extraction

Reactive extraction experiments were conducted in sealed, chemically resistant Schott bottles (250 ml). The bottles were filled with methanol to achieve the required pre-determined molar ratio and relevant amount of catalyst (NaOH) then was dissolved in the methanol. The mixture was then pre-heated to the reaction temperature in a programmable incubator shaker (IKA, Germany). Jatropha seeds (10 g) were put into the bottles after the mixture reached previously set temperature. The reaction was conducted in an incubator shaker with controlled temperature, agitation speed and time.

The bottle was taken out from the incubator and vacuum-filtered after the reaction. This will separate the meal and the liquid mixture. To ensure that the transesterification reaction stopped, the catalyst was neutralised using a small amount of glacial acetic acid. 2 mL of the sample was then analysed by gas chromatography (GC) to determine the yield. The remaining mixture was put in an evaporator flask, to evaporate the methanol. The rotary evaporator (Buchi, Switzerland), which was set at 55°C - 60°C evaporated the methanol and collected it in another connecting flask.10 mL of hexane then added to the remaining mixture. The solution then was transferred to separating funnel and was allowed to separate gravitationally. After 1 hour, two distinct layers, the biodiesel-rich upper layer, and the glycerol-rich lower layer appeared. The upper layer was yellowish in colour the lower layer, dark brown. The mixture was then washed with warm water to remove impurities from the upper layer. After the glycerol layer was drained, the hexane in the upper layer was evaporated using a hot plate in fume hood. The mass of the purified biodiesel was then recorded, and analysed by GC to determine percentage methyl ester. The solid part of the reaction (the meal) was heated in the oven at 100°C to remove the methanol. It was then weighed and recorded to determine the percentage of weight loss.

Table 1 listed the setting of all parameters studied in the experiments. The coloured boxes indicate the parameter which was varied for the particular experiment set.

Table 1: List of experiments conducted with respective parameters setting

Experiment	Parameters					
	Seed size (mm)	Mixing speed (rpm)	Reaction temperature (°C)	Reaction time (min)	Catalyst concentration (N)	Alcohol:oil molar ratio
Effect of particle size	<0.5-4	400	60	60	0.1	400
Effect of mixing speed	<0.71	100-400	60	60	0.1	400
Effect of reaction temperature	<0.71	400	30-60	60	0.1	400
Effect of reaction time	<0.71	400	60	10-60	0.1	400
Effect of catalyst concentration	<0.71	400	60	60	0.1-0.2	400
Effect of methanol to oil molar ratio	<0.71	400	60	60	0.1	200:1-600:1

Analysis

A flame ionisation detector (FID)-equipped HP5890 Series II (Hewlett Packard, USA) gas chromatograph fitted with BPX70 column, 30 m long x 0.32 mm ID x 0.25 μm film thickness (SGE, Australia) was used to analyse the samples. Helium was used as carrier gas at a pressure of 7 psi. The oven was maintained at 230°C for 30 minutes.

The data was acquired and processed using Clarity Chromatography Station for Windows (DataApex, Czech Republic). The software allowed integration of peaks to be performed on the chromatogram. For the calculation of FAME conversion as well as methyl ester content, the solvent peak was excluded from the chromatogram, and this was done by integrating the peaks eluded from 10 – 30 minutes.

The internal standard, methyl heptadecanoate solution was prepared prior to the sample preparation. 250 mg of sample was then weighed and placed in the vial. After that, 5 ml of methyl heptadecanoate solution was added to the sample. The mixture was mixed thoroughly using the MS1 Minishaker (IKA, Germany).1 μL of sample was then injected to the GC using a 5 μL microsyringe (SGE, Australia).

Yield and FAME Percentage Calculation

The yield was calculated by modifying the standard for calculating FAME (BS EN 14103:2003). Firstly, the mass of the liquid mixture after the separation of solid (meal) and liquid (mixture of methyl ester, methanol, sodium hydroxide, glycerol and unreacted triglycerides) was taken and recorded (w). Methanol was used to prepare the internal standard. After adding internal standard to the sample, the sample was then injected to the GC.

The ester content in the sample, C, expressed as a mass fraction in percent, can be calculated using Equation 1 below.

$$C = \frac{(\Sigma A) - A_{Ei}}{A_{EI}} \times \frac{C_{EI} \times V_{EI}}{m} \times 100\% \qquad \textbf{Equation 1}$$

where

$\sum A$	= total peak area from the methyl ester $C_{14} - C_{24:1}$
A_{EI}	= peak area corresponding to methyl heptadecanoate
C_{EI}	= the concentration, in mg/ml of the methyl heptadecanoate solution
V_{EI}	= the volume, in ml of the methyl heptadecanoate solution being used
M	= mass of the sample (mg)

The yield then was calculated by multiplying C with mass of the sample, w, as in Equation 2.

$$Yield(\%) = C(\%) \times w(g) \qquad \textbf{Equation 2}$$

FAME content is the percentage of methyl ester in the biodiesel rich phase of the sample. As in the yield calculation, the method and calculation was based on the BS EN 14103:2003 standard. In the sample preparation, heptane was used as a solvent to dissolved methyl heptadecanoate, the internal standard.

Results and Discussion

The results of all the experiments are as in Figure 1. In this study, the molar ratio of methanol to oil was varied from 100 to 600 (Figure 1a). No methyl ester was produced at the lowest molar ratio (100), even though 18.1% of mass was extracted from initial 10g of seeds. The yield of methyl ester was significantly higher at a ratio of 200, although there had been little increase in % mass loss (21.9%). The yield then increases steadily as the ratio increases. The yield at a molar ratio of 300 is 74.7% and the yield for 400, 500 and 600 are 81.9%, 85.7% and 86.9%, respectively. The mass loss levels off when the ratio reaches 400.

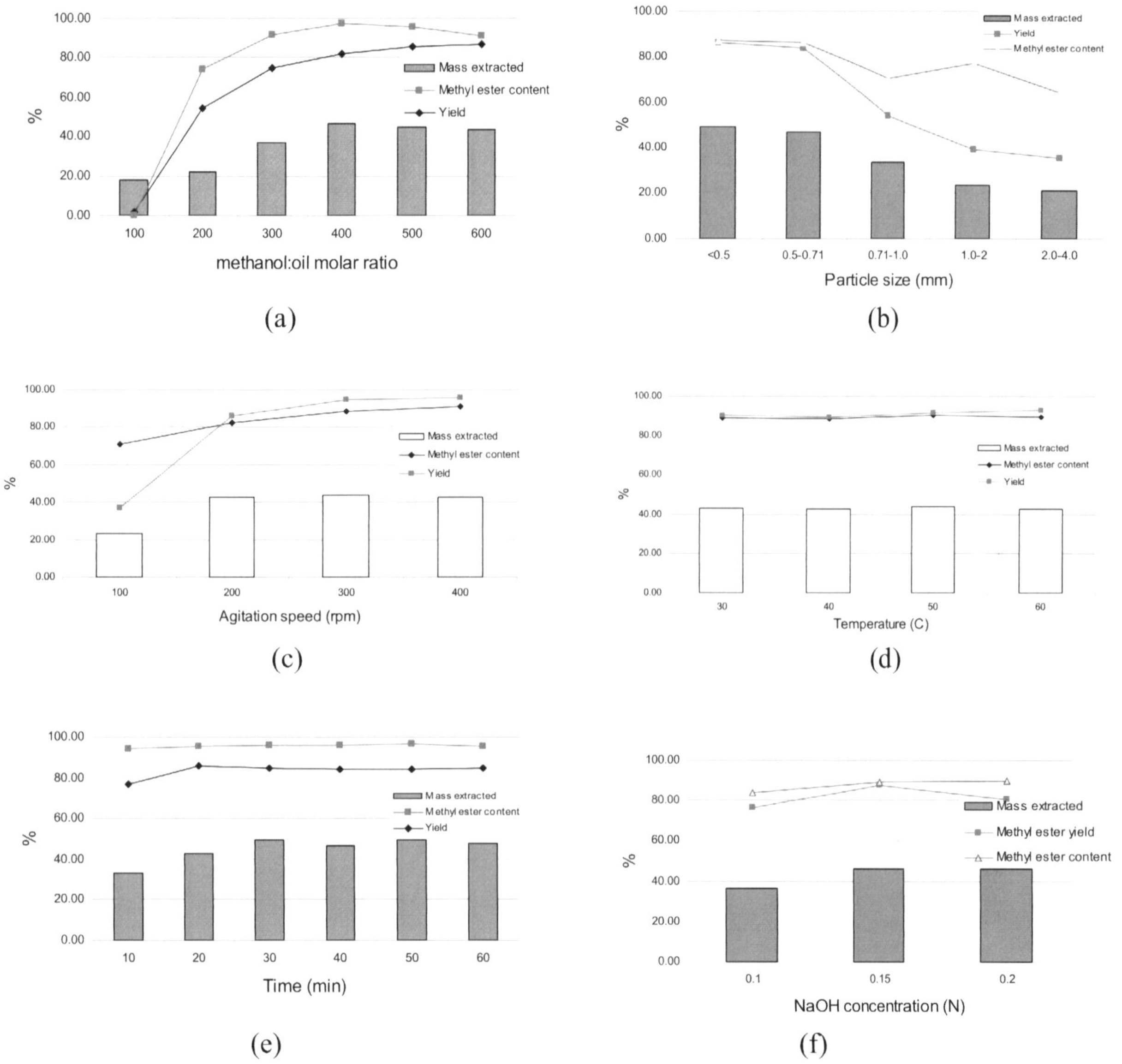

Figure 1. Methyl ester yield, methyl ester content and mass extracted on various experiments parameters (a) 100 to 600 molar ratio of methanol to oil.; (b) <0.5 mm to 2-4 mm particle size; (c) 100 to 400 rpm agitation speed; (d) 30°C to 60°C reaction temperature; (e) 10 minute to 60 minute reaction time and (f) 0.1N to 0.2N NaOH concentration. The other parameters set as in Table 1.

The results suggested that the amount of methanol must be very high to achieve an appreciable rate. This was presumably to drive the penetration of alkaline methanol into the seed, as observed in Ren *et al* [1]. Further increasing the excess of methanol (e.g.600) does not greatly increase the yield, but is undesired, as it will increase the load on downstream separation processes. This plateau has also been observed by Qian *et al* [2]. They also found, however, that the amount of methanol required was a function of the catalyst concentration. Haas *et al* [3] also found that the greater amount of alcohol used in the process, the lower the catalyst concentration needed to give a good yield. Above a molar ratio of 300, the methyl ester content is above 90%, but it exhibits a maximum around 400, indicative of significant amounts of triglyceride had been converted into methyl ester.

In Figure 1(b), the yield of methyl ester decreases with increasing particle size, when the particle size is larger than some threshold value of around 0.71 mm. No significant difference was determined between the yield for the smallest two particle size ranges, <0.5 mm and 0.5 – 0.71 mm, with the former yielding 86.1% conversion and the latter 83.7%. The largest particle size range (2–4 mm) produced the lowest yield, at 35.5%. This finding is consistent with those reported by Qian *et al* [2]. In terms of methyl ester content, only the two smallest particle sizes produced more than 80% of methyl ester in the sample. The other particle sizes produced lower methyl ester contents; with size 0.71-1 mm has 70.5% methyl ester, size 1-2 mm; 77.0% and 2-4 mm, 64.0%. No further analysis was done to determine the other compounds in the sample, but the most likely suggestion would be the presence of unreacted glyceride in the form of mono-, di- and triglycerides [4.5]. The chances of other compounds such as glycerol and polar lipids being present in the sample are small, as they will have been removed by settling and hexane washing.

It is noticeable for Figure 1(c) that the mixing intensity is not a mass transfer limiting factor once it reaches 300 rpm. The finding suggests that there is no point increasing the agitation beyond this level, as it will not lead to a significant improvement in yield. However, decreasing the speed to 200 rpm, and further to 100 rpm, decreased the yield (from 94.8%) to 85.7% to 37.2%. A similar pattern was also observed by Meher and co-workers in their work on alkali-catalyzed transesterification of *Pongamia pinnata* oil [6]. They attributed the low yield (< 85%) achieved with 180 rpm mixing intensity to incomplete reaction as opposed to 97% yield when 360 rpm and 600 rpm mixing intensity were applied.

There was no significant difference between the masses extracted at the 4 different temperatures (Figure 1d). There was also minimal change in the yield and methyl ester content. The results obtained, are in agreement with observations by Haas and co-researchers [3]. Both observed that triglyceride can be converted to FAME at both low and high temperature (23°C and 60°C in the case of Haas' group). However, they also found that at the lower temperature 2.4 times more methanol was required to achieve the same conversion as at 60°C. This phenomenon also has been reported elsewhere [2]. In contrast, temperature influences both the reaction rate and yield in conventional transesterification [7].

The results obtained from the reaction time experiments are presented in Figure 1(e). The mass extracted from the seeds during reactive extraction were similar (around 40%) starting from 30 minute onwards. This result mirrors the methyl ester yield, which also reached a plateau after 30 minutes. It is therefore very likely that the reactive extraction itself completed between 20 and 30 minutes. It also can be observed that reactions less than 20 minutes did not achieve high yields. This is in agreement with other findings, which report rapid increases in yield within the first 30 minutes of reaction [2,3].

The addition of NaOH, although in small quantity (0.1 N) give a significant effect in term of converting oil to methyl ester. Increasing the concentration to 0.15 causing the yield to increase from 76.2% to 87.8%. However further increment to the concentration (0.2 N) decrease the yield to 80.8%. It is also interesting to note that, with a further increase in NaOH concentration, an emulsion started to form, consequently reducing the yield. The most likely cause of the emulsion is the formation of soap which is a competing reaction in alkali-catalysed transesterification process. Interestingly, it is apparent from Figure 1(F) that through reactive extraction, the high free fatty acids content of jatropha oil had no negative effect to the yield.

Conclusion

This study was designed to determine the effect of various parameters on the reactive extraction of jatropha. Of the 6 parameters investigated (particle size, mixing speed, reaction temperature, reaction time, NaOH concentration and molar ratio of methanol to oil), 5 parameters have some effect on the yield of FAME. The only parameter which did not exhibit any significant change in 60 minute is the reaction temperature. Particle size and molar ratio of methanol to oil strongly affect the yield of methyl ester: it increases with increasing molar ratio to an optimum, and increases with decreasing particle size to a plateau. Investigation of the effect of mixing speed indicated that a certain minimum speed must be provided to minimise the immiscibility between oil and alkaline methanol, otherwise the FAME yield decreases. The reaction was shown to be complete after 20-30 minutes. This study demonstrated that for jatropha, as for other oilseeds, some form of alkaline catalyst, in this case NaOH, is required to achieve a meaningful yield when performing extraction with. Finally, there is an important limitation that must be considered. It is impossible to rule out the interaction between parameters as a consequence of the involvement of 6 different parameters in this reactive extraction reaction. Therefore, this one-at-a-time series of experiments is a precursor to a full design of experiments, in which the interaction between parameters can be identified and quantified.

References

1. Ren Y, Harvey A. and Zakaria R. (2010), "Biorefining based on Biodiesel Production: Chemical and physical Characterisation of Reactively Extracted Rapeseed," *Journal of Biobased Materials and Bioenergy*, 4(1), pp. 79-86.

2. Qian J., Wang F., Liu S. and Yun Z (2008) "In Situ Alkaline Transesterification of Cottonseed Oil for Production of Biodiesel and Nontoxic Cottonseed Meal," *Bioresource Technology*, 99(18), pp. 9009-9012.

3. Haas MJ., Scott KM., Marmer WN. and Foglia TA. (2004), "In Situ Alkaline Transesterification: An Effective Method for the Production of Fatty Acid Esters from Vegetable Oils", *Journal of the American Oil Chemists' Society*, 81(1), pp. 83-89.

4. Gerpen JV., Shanks B., Pruszko R., Clements D. and Knothe G. (2004), "Biodiesel Production Technology". *National Renewable Energy Laboratory (NREL) Report*, Colorado.

5. Ma F. and Hanna MA.(1999), "Biodiesel Production: A Review," *Bioresource Technology*, 70(1), pp. 1-15

6. Meher LC., Dharmagadda VSS. and Naik SN., "Optimization of Alkali-catalyzed Transesterification of Pongamia Pinnata Oil for Production of Biodiesel," *Bioresource Technology*, 97(12), pp. 1392-1397

7. Meher LC., Vidya Sagar D. and Naik SN., "Technical Aspects of Biodiesel Production by Transesterification - a Review," *Renewable and Sustainable Energy Reviews*, 10(3), pp. 248-268

(467d) A Novel Approach for Carboxylic Acid Recovery From Fermentation Broths

Alvaro Orjuela, Chemical Engineering & Material Science, Michigan State University & Universidad Nacional de Colombia, East Lansing, MI, Abraham J. Yanez, Chemical Engineering & Material Science, Michigan State University, East Lansing, MI, Carl T. Lira, Chemical Engineering and Material Science, Michigan State University, East Lansing, MI and Dennis Miller, Michigan State University, East Lansing, MI

Biobased fermentation processes uniformly produce dilute aqueous solutions of the carboxylic acid as the primary products. Many of those fermentation systems yield a mixture of acids and other products, depending on the intended metabolic pathways and conditions within the bioreactor. The recovery and purification of individual species from these fermentor effluents can be expensive, accounting for 50 to 80% of the overall production cost. For succinic acid fermentation, significant amounts of acetic, formic, pyruvic and/or lactic acid can be formed as byproducts and are present in the broth in addition to non-converted sugars, biomass (cell growth), salts, and supplements. Because fermentation is most efficient at intermediate pH (5-7), base (NaOH, Ca(OH)2, ammonia) is used to control pH that naturally declines as the acid accumulates. The carboxylic acids produced are thus typically obtained as salts of the base used. Several alternatives have been proposed to recover succinic acid from the acid salt broths; these can be classified according to the approach as direct precipitation (as salt or acid), extraction (solvents, amines), ion exchange, membrane recovery, or esterification. In the traditional process, the aqueous broth containing succinate salt is acidified with H2SO4 at a pH below the pKa1 of succinic acid (pH < 4). With calcium as the cation, calcium sulfate is precipitated; the sulfate is removed by filtration and the SA is subjected to several crystallization steps. Because succinic esters are useful as platform species to produce a variety of value-added commodity and specialty derivatives, we have developed and present here a novel approach to directly recover succinic acid esters from fermentation. Following partial water removal, the process performs acidification with sulfuric acid in ethanol. Acidification of salts in EtOH releases succinic acid which simultaneously reacts to form mono- and di-ethyl succinate, while inorganic salts (sodium or calcium sulfate) precipitate out of the solvent. Other carboxylic acids removed in traditional processes are also recovered as ethyl esters which can be separated easily than original mixed acids. Experiments performed using pure and mixed succinate and acetate salts and actual fermentation products show high recovery efficiency and rapid conversion rates.

See more of this Session: Innovations in Bio-Refinery Processes
See more of this Group/Topical: Process Development Division

A TECHNO-ECONOMIC SIMULATION OF ALTERNATIVE FEEDSTOCK STRATEGIES FOR AN INTEGRATED FOREST BIOREFINERY AIChE 2010 ANNUAL MEETING

Anna Kalliola, VTT Technical Research Centre of Finland, Espoo, Finland
Juha Hakala, VTT Technical Research Centre of Finland, Espoo, Finland
Pertti Koukkari, VTT Technical Research Centre of Finland, Espoo, Finland

Abstract

A simulation based feasibility study of a paper mill producing simultaneously printing paper and mechanically pre-treated biorefinery feedstock was performed. A process model of a modern paper mill served as a reference for the new process alternatives, in which a TMP reject fraction (feedstock) was conducted out of the papermaking process. The share of the removed pulp fraction in the paper was compensated either by using other raw materials (*e.g.* recycled pulp) or alternative process/paper machine operation practices. Manufacturing costs for each new concept were estimated by computing the variable costs. The cost effects were allocated both for tonne of paper produced and for tonne of TMP reject pulp produced. Thus, a unit cost for the biorefinery feedstock was obtained for each alternative. The unit cost varied from 60 to 360 €/ad-t feedstock.

The simulation method developed can be utilized for the early evaluation of alternative concepts of process renewal.

Keywords: Integrated Forest Biorefinery, Mechanical pulping, Production economics, Simulation Based engineering

Introduction

Economic pressures and more stringent legislation on environmental issues have enforced the renewal of the forest industry. The pulp and paper making (P&P) companies are pursuing new revenues from the emerging lignocellulosic biorefining industry by producing biofuels alongside with their conventional products. The development of sustainable wood-based non-fuel products is also widely investigated. However, the multitude of process and product options, combined with the raw material base in the mill region, makes the industry renewal a difficult task. The current integrated biorefineries are mainly utilizing the P&P side streams or by-products as feedstock. However, it is anticipated that in the future multiple volumes of feedstock will be needed to meet the increasing supply of bioproducts.

The use of mechanical pulp is an established part of current papermaking. In year 2009 the volume of all mechanical and semi-chemical pulp produced in Finland was 3.2 Mt being 37 % of total pulp produced [1]. In thermomechanical pulping (TMP) wood chips are refined at elevated temperatures and under pressure into fibrous pulp. A significant part (25-40%) of the pulp produced at the main refining stage belongs to a sc. reject fraction which needs to be further refined at a reject refining stage to reach the desired fineness for the final paper products. Reject refining is energy intensive. With a successful choice of new uses for this raw material fraction, the process could be simplified and the profitability of the production site could be improved.

The average production of the reject fraction in a modern TMP department is 100-200 t/day. This can be considered as a high volume source of mechanically pre-treated feedstock for new bioproducts. A paper mill producing simultaneously its conventional product *i.e.* paper and 'TMP feedstock' for biorefining then offers a pertinent industrial case for a study with a concept model. In this paper five different process concepts without the reject refining stage were compared for evaluating the techno-economic feasibility of the simultaneous production of printing paper and biorefinery feedstock. The processes were simulated with the steady state process simulator Balas (a software product of VTT) [2] to assess material and energy balances for each concept. Their manufacturing costs were compared by computing the variable costs. The potential bioproducts refined from the TMP reject fraction were yet excluded from this study.

Approach for techno-economic process simulation

A process model of modern paper mill [3] consisting of TMP department, recycled fiber deinking process, paper machine, power plant, and effluent treatment served as a reference (Figure 1) in this comparative study. The reference mill was producing 1335 tons of light weight coated (LWC) printing paper on daily bases. The main raw materials used in the paper production were mechanical fiber (TMP), chemical fiber (bale kraft), filler and coating pigments. Recycled fiber (RCF) was not a paper component in the reference case (RCF line was not operational). However, the RCF line was part of the model to enable the simulation of alternative process concepts.

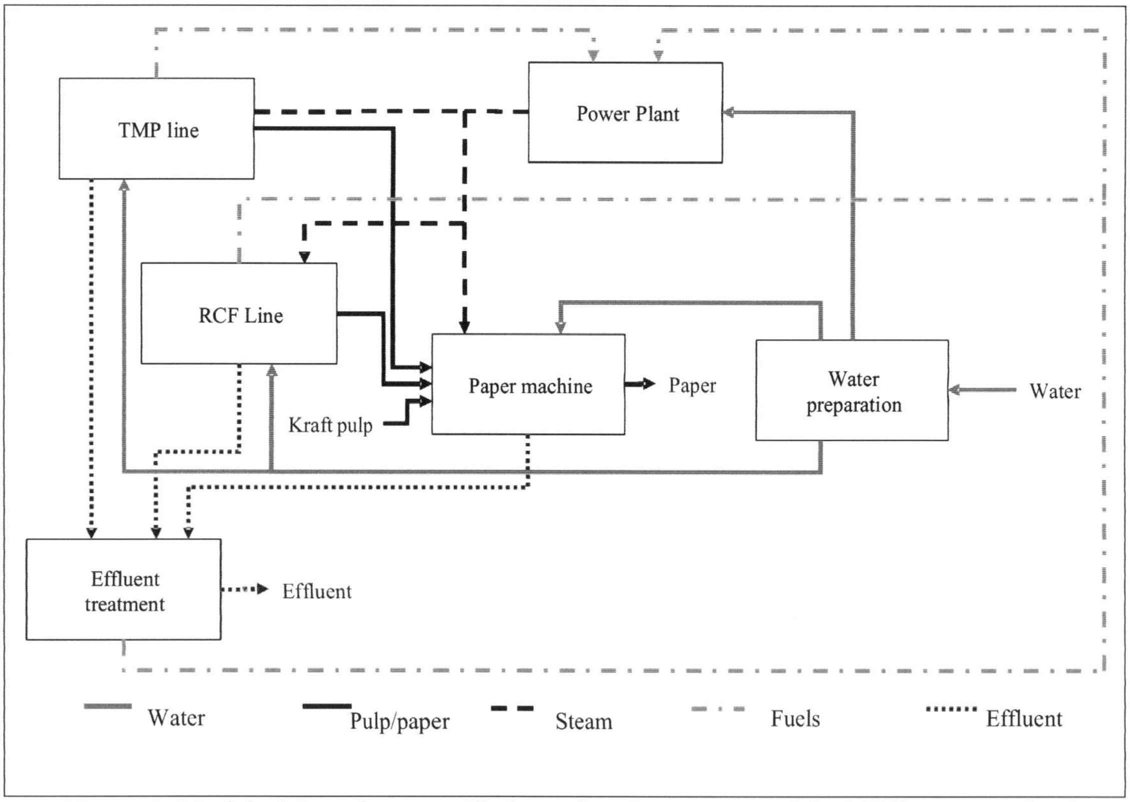

Figure 1. Model of the reference mill: the main sub-processes of the LWC paper production.

A general problem of process analyses, particularly when evaluating emerging technologies, is the confidentiality of process and price data. Yet, by using generic key data from available sources this problem may be circumvented and at least a consistent comparison of alternatives can be produced. For considering the economic aspect in this study, unit prices for the variable costs (raw materials, electricity, oil etc.) were obtained from public databases such as Foex and Eurostat (Table 1), which were applied on the simulated process values by using MS Excel interface. An estimated average price level of January 2009 was then adapted for all simulation alternatives.

Table 2. Unit prices for the variable costs considered in the simulations, based on average price level on January 2009 [4,5,6,7,8,9].

Cost component	Cost	Unit	Notes
Northern bleached kraft pulp (softwood, long fibre)	454	[€/ ad-t]	90% d-c
Bleached kraft pulp (hardwood, short fibre)	418	[€/ ad-t]	90% d-c
Wood (softwood) at the factory	33.7	[€/ t]	42% d-c
Sorted graphic paper for deinking	80.4	[€/ad- t]	
Filler	110	[€/ dry-t]	
Water	0.05	[€/ m3]	
Electricity	50.0	[€/ MWh]	
Natural gas	23.4	[€/ MWh]	
Oil (heavy fuel oil)	0.36	[€/ kg]	
CO2 emission trading	15	[€/ ton CO2]	
Ash treatment cost	0.002	[€/t]	
Effluent treatment costs:			
COD in	0	[€/t COD]	
Flow	0.30	[€/m3]	
COD released	0	[€/t COD]	

Process alternatives

The share of the reject fraction of TMP fibers is 35.9 % and the consistency 35 % (after reject screw press). Because there is no pulp entering the reject refining in the new concept, this stage can be uninstalled. The process model of the TMP line is presented in figure 2. Specific energy consumption of the main and reject refining as well as the option for the new feedstock concept are shown.

The fiber loss in the paper product caused by the removal of TMP reject fraction corresponds to 19 % of total fiber loss. This loss can be compensated using other raw material components or could be ignored (hypothetical and causing deterioration in the strength properties of the paper without new improved paper forming technology).

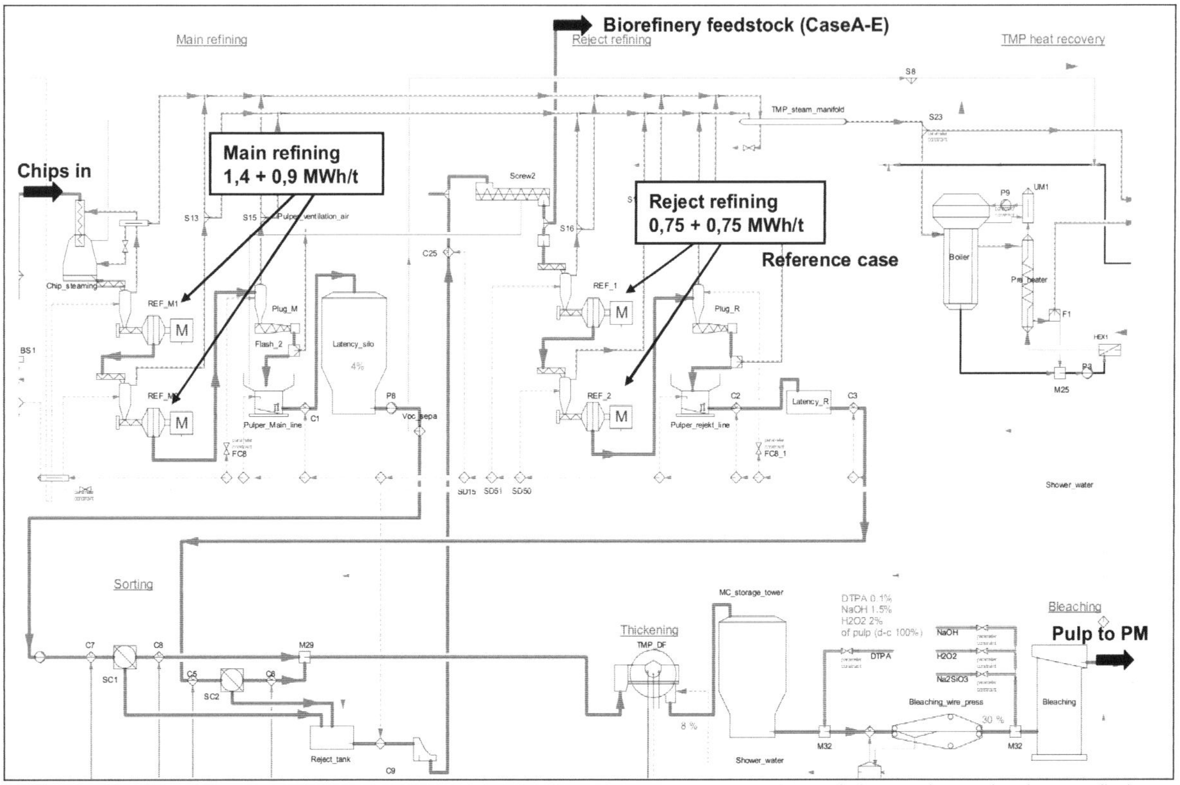

Figure 2. Process model of the TMP line. Specific energy consumption of the main and reject refining, and the option for the new feedstock concept are shown in the figure.

Altogether five different process concepts without the reject refining stage (cases A-E) were simulated as listed below. The TMP reject pulp was conducted out of the process in each alternative. In three cases (B, D, E) the per tonne based paper production was kept constant (the same as with the reference) and the pulp removed was compensated either with increased wood intake or with recycled or chemical kraft pulp. In two cases (A, C) the paper production was assumed lower than in the reference (either thinner paper or less paper) to adjust with the decreased share of TMP pulp.

The net changes in the cost structures (reference mill as comparison basis) were computed for each alternative as cost effects. They were allocated both for tonne of paper produced and for tonne of TMP reject pulp produced. Thus, a unit cost for the biorefinery feedstock was obtained for each process option. For the reference LWC paper mill the costs considered here covered ca. 75% of manufacturing costs even though labor and chemicals were not included.

Also, a process concept (case F) with the acceptance of reject fraction as such for the paper production was simulated. In this alternative the reject refining was not needed and the mill integrate was producing only paper without the feedstock production. The case F is rather hypothetical and describes a condition with lower quality requirements of the paper product.

Reference mill

- Intake of spruce chips to TMP line ca. 1360 t /day (at dry content of 42 %)
- 35,9% of TMP is refined at reject refining stage
- Specific energy consumption (SEC) at reject refining 0.75+0.75MWh/t
- Daily based paper production 1335 t, grammage 70 g/m^2

Case A: Less paper

- Daily based paper production (tons) is decreased to adjust with the decreased share of TMP pulp (reduction of PM speed, paper grammage 70 g/m^2)
- The share of the other paper making raw materials such as kraft, filler and coating pigments is decreased to adjust the decreased paper production, and *e.g.* the flow rate of PM shower waters is connected the production

Case B: More wood

- Increased wood intake to compensate the share of TMP reject fraction (feedstock)
- The share of the other paper making raw materials were kept constant
- No changes in paper machine (same speed and production as with the reference)

Case C: Thinner paper

- Paper grammage decreases to correspond the removed TMP share
- No change in PM speed
- Wood, bleached kraft pulp and filler intake remains constant in tonnage basis, but due to reduced TMP pulp share in the paper, their specific intake values increases

Case D: More kraft

- Chemical fiber (bale kraft) usage is increased to correspond the removed TMP share

Case E: Added recycled paper

- Recycled fiber is used to correspond the removed TMP share
- RCF process line is a part of the mill integrate

Case F: TMP reject to PM

- TMP reject pulp is accepted to the paper production
- No simultaneous biorefining feedstock production

Results and discussion

A unit cost for the TMP reject pulp (feedstock), corresponding the per tonne income needed from the biorefinery products (here unspecified) was obtained for each process alternative (case A-E). Figure 3 indicates the amount of feedstock available in each case - grey squares, while the cost effect to be covered with the new products are shown in black. The unit cost varied from 60 to 360 €/ad-t feedstock. True feedstock production cost in the value chain is higher due to other possible manufacturing costs (direct labor, manufacturing overheads) and fixed costs (*e.g.* investments). Also, feedstock selling price needs to cover the requested profit margin.

Figure 4 presents the paper production in each case, and the cost effect (variable costs) of the reject pulp computed per tonne of paper produced. The distribution of the variable costs per tonne of paper is illustrated in Figure 5. Main cost factors in each alternative are discussed below.

Case A: Less paper

The specific purchased electricity per tonne of paper increased since less paper was produced while the main line refining consumed as much electricity as before (in the reference case). By-passing the reject refining did not cover this effect. Wood purchasing did not change, which led to increase in wood specific consumption per tonne of paper since less paper was produced. Usage of the other raw materials per tonne of paper followed the paper production rate, and did not cause changes in their specific consumptions.

Case B: More wood

Paper production rate was kept constant, but the wood intake was increased. The greater flow through the main line refining was needed to produce the same amount of paper. The specific electricity consumption was thus increased. By-passing the reject refining was not enough to cover this effect. Also, more process steam was generated and it was utilized at the paper machine, so that the need of low pressure steam produced at the power plant was diminished. Thus, in the power plant more steam was conducted to a condensing turbine causing little compensation on the electricity purchase. As in case *Less paper*, there were no changes in the specific consumptions of other raw materials.

Case C: Thinner paper

Production of thinner paper saves raw materials and utilities. Case *Thinner paper* was evaluated both on tonnage and on surface area basis. The latter considers the functionality of paper surface area.

Tonnage basis: Specific electricity savings were obtained since the share of pulp and filler in the paper recipe were increased, replacing the energy intensive TMP pulp.

Surface area basis (not presented in Figure 4): There were no significant changes in the amount of raw materials per surface area of paper, since the produced surface area remained almost constant (19.15 km^2/day). The paper was just thinner. Remarkable savings were obtained in purchased electricity since the same surface area of paper was produced with less energy intensive paper recipe (less TMP pulp) and without the reject refining. Savings in refining electricity reduced the process steam generation, increasing the need of low pressure steam from power plant, and thus also increased the oil usage (as auxiliary fuel).

An important factor is the pricing of the thinner paper. If the price can be kept at the same level as for the original reference paper, assuming that the quality is maintained, savings can be reached despite the fact that biorefinery feedstock is produced simultaneously from the same wood intake as with the reference mill. However, reduction in paper price should be expected, due to the quality decrease. If pricing remains on tonnage basis, price increase is required since more functional area is provided per tonne of paper.

Case D: More kraft

The share of TMP reject pulp was compensated with the bleached kraft pulp to maintain the paper production constant. Purchase of kraft pulp is clearly seen in Figure 5. Specific electricity consumption decreased due to the increased share of bale kraft pulp in the paper recipe, thus providing savings in electricity. Again, process steam generation decreased along with the decreased electricity consumption in refining, however finally increasing the additional usage of auxiliary fuel (oil). Cost effect was quite negative mainly due to purchase of kraft pulp.

Case E: Added recycled paper

The share of TMP reject pulp was compensated with recycled pulp produced from sorted graphic paper at RCF plant. Purchased electricity and auxiliary fuel consumption behaved similarly as in the case D. Main cost effect is caused by purchase of sorted graphic paper. Total cost effect was competitive when compared with other alternatives.

Case F: TMP reject to PM

TMP reject pulp was accepted to paper production. Purchased electricity reduced due to by-passing the reject refining. This effect was comparable with alternatives *More kraft* and *Added RCF*.

Reduced electricity consumption caused additional usage of auxiliary fuel. Total cost effect was favorable.

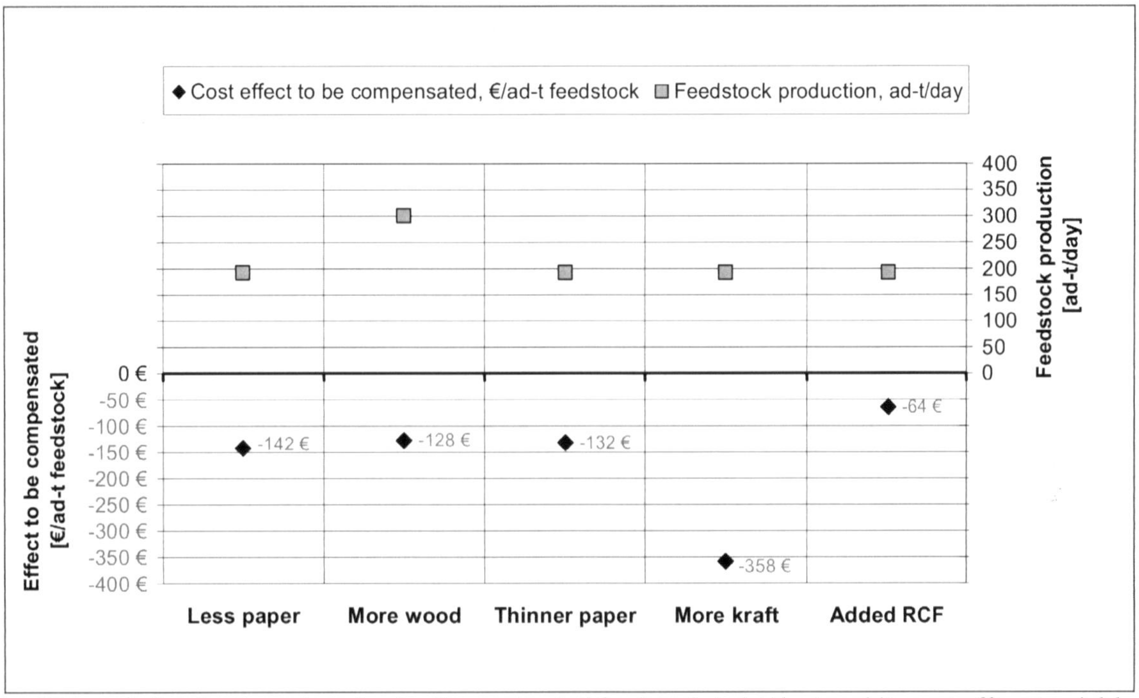

Figure 3. Production of mechanically pre-treated feedstock (ad-t/day) and its cost effect (variable costs, - sign means cost increase) to be covered (€/ad-t feedstock) in different process alternatives.

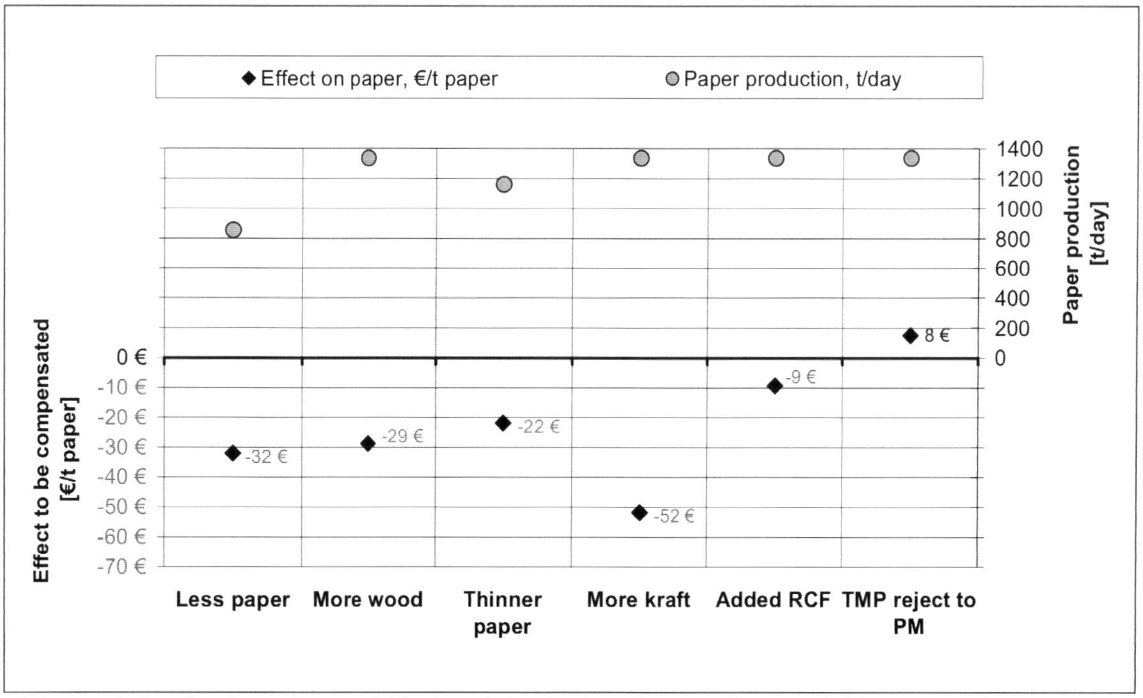

Figure 4. Production of paper (t/day) and the feedstock cost effect (variable costs, - sign means cost increase) to be covered expressed as per tonne of paper in different process alternatives.

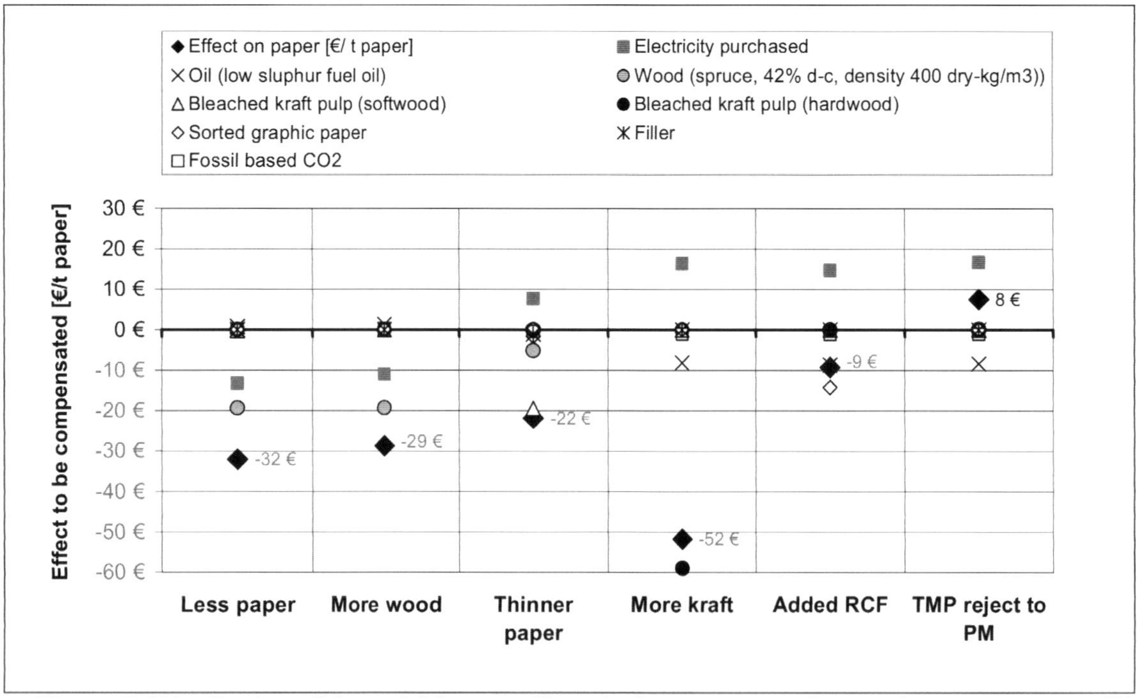

Figure 5. Distribution of the variable cost effects (- sign means cost increase) per tonne paper (€/t paper) in different process alternatives.

Conclusions

A simulation based techno-economic feasibility study of a paper mill producing simultaneously printing paper and mechanically pre-treated feedstock for new bioproducts was performed. Removing the reject refining stage allows for a major feedstock stream from the thermomechanical fraction of papermaking pulp for optional biorefining. The fraction removed may be compensated by reducing paper grammage, its production volume or by increasing other feedstock fractions. The viability of the process renewal alternatives depends on their technical feasibility, price levels and market conditions. The simulation suggests that addition of feedstock, either as wood or recycled fiber is necessary to compensate the lack of the TMP reject in paper production under the prevailing market conditions. With this compensation the combined effect of cost changes and biorefinery feedstock sales appear clearly positive and provide resources for additional investments. Yet, with a changing market situation the production of less or thinner paper together with new bioproducts may emerge as interesting alternatives in the future. For such cases, further studies on the aspects of both paper - and bioproduct qualities will be necessary.

References

1. Finnish Forest Industries Federation. 2009. Metsäteollisuustuotteiden tuotantomäärät Suomessa (Forest industry production rates in Finland) Metsäteollisuuden vuosi- ja nelj.vuositilastot (annual and quarter statistics). (http://www.metsateollisuus.fi) (http://www.forestindustries.fi)
2. www.balas.fi
3. Hakala, J., Manninen, J. Ruiz, J., Kamml. G, Lundström, A., Samuelsson, Å. 2005. D6.1.1 Description of the reference paper mills for ECOTARGET project. EU project Ecotarget deliverables 500345. (http://www.ecotarget.eu/news/ECOTARGET%20Technical%20deliverable%20D6.1.1.pdf)
4. Ekorosk Oy. 2009. Käsittelymaksut 1.1.2008 alkaen -ash (handling fees starting 1.1.2008 -ash (http://www.ekorosk.fi/en/default.html)
5. Eurostat. 2009. Natural gas prices for large industrial standard consumers. (http://epp.eurostat.ec.europa.eu/portal/page/portal/eurostat/home)
6. Eurostat. 2007. Electricity prices for large industrial standard consumers (http://epp.eurostat.ec.europa.eu/portal/page/portal/eurostat/home)
7. Finnish Oil and Gas Federation. 2009. Öljytuotteiden kuluttajahintaseuranta (Oil based products consumer price statistics) (http://www.oil-gas.fi/eng/)
8. Central Union of Agricultural Producers and Forest Owners (MTK). 2009. Kuitupuun hankintahintakehitys, viimeiset 52 viikkoa (pulpwood purchasing price development, the latest 52 weeks). (http://www.mtk.fi/en_GB/)
9. FOEX Indexes Ltd. 2009. (http://www.foex.fi)
   PIX recovered paper indexes Europe
   THE PIX pulp Europe benchmark indexes
   THE PIX paper Europe benchmark indexes

(467f) Integrated Production of Succinic Acid with the Separative Bioreactor

Yupo J. Lin, Cynthia S. Millard, Saurav Datta, Seth W. Snyder and Edward J. St.Martin, Argonne National Laboratory, Lemont, IL

Ion-exchange resins (IXR) have been widely used by industry for catalysis as well as an absorbent for removal of dilute salts from process streams. Electrodeionization (EDI), a combination of ion-exchange resin technology and electrodialysis membrane separation, has further enhanced the utilization of IXR and can significantly reduce capital and operating costs compared to conventional ion-exchange column. Currently, commercial EDI processes are used primarily to produce ultra-pure water for the semiconductor and pharmaceutical industries. ANL researchers have successfully improved on conventional EDI technology by developing "IXR wafers" to immobilize the IX resins into a porous substrate. The novel resin wafer (RW) enables application of EDI to a wider range of applications compared to conventional loose resin EDI processes. The resin wafer is a porous matrix of resin beads immobilized by polymer. Using this technology, ANL has developed several process implementations that take advantage of EDI for bioprocessing as well as bioconversion. The chemical industry is actively seeking bio-catalytic replacements for traditional chemical catalysis reactions. In most cases, separation represents a significant fraction of the overall cost of the process. Therefore, bio-catalytic process designs that focus on maximizing product concentration, limiting impurities, and reducing downstream processing steps are needed. To help meet these needs we developed the EDI Separative Bioreactor to integrate up-stream bioreactor and down-stream product separation into one unit. This membrane-based Separative Bioreactor enables simultaneous bio-catalytic production (fermentations or enzymatic reactions) of organic acids and immediate product separation. Direct capture of organic acid products from bio-transformations offers two immediate benefits: prevention of reactor acidification and avoidance of product inhibition. Argonne also observed that rapid organic acid capture during production offered a new way to control reactor pH without the addition of buffers for pH control. In this presentation, we will discuss an integrated separative bioreactor designed for pure succinic acid production. This system consist a CO2 capture reactor, a succinic acid fermentation and in-situ pure succinic acid recovery using resin-wafer electrodeionization (RW-EDI). The fermentation is setup in a temperature-controlled, stirred bioreactor using a succinic acid producing E.coli mutant developed at Argonne which converts glucose to succinic acid using one mole CO2 per mole of glucose to produce one mole of succinic acid and other products. CO2 is provided from the CO2 capture reactor in the form of carbonate which is formed by bubbling CO2 gas through a basic solution. This solution is used to neutralize the succinic acid fermentator and provide the CO2 for the reaction. The fermentation broth was constantly sent to the RW-EDI device to separate succinic acid from the solution and recycle the base to the CO2 capture reactor. This work was supported by DOE Office of Biomass Program.

See more of this Session: Innovations in Bio-Refinery Processes
See more of this Group/Topical: Process Development Division

(517a) Modeling of Steam-Air-Blown Gasification for Biomass in a Dual Circulating Fluidized Bed (CFB) Gasifier

Son Ich Ngo[1], Thanh D. B. Nguyen[1], Young - Il Lim[1], Won Yang[2], Uen-Do Lee[2] and Byung-Ho Song[3], (1)Department of Chemical Engineering, Hankyong National University, Anseong, South Korea, (2)High temperature processing R&D, Korea Institute of Industrial Technology, Cheonan, South Korea, (3)Department of Chemical Engineering, Kunsan National University, Gunsan, Jeonbuk 573 – 701, South Korea

A three-steady-stage thermodynamic equilibrium model including mass and energy balances was applied for steam-air-blown biomass gasification in a dual fluidized bed (DFB) to calculate the product gas composition, the carbon conversion, the heat recovery of biomass, and the heat demand needed in bubbling fluidized bed (BFB) gasifier. A complete combustion of both unconverted char and additional fuel was assumed in the combustion reactor (riser). The heat required for gasification reaction was provided by the circulating bed material (silica sand).

The final composition of the producer gas is obtained from two-stage equilibrium model incorporated with biomass pyrolysis. For the composition of H_2 and CH_4 of the producer gas, the equilibrium model shows good agreement with experimental data collected from others literature. The effects of reaction temperature, steam to fuel ratio, and air to fuel ratio on the composition of producer gas and overall performance of a DFB gasifier are evaluated.

See more of this Session: Biorefinery – Thermochemical Conversion of Biomass II
See more of this Group/Topical: Forest and Plant Bioproducts Division

(517c) Thermal Deoxygenation of Levulinic Acid

Thomas J. Schwartz, Paige Case, Adriaan van Heiningen, G. Peter van Walsum and **M. Clayton Wheeler**, Chemical and Biological Engineering, University of Maine, Orono, ME

Levulinic acid can be produced in 70% of theoretical yield from cellulosic biomass, but levulinic acid and its derivatives, such as ethyl-levulinate, have low heating values associated with high oxygen-to-carbon ratios. This talk presents results of thermal deoxygenation (TDO) of levulinic acid. TDO involves neutralizing the acid with calcium hydroxide and heating the resulting salts to temperatures between 350°C and 450°C. Aliphatic acids subjected to TDO are dimerized to form longer-chain ketones along with calcium carbonate. The TDO products of levulinic acid have been characterized using a combination of techniques, and more complicated chemistry was observed which results in significantly more deoxygenation than would be predicted. Because of the low oxygen to carbon ratio and the cyclic nature of the products, they are good candidates for upgrading by mild hydrocracking to form C5-C12 hydrocarbons for use as a blending agent in conventional jet fuel.

See more of this Session: Biorefinery – Thermochemical Conversion of Biomass II
See more of this Group/Topical: Forest and Plant Bioproducts Division

(517d) Characterization of a 200 Kw Fluidized Bed Biomass Gasifier

Daniel J. Sweeney, Institute for Clean and Secure Energy, The University of Utah, Salt Lake City, UT, Brett Christensen, Institute for Clean & Secure Energy, The University of Utah, Salt Lake City, UT and Kevin J. Whitty, Institute for Clean & Secure Energy, Dept. of Chemical Engineering, The University of Utah, Salt Lake City, UT

Utilization of biomass waste streams is critical for efficient and cost-effective operation of pulp and paper mills. Gasification of e.g. black liquor, forest residues and nearby available agricultural residues is a promising alternative for processing such waste streams. Although gasification is a well established process in many different industries (petroleum refining, coal power generation, etc.), its use in the forest products industry is so far relatively limited.

The University of Utah operates a 200 kW fluidized bed gasifier capable of feeding up to 65 lbs/hr of solid biomass. This presentation describes experience with this system including the influence of different gasifier operating conditions on synthesis gas composition and carbon conversion. In addition, qualitative "lessons learned" relating to operation of the pilot scale biomass gasifier are presented. The information provides a characterization of the pilot scale fluidized gasifier for future research on syngas cleanup, including reduction of tar and sulfur species.

See more of this Session: Biorefinery – Thermochemical Conversion of Biomass II
See more of this Group/Topical: Forest and Plant Bioproducts Division

Biomass Co-Firing for CO2 Management: Full-Scale Field Test

Jacob Beutler, Larry Baxter
Chemical Engineering Department
Brigham Young University
Provo, UT, USA 84602
Sønnik Clausen & Alexander Fateev,
Plasma Physics and Technology Programme
Risoe DTU National Laboratory for Sustainable Energy
Frederiksborgvej 399
4000 Roskilde, Denmark
Søren Hvid
Dong Energy A/S
Kraftværksvej 53
7000 Fredericia, Denmark

Abstract

This document reports experimental, laser-based velocity and velocity fluctuation measurements in the near-burner region of an operating, utility-scale boiler firing a blend of coal and straw. These measurements are part of a larger campaign to characterize the near-burner flame structure with experimental and theoretical velocity, temperature, gas composition, and particle measurements at utility scale. With the exception of one set of LDA measurements obtained from a coal-fired flame in a commercial boiler, these may be the first full data set of this type describing cofired flames. As with any commercial-scale measurement, flow access and maintaining steady operation present major challenges in this work, though this investigation went to great lengths to address both. The data clearly show the existence of a central recirculation zone with surrounding swirling air. Up to four replicate measurements were made at some locations at different times during the campaign. Variation among these replicates is small compared to the trends in the data and the measured velocity fluctuation at that point, indicating that the data are at least semi-quantitative.

Introduction

An increasing emphasis on mitigating global climate change (global warming) over the last few decades has created interest in a broad range of sustainable or alternative energy systems to replace fossil fuel combustion. Cofiring biomass with coal in traditional large-scale coal power plants represents one of the lowest risk, least costly, near-term methods of CO2 mitigation. Simultaneously, it is one of the most efficient and inexpensive uses of biomass(Baxter 2005).

There are some challenges to cofiring biomass with coal. The majority of cofiring demonstrations have focused on the challenges of preparation, storage, and handling. This is due to the differences in physical properties of biomass and coal. Biomass has a bulk energy density 10% or less of that of coal, requiring much larger storage and handling facilities than an equivalent amount of coal by heat content. Biomass has a relatively high moisture content, low friability, and high abrasiveness making it much harder to pulverize than coal. Coal is normally pulverized in pf boilers such that 70% of the particles pass through a 200 mesh sieve. Tests have shown woody biomass mass fraction must be <5% for pc boilers and <20% in cyclone boilers when preblended because milling limitations. When blending herbaceous biomass or bark, pluggage can occur even at these low levels of blending (Tillman 2000). A separate feed system should be used when cofiring larger fractions in PC boilers to avoid derating of the boiler and carbon-in-ash caused by pulverizer performance limits (Hughes 1998).

Over 200 cofiring tests or demonstrations have been carried out on full-scale boilers in the US alone. Many of these investigations have focused on the feasibility of fuel handling and supply, with the majority of the others concerned with the effects on emissions. In all cases, emission measurements have been obtained well down stream of the fire box (e.g. stack). These tests have shown a reduction in SO_x and often in NO_x. They have also proven the concept of cofiring in all types of boilers, albeit optimization of the fuel handling and supply system is needed and most of the tests have been short term. While these findings are significant, research is lacking in other areas, such as ash deposition, corrosion, burnout, and other fireside issues.

Several investigations have been done with measurements in the near-burner region of pf flames. These investigations include some or all of the following information: velocity fields, gas temperatures, particle temperatures, gas composition data, and carbon burnout. Most of these investigations have been on coal flames(Godoy, Ismail et al. 1989; Costa, Costen et al. 1991; Truelove and Holcombe 1991; Weber, Dugué et al. 1992; Jensen, Ereaut et al. 1994; Smart, Woycenco et al. 1996; Nazeer, Pickett et al. 1999) and most of these have been in pilot-scale systems. There are relatively few biomass or biomass-coal blend flame investigations and these are almost entirely in pilot-scale systems. Some important flame characteristics, such as the stoichiometry and intermittency of oxygen near the flame center, quite likely change with increasing scale. Measurements from within a biomass or co-fired flame are more rare. Several parametric investigations have been performed with emissions analyzed. These data have shown co-firing with lower-S and N biofuels decreases SOx and usually NOx, but not in all cases. Some in-flame measurements are listed below.

One cofiring investigation has been performed where the fuels were not pre-blended before firing. This investigation burned sawdust-coal and sewage sludge-coal. However, only emission data was obtained (NOx and particle burnout). It was found that injecting sawdust in center was better and ignited sooner and enhanced combustion intensity with lower NOx and higher efficiency (max at 30% by energy). Not much difference with firing location of sewage sludge (Abbas, Costen et al. 1994).

A more recent investigation of single fuel flames of a lignite, bituminous coal, and oak sawdust has been performed. It was found that biomass flame is much more intense near the burner (higher CO and UHC) and also a longer flame (combustion zone). The conclusion was 2-stage combustion where volatile combustion in the 2nd stage comes from large biomass particles (Ballester, Barroso et al. 2005). This is consistent with findings in this research group as explained below.

A new investigation in a 3 MW_{th} test facility did a parametric cofiring investigation with increasing biomass (sawdust 80% oak/20% pinewood). However, fuel was screw fed (by volume) so heat input differed. Secondary air flow was necessarily increased with increasing biomass to move the flame from the burner so conclusions cannot directly be based on biomass fraction alone. Coal (South African) and wood were pre-blended and milled together. $T_{particles}$ was measured by 2-color pyrometry imaging in flame zone. Gas temperature and NO were reported at 2 axial positions in the flame. Also, emissions (O_2, CO_2, CO, NO, and SO_2) were reported in flue along with fly ash properties (unburnt carbon, volatile matter, and inert composition). This study showed an increase in NO with biomass increase. However, cofired flames were less stable and the increase in particle size may contribute to greater NO_x. Biomass addition significantly reduced unburnt carbon, likely because of quicker ignition and more intense flames due to the higher volatile content of biomass. However, alkali content in ash increases with biomass fraction and fouling problems visually appeared on suction pyrometer probe (Molcan, Lu et al. 2009).

This research is a continuation of the work done by Brad Damstedt and Chunyang Wu at BYU. Their work included detailed flame composition maps of pure coal, pure biomass, and cofired flames in a 150 kW_{th} swirl-stabilized burner and detailed particle combustion model and data development. These investigations developed the first detailed gas composition maps of a non-pre-blended dual fed burner and significantly changed perceptions about the structure of cofired, pure biomass, and many coal-based flames. Although these analyses involved among the largest scale flames that can be generated at laboratory scale, some of the features noted, specifically the presence of oxygen at least momentarily essentially everywhere in the flame region, may not be as significant at the much larger flames of commercial scale. In addition, although the measured and predicted velocity profiles measured in the small-scale system reached a new level of agreement in this research group, no reacting flow velocity results were available for comparison (Damstedt 2007; Wu, Tree et al. 2007).

While some of these investigations were performed on close to commercial-scale burners, none were from a burner in a multi-burner combustor. Because factors affecting operation of a swirl-stabilize burner scale in different ways, there is uncertainty as to whether results from bench-scale measurements accurately represent commercial scale burners. For example, it is necessary to burn smaller biomass particles in the bench-scale burners than in commercial scale to achieve stable flames. Also heat transport by radiation cannot be accurately scaled from commercial boilers to a single burner facility. Mathematical models suggest that these differences are unlikely to significantly affect test furnace results (Abbas, Costen et al. 1996). However, no commercial near-burner tests exist up to this point.

This research focuses on the flame composition and structure from a single burner in a 24-burner, 350 MW_e power plant. Both traditional extraction methods and in situ spectroscopic methods were employed to simultaneously determine gas composition, gas temperature, and particle temperature in the near-burner region. The gases measured include major components: H_2O, CO_2, CO, and O_2; minor components: CH_4, C_2H_2, other UHC, and PAH; and nitrogen species: NO, HCN, and NH_3. Gas velocities in the axial and vertical direction were obtained through laser Doppler anemometry (LDA). Particles were sampled throughout the flame and are being analyzed on a basis of composition and size distribution. These field tests and accompanying model predictions were developed and carried out in coordination with partners at Risø DTU, Dong Energy, and Aalborg University, namely Sønnik Clausen, Alexander Fateev, Karsten Nielsen, Søren Hvid, and Søren Kær.

This investigation may be the most detailed attempt to date to characterize the near-burner region of a utility-scale boiler operating at full load. The work is undertaken to better characterize the flame structure and other features of this boiler. This paper focuses on the LDA velocity results. Data analysis of other techniques is ongoing and will be included in subsequent papers.

Experimental Setup

A Dantec Dynamic FiberFlow LDA system was used to measure two orthogonal velocity components, in this case arranged to represent the axial flow with respect to the burner axis of symmetry and the vertical or lift velocity.

A 27 mm diameter probe with a focal distance of 160 mm was chosen and purchased by Risø. This probe fits just inside the core of the 5-meter (60 mm diameter) probe so its temperature can be maintained below 50°C. A glass window sits in front of the LDA optics and a non-reflective cavity that extends 8 cm from the LDA focusing lens is screwed on the end to minimize radiative heat flux onto the lens. Therefore, the focal distance is only 8 cm in front of the probe tip. This is advantageous to minimize error due to beam steering from a particle laden flow, but may cause some disturbance in the flow field.

The LDA system at Risø has been refurbished and updated. The refurbished laser (6 watt Coherent Innova 306 Argon Ion Laser) is again capable of running near 6 watt power and a more modern, smaller probe—Fiberflow from Dantec Dynamics—has been purchased that can fit into the probe. The laser was only run at 160 mW power because backscatter intensity was high enough to give a sampling rate to ensure the detection of very small particles. The two signal processors used are BSA series 57N running the BSA version 2.1 software, also available from Dantec Dynamics. The two lines used are the blue light at 488 nm and the green light at 514.5 nm.

Parametric tests were done on the settings for the LDA system prior to the actual measurement campaign. These tests were run to maximize the sampling rate. This would ensure that the velocity of the large particles that do not follow gas flow is diluted out by the velocities of the much more abundant small particles. Thus the gas velocity can be approximated by the LDA particle velocities measured. Sampling rates exceeded 5000 Hz at all positions.

Dong Energy's SSV plant

Measurements were taken in the flame at Unit 4, Burner 46 at Dong Energy's Studstrup Power Station (SSV) in Studstrup, Denmark (see Figure 1). This pulverized-fuel, wall-fired boiler operates on approximately 10% wheat straw by energy and 90% coal. The boiler contains 24 total burners, six on each row with two rows on each of the front and back wall (see back wall burner layout in Figure 2). All burners are fed pure coal except the inner 4 on row 40: burners 42, 43, 44, and 45. These four burners have been modified such that the oil lance has been moved from the center of the burner to allow straw feeding. This modification to allow injection of high volatile fuel in center is consistent with literature findings.(Abbas, Costen et al. 1993) The oil lance is angled in through the tertiary air swirl so it can still be inserted for startup and retracted during normal operation. These four burners are fed equivalent amounts of wheat straw and coal by energy. For this project, burner 46 was modified in a similar fashion to allow for cofiring while burner 45 is fed pure coal. The selection of burner 46 results from consideration of probe access, boiler superstructure, and feed system routing.

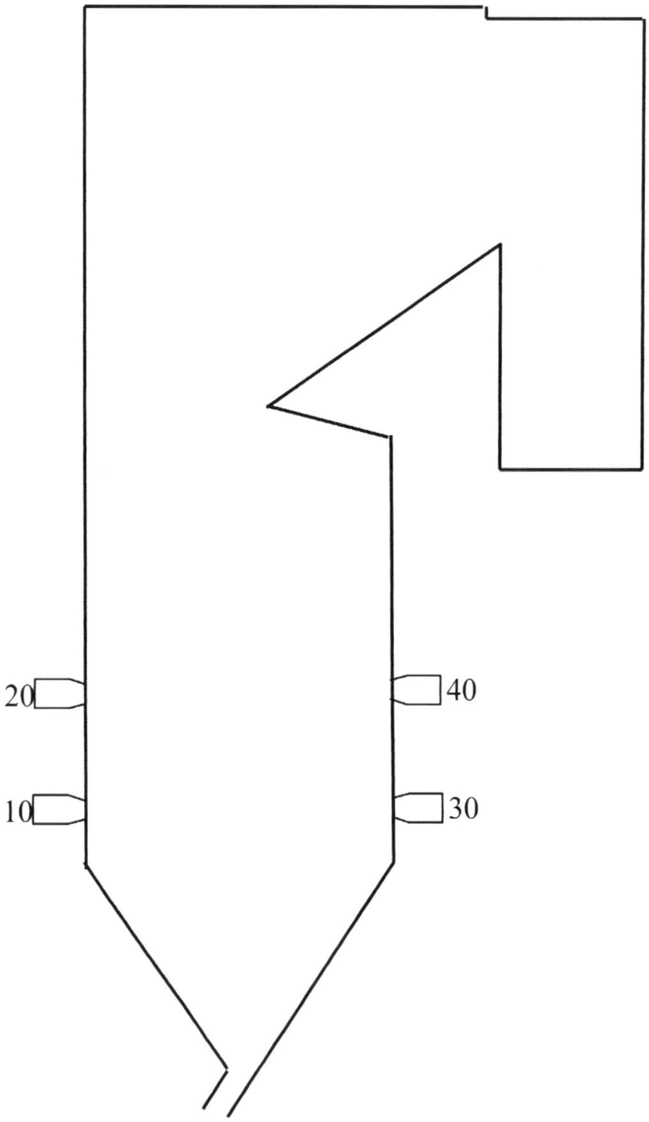

Figure 1. SSV general layout showing burner row positions. Measurements taken at burner 46 on row 40.

The boiler is normally cofired on wheat straw that can pass through a 50 mm sieve. However because of expected bias in the particle collection probe (16.6 mm diameter), smaller sieves were used in these tests. A 16 mm sieve was used in this campaign. Since most straw particles have high aspect ratios, much longer straw particles escaped through the sieve and were seen in this investigation. It is assumed that this reduction in sieve size will not significantly affect the flame structure.

Figure 2. Burner set-up on SSV back wall. Front wall has identical setup with burners aimed directly towards each other.

Burner dimensions and measurement port positions appear in **Error! Reference source not found.**. The four measurements ports (labeled A-D with A being closest the burner) are elongated as shown in Figure 3, which allows angling the probe to get measurements from any axial position. The center of the measurement ports is approximately 15 cm above the burner centerline because of flame lift. There is a structural pillar directly in front of Port A limiting flame access by requiring the probe be inserted on a 5-10° angle towards the flame and only positions in the bottom half of the flame can be measured.

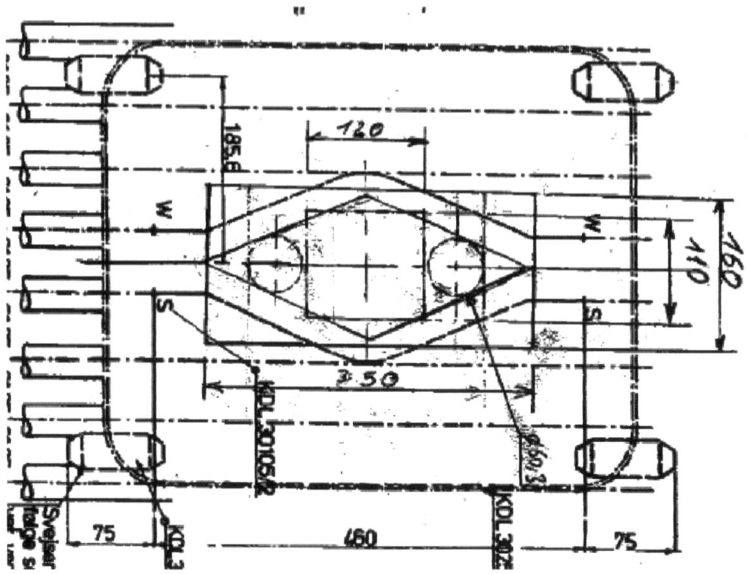

Figure 3. Dimensions of measurement port. The elongated port allows for several axial and radial positions to be measured from single port.

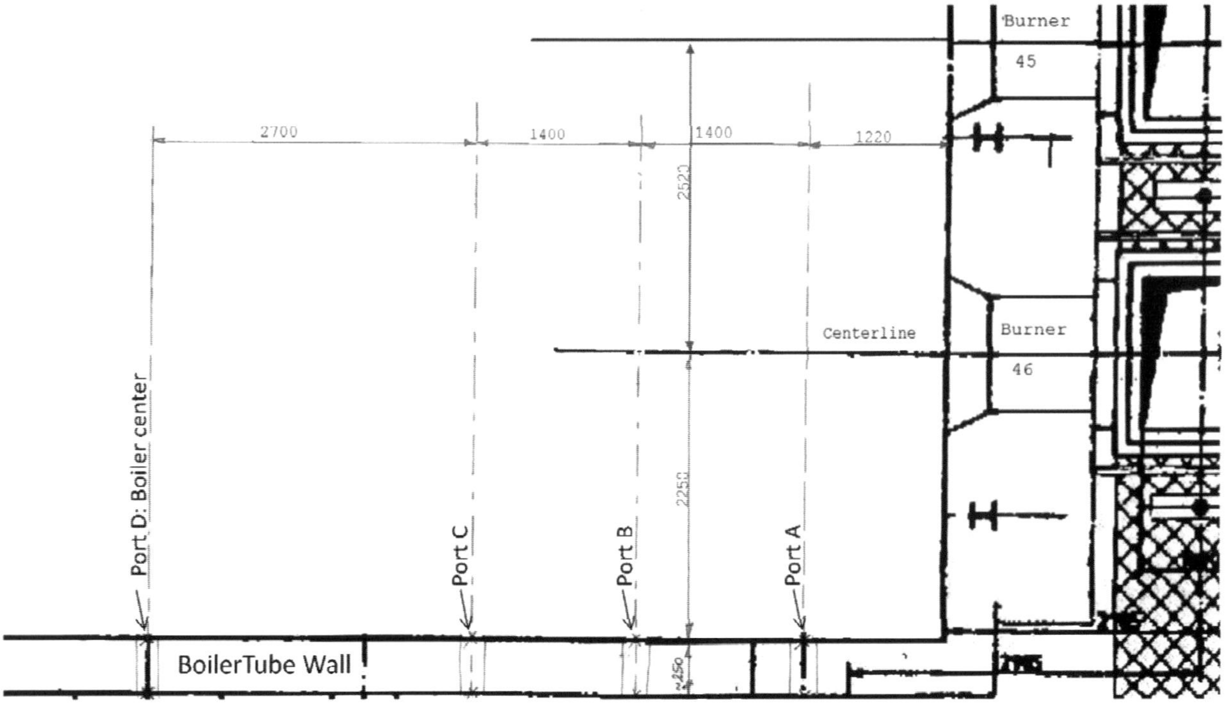

Figure 4. Measurement port positions from top view of the near-burner region. All dimensions are in mm.

Measurements on this particular flame spanned five different days. LDA velocity measurements were obtained on the same day but still spanned approximately 16 hours of operation. Operation on row 40 was held at a constant full-load during the whole campaign. Flows on the other three rows of burners were allowed to change some in order to more closely follow demand curves.

Results

The results of the measurements are difficult to display and even more difficult to understand in static figures, as appear here. Figure 5 illustrates the sampling grid for the velocity measurements. The points represent locations away from the wall (nearly horizontal axis), across the burner face (nearly vertical axis), and in the direction of flow from the burner (axis nearly into the page) where velocities were measured, displayed in the burner flow direction. The burner is located approximately in the center 100 cm of this measurement block. Probe access was from the left (as seen here) side of the flow through a narrow slit installed between steam tubes. The points nearest the burner are about 1.5 m from the burner outlet, or about 1 m from the boiler front wall, with the burner quarl recessed about 0.5 m into the wall. The points furthest in the flow direction are approximately at the center of the boiler (about 700 cm from the front wall). The generally fan-shaped pattern of points results from the physically limited range of angles at which the probe could be placed because of structural components outside the sampling window. The measurements along the side of the figure are in cm.

A total of 235 measurement locations are included in the grid, with replicate or near-replicate measurements at 22 of these points. Most commonly, only two replicates are included, though there are some points with three and four replicates each. The measurements at replicated points occurred at significantly different times and provide some measure of data precision. Table 1 summarizes this analysis for the case with four replicated results, which is located at position (272.32, 38.27, 262.00) in the plots, or at the approximate middle of the diagnostic area. As seen in the table, the standard deviation of the mean velocities is much smaller than mean of the measured fluctuations in the flow and typically is about 0.7 m/s for the velocities and 0.25 m/s for their fluctuations. The average ratio of standard deviations of replicated measurements to velocity-rms values is 0.25 with a maximum ratio of 0.55, seen at a position of low velocities. This indicates that the natural velocity fluctuations are greater than deviation of average values of repeated measurements.

Table 1. Statistical analysis of measurements based on four replicated results

	Axial Vel	Vert Vel	Axial rms Vel	Vert rms Vel
mean	8.77	3.67	5.05	5.30
Std. Dev.	0.60	0.78	0.21	0.35
Coef. Variation	0.07	0.21	0.04	0.07

LDA measurements were taken at each axial position corresponding to the port location (1.22 m, 2.62 m, 4.02 m, and 6.72 m). Figure 5 shows the 3-D plot of all measurement points. The probe was tilted up and down and extended into the boiler as much as possible to fill out the velocities in the flame region. As seen it was not possible to get a lot of information from the half of the flame region nearest the boiler wall. Figure 6, Figure 7, Figure 8, and Figure 9 display the axial velocities on a plane corresponding to the port locations. Figure 10, Figure 11, Figure 12, and Figure 13 display the vertical velocities at these same planes. In each instance a circle is super-imposed upon the figure to show the diameter and position of the quarl exit.

The general structure of a flame region from a low-NO_x burner is captured in the following figures. A distinct internal recirculation zone is seen in the displays of the axial velocities at 1.22, 2.62, and 4.02 meters from the quarl exit (Figure 6, Figure 7, and Figure 8). Large negative velocities (~-14 m s^{-1} maximum) are seen in the measurements nearest the flame. These large negative velocities dilute down moving away from the burner until they are nearly 0 m s^{-1} approximately 4 meters into the boiler. Measurements taken at the center of the boiler, 6.7 meters from the burner, clearly show the influence of the opposing burner with low velocities throughout the whole measurement region. The effect of the neighboring flame (to the right of displayed velocities is also clearly seen with axial velocities to the right of the quarl being higher than velocities to the left of the quarl in each case.

The swirl from the secondary and tertiary air is captured in the vertical velocity figures. The swirl is counter-clockwise which is shown most dominantly in measurements near the burner as shown in Figure 10. This swirl quickly dies out and is not seen further downstream. Also the influence of the recirculation air entering the bottom of the boiler and the inlets from the row of burners directly below this burner is also captured since the average vertical velocity is positive in each case. Flame lift is captured moving away from the burner indicated by increasing average vertical velocities. This phenomena was verified with IR video taken just prior to LDA measurements.

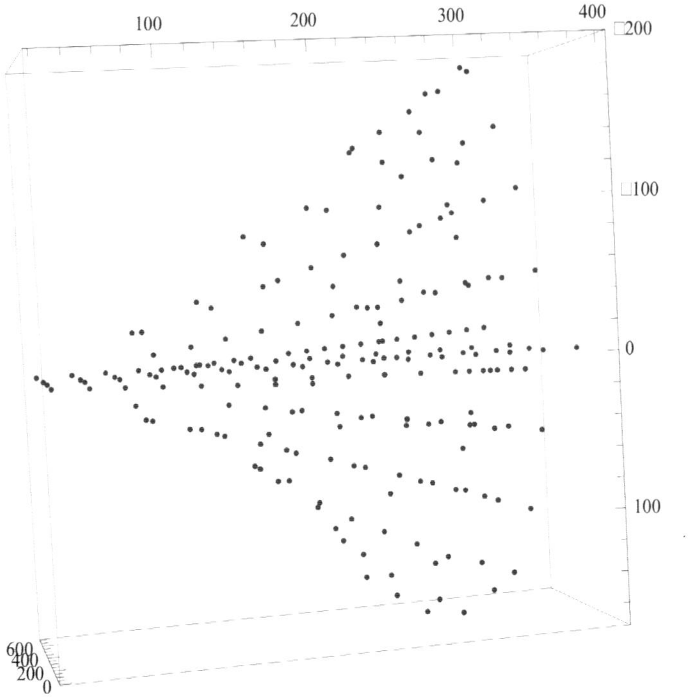

Figure 5. 3-D pattern of sampling points from the wall (left) into the boiler. The measurements represent distances away from the wall (nearly horizontal axis), across the burner face (nearly vertical axis), and in the direction of flow from the burner (axis nearly into the page). The measurements along the side of the figure are in cm.

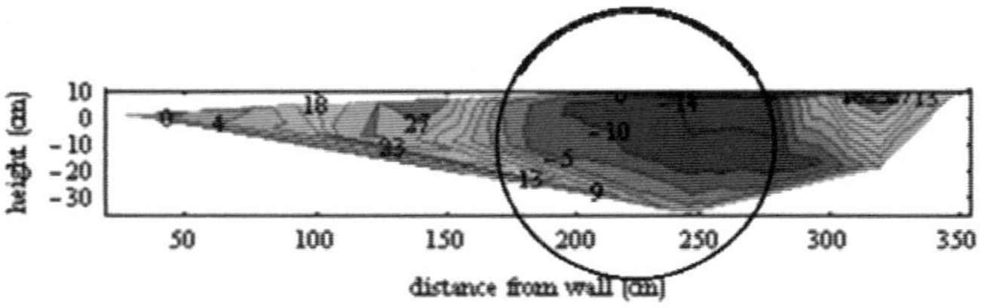

Figure 6. Axial velocities at 1.22 meters from quarl exit. Notice high reverse velocities of -14 m s^{-1} in the internal recirculation zone (IRZ). Also high velocities seen from secondary/tertiary air around IRZ.

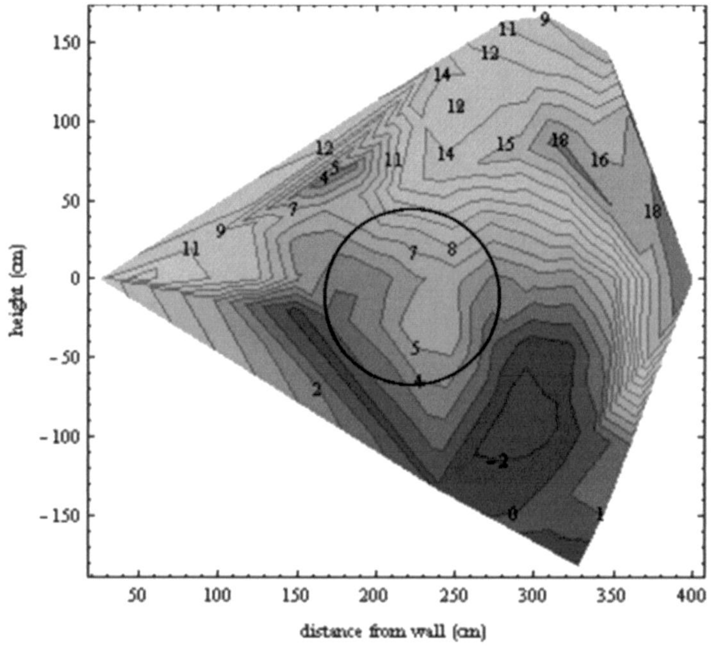

Figure 7. Axial velocities at 2.62 meters from quarl exit. Higher velocities are still evident in secondary/tertiary air region with lower velocities in the center.

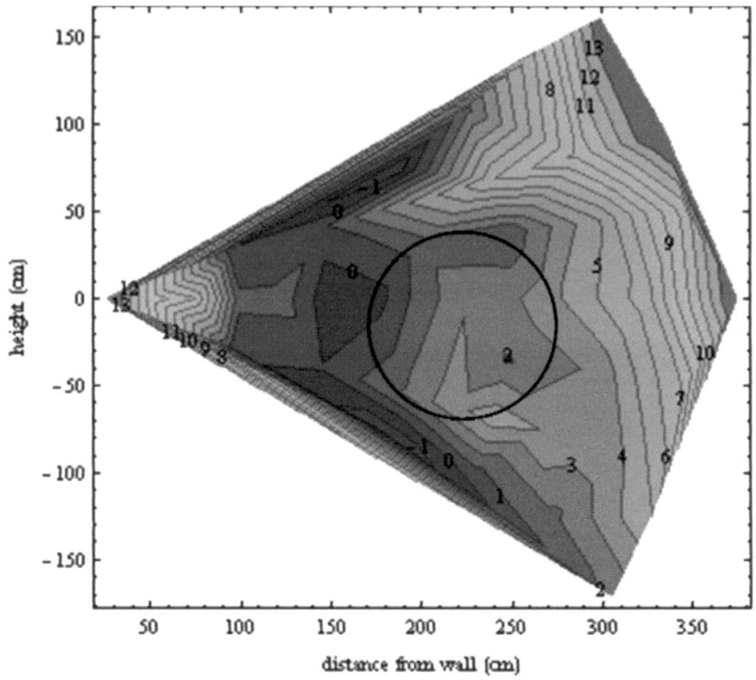

Figure 8. Axial velocities 4.02 meters from quarl exit. This appears to be the end of the IRZ.

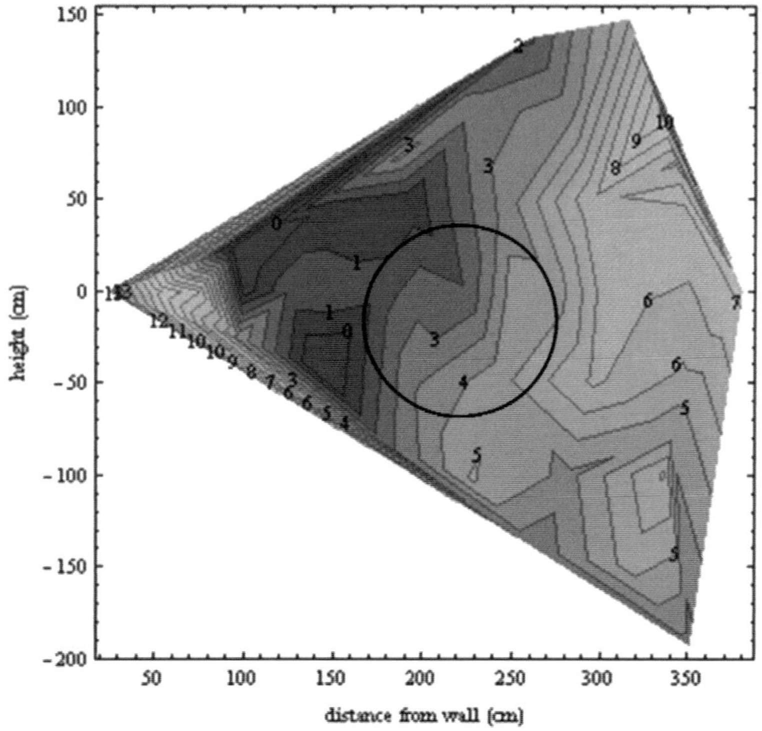

Figure 9. Axial velocities at 6.72 meters from quarl exit (center of the boiler). Low velocities expected since opposing flame regions meet at center of boiler.

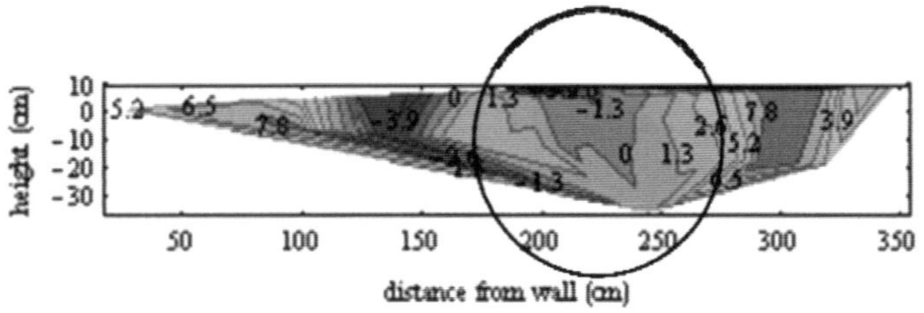

Figure 10. Vertical velocities 1.22 meters from quarl exit. Secondary/tertiary air swirl is counter-clockwise and clearly seen from data.

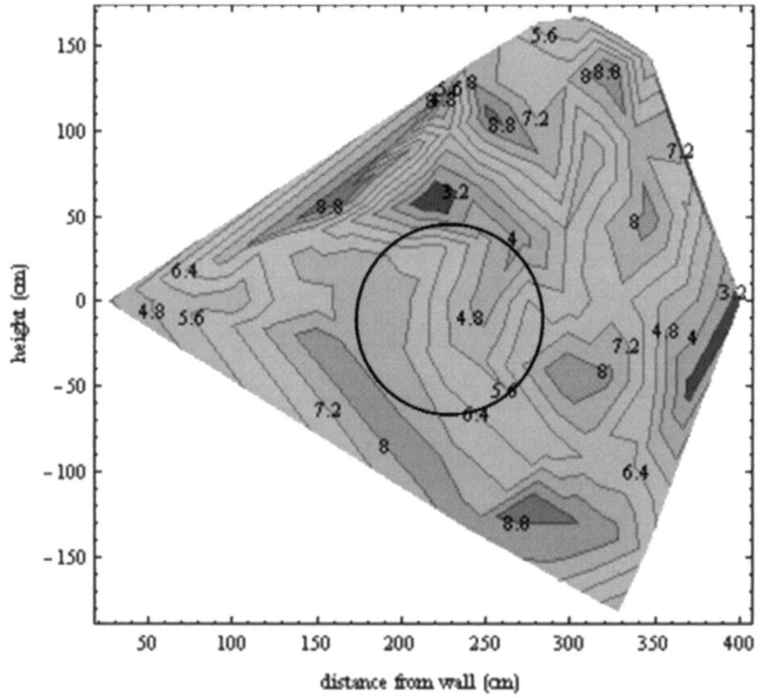

Figure 11. Vertical velocities 2.62 meters from quarl exit.

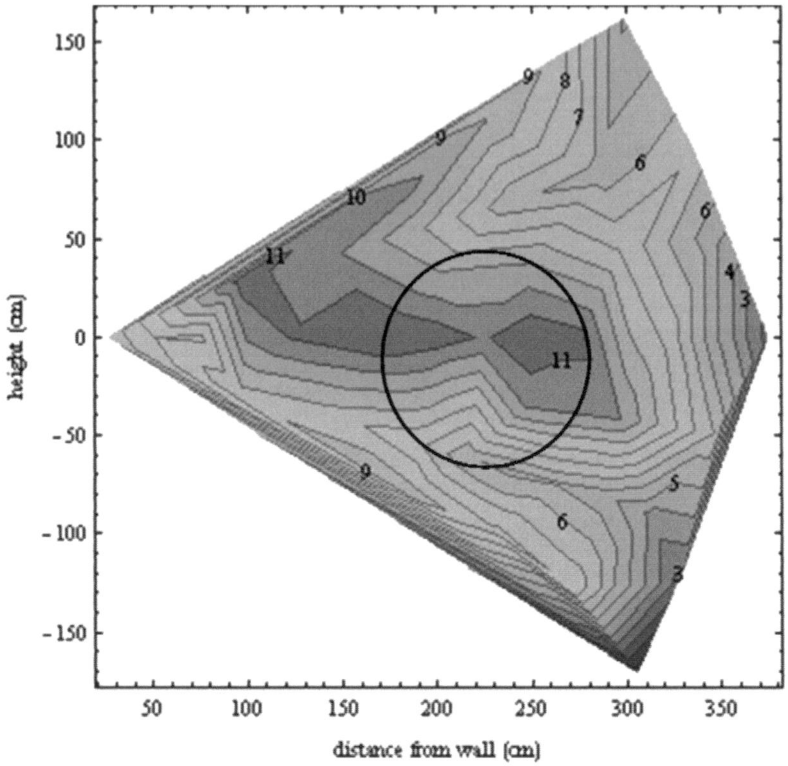

Figure 12. Vertical velocities 4.02 meters from quarl exit. Average vertical velocity continues to increase as flame lifts.

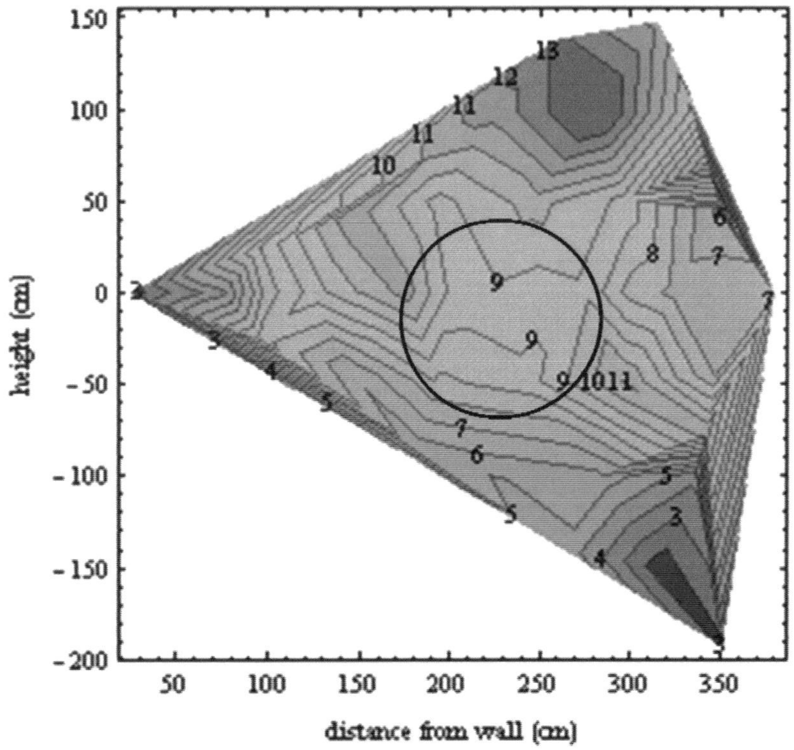

Figure 13. Vertical velocities 6.72 meters from quarl exit.

Conclusions

Experimental, laser-based velocity and velocity fluctuation measurements in the near-burner region of an operating, utility-scale boiler firing a blend of coal and straw show clear indications of internal recirculation zones, toroidal axial and tangential velocities around this recirculation zone, relatively rapid decay, and significant burner-burner interaction. All of these features agree with theoretical and design features of these flows. These measurements are part of a larger campaign to characterize the near-burner flame structure with experimental and theoretical velocity, temperature, gas composition, and particle measurements at utility scale. These may be the first data of this type describing cofired flames and possibly any near-burner flame from a utility-scale boiler. As with any commercial-scale measurement, flow access and maintaining steady operation represent major challenges in this work, though this investigation went to great lengths to address both. Up to four replicate measurements were made at some location at different times during the campaign. Variation among these replicates is small compared to the trends in the data and the measured velocity fluctuation at that point, indicating that the data are at least semi-quantitative.

Bibliography

Abbas, T., P. Costen, et al. (1993). "The Effect of the Near Burner Aerodynamics on Pollution, Stability and Combustion in a pf-Fired Furnace." Combustion Science and Technology **93**(1-6): 73-90.

Abbas, T., P. Costen, et al. (1994). "THE INFLUENCE OF BURNER INJECTION MODE ON PULVERIZED COAL AND BIOMASS COFIRED FLAMES." Combustion and Flame **99**(3-4): 617-625.

Abbas, T., P. G. Costen, et al. (1996). "Solid fuel utilization: From coal to biomass." Symposium (International) on Combustion **26**(2): 3041-3058.

Ballester, J., J. Barroso, et al. (2005). "Comparative study of semi-industrial-scale flames of pulverized coals and biomass." Combustion and Flame **141**(3): 204-215.

Baxter, L. (2005). "Biomass-coal co-combustion: opportunity for affordable renewable energy." Fuel **84**(10): 1295-1302.

Costa, M., P. Costen, et al. (1991). "Detailed measurements in and modelling of an industry-type pulverised-coal flame." Symposium (International) on Combustion **23**(1): 973-980.

Damstedt, B. (2007). Structure and nitrogen chemistry in coal, biomass and cofiring low-nitrogen oxide flames. United States -- Utah, Brigham Young University. **Ph.D.:** 261.

Godoy, S. G., M. Ismail, et al. (1989). "Measurements in a Cylindrical Pulverized Coal Furnace. Influence of Swirl Number on Combustion Performance." Combustion Science and Technology **67**(1): 59 - 72.

Hughes, E. E. (1998). Utility Coal-Biomass Cofiring Tests. Proceedings to the Department of Energy.

Jensen, P. A., P. R. Ereaut, et al. (1994). "LOCAL MEASUREMENTS OF VELOCITY, TEMPERATURE AND GAS-COMPOSITION IN A PULVERIZED-COAL FLAME." Journal of the Institute of Energy **67**(470): 37-46.

Molcan, P., G. Lu, et al. (2009). "Characterisation of biomass and coal co-firing on a 3 MWth Combustion Test Facility using flame imaging and gas/ash sampling techniques." Fuel **88**(12): 2328-2334.

Nazeer, W. A., L. M. Pickett, et al. (1999). "In-Situ Species, Temperature and Velocity Measurements in a Pulverized Coal Flame." Combustion Science and Technology **143**(1): 63 - 77.

Smart, J. P., D. M. Woycenco, et al. (1996). "Studies on scale-up of swirl-stabilised pulverised-coal burners in the thermal input range 2.5-12 MW." Journal of the Institute of Energy **69**(480): 131-143.

Tillman, D. A. (2000). "Biomass cofiring: the technology, the experience, the combustion consequences." Biomass & Bioenergy **19**(6): 365-384.

Truelove, J. S. and D. Holcombe (1991). "Measurement and modelling of coal flame stability in a pilot-scale combustor." Symposium (International) on Combustion **23**(1): 963-971.

Weber, R., J. Dugué, et al. (1992). "Quarl zone flow field and chemistry of swirling pulverized coal flames: Measurements and computation." Symposium (International) on Combustion **24**(1): 1373-1380.

Wu, C., D. Tree, et al. (2007). "Reactivity of NH3 and HCN during low-grade fuel combustion in a swirling flow burner." Proceedings of the Combustion Institute **31**: 2787-2794.

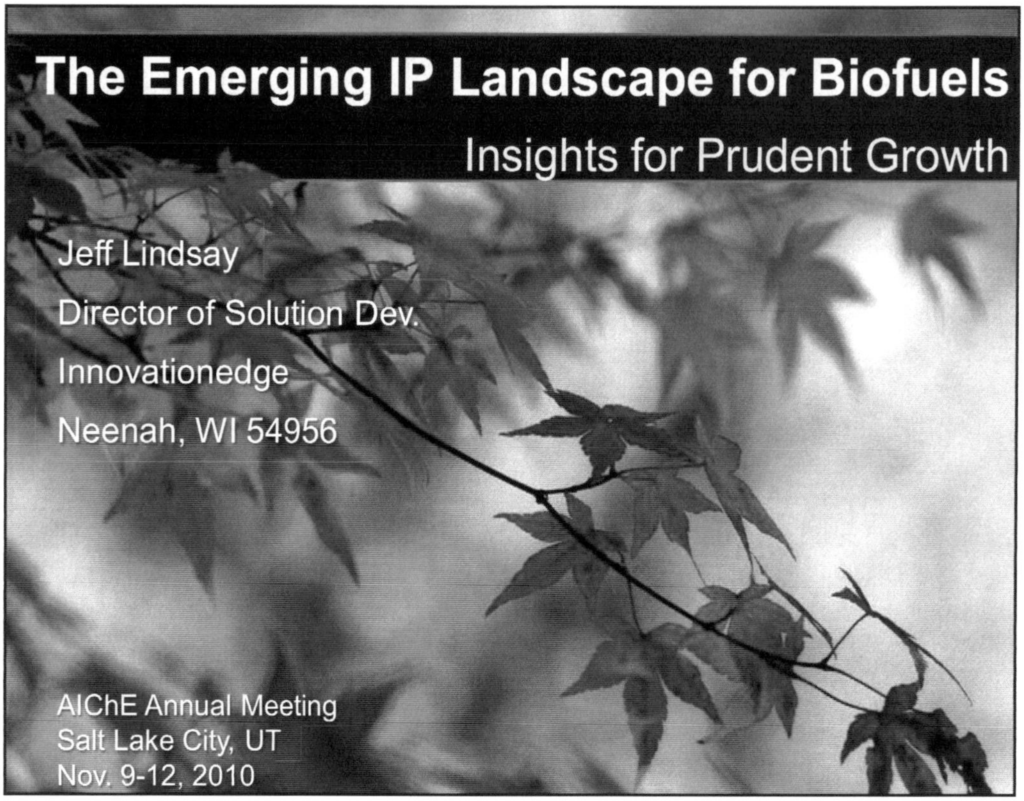
The Emerging IP Landscape for Biofuels
Insights for Prudent Growth
Jeff Lindsay
Director of Solution Dev.
Innovationedge
Neenah, WI 54956
AIChE Annual Meeting
Salt Lake City, UT
Nov. 9-12, 2010

About Innovationedge
ie
• Innovation and business consultancy
• Founded by Cheryl Perkins, former Chief Innovation Officer of Kimberly-Clark
• Projects and guidance for corporations & startups
– Open innovation, innovation strategy
– New product development, commercialization
– Intellectual asset strategy & creation
• Education mission also
nnovation

Who Claims the 1st Biofuel Patent?

Quiz Time

- What company's IP recognizes a possible role of blimps in the world of biofuels?
  - Hint: they also recognize the need for mobile ethanol plants to go where the biomass is.
- What company is using biotech to create inorganic catalysts for next-generation biofuels?
  - Siluria (e.g., for methane to ethanol)
- Who recently won the contract to build the world's largest biomass-only power plant – and where is it?
  - Foster-Wheeler, Poland (190 MW), *Forbes*, 6-7-2010.

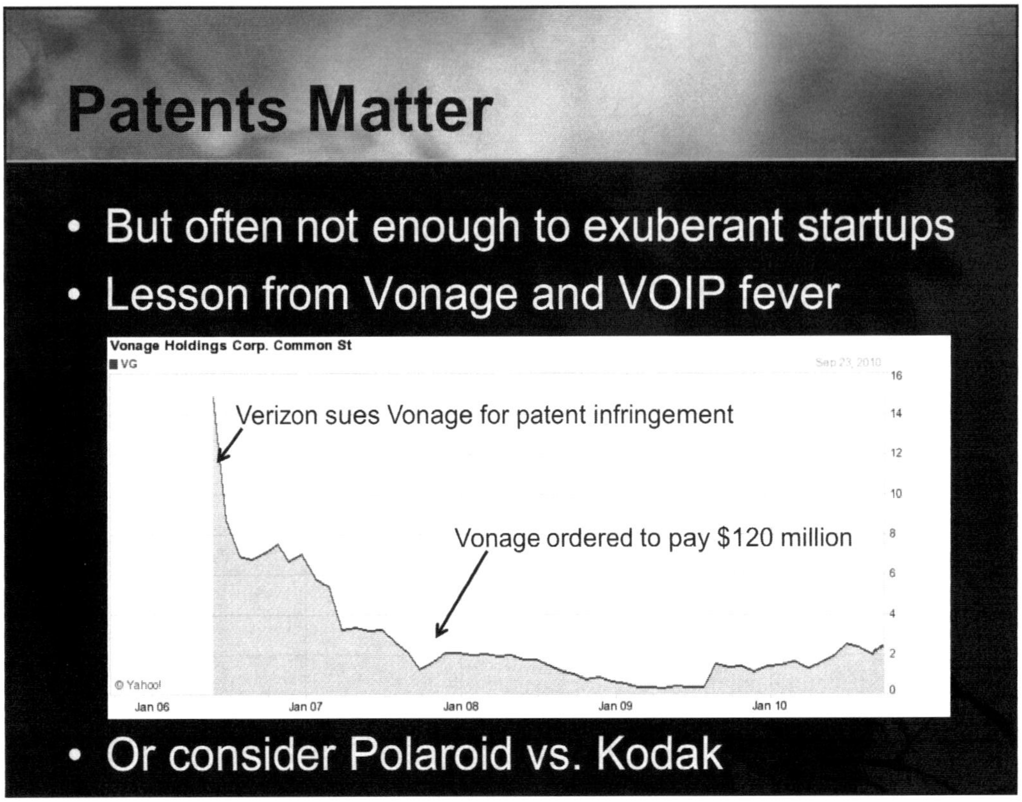

Patents Matter
• But often not enough to exuberant startups
• Lesson from Vonage and VOIP fever
Vonage Holdings Corp. Common St
VG
16
14
12
10
8
6
4
2
0
Verizon sues Vonage for patent infringement
Vonage ordered to pay $120 million
© Yahoo!
Jan 06
Jan 07
Jan 08
Jan 09
Jan 10
• Or consider Polaroid vs. Kodak

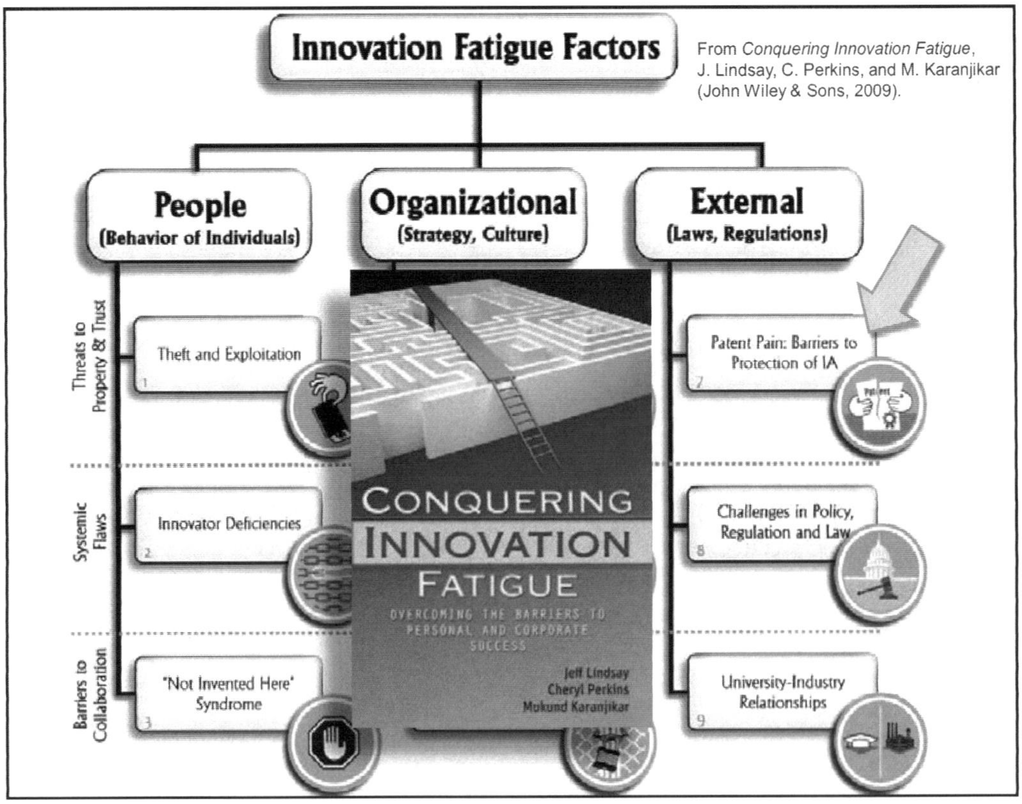

Innovation Fatigue Factors
From *Conquering Innovation Fatigue*, J. Lindsay, C. Perkins, and M. Karanjikar (John Wiley & Sons, 2009).
People
(Behavior of Individuals)
Organizational
(Strategy, Culture)
External
(Laws, Regulations)
Threats to Property & Trust
Systemic Flaws
Barriers to Collaboration
Theft and Exploitation
1
Innovator Deficiencies
2
"Not Invented Here" Syndrome
3
Patent Pain: Barriers to Protection of IA
7
Challenges in Policy, Regulation and Law
8
University-Industry Relationships
9
CONQUERING INNOVATION FATIGUE
OVERCOMING THE BARRIERS TO PERSONAL AND CORPORATE SUCCESS
Jeff Lindsay
Cheryl Perkins
Mukund Karanjikar

The Study

- Ongoing analysis of biofuels IP using a variety of tools (PatBase, other systems)
- Combining IP trends with other information
- Several groupings
  - Focused biofuels companies
  - Big oil and auto
  - Paper and wood products companies
- Technology areas: gasification, fermentation/biological, chemical
- Emphasis on cellulosic biomass

Key Messages

- Much of the promise in biofuels is found in startups &young companies
  - Anxious to bring products to the market
  - At risk of overlooking IP barriers
  - At risk of missing IP opportunities
- Much work is being done in organi-zations where IP is not a priority
- IP will matter and may determine the future landscape of the industry.

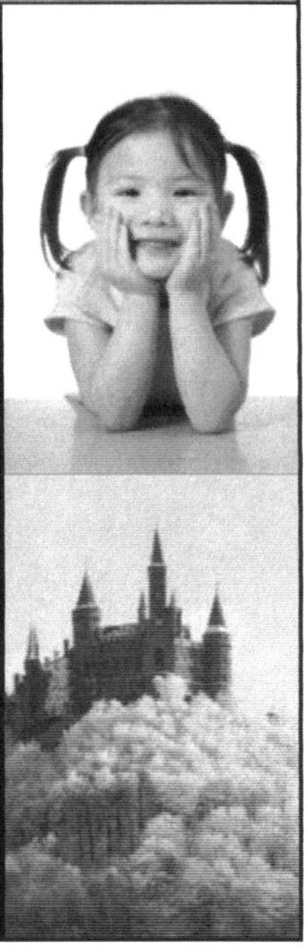

Key Trends

- Explosion in biofuels technologies, from gasification to fermentation (intl. class C10, C12)
- Very large recent boom in biological conversion of biofuels (e.g., intl. class C12)
  - Synthetic biology
  - New microbial strains
  - New uses of microbes
  - New enzymes
  - Fermentation methods

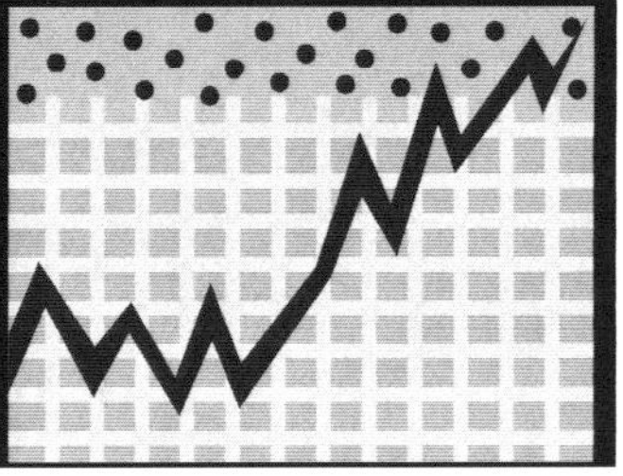

Observations

- There is a move toward protecting value-added products (e.g., THF, levulinic acid, or chemicals derived from ethanol)
- Advances in genetic engineering methods to produce microbes and enzymes for value-added products
- Supply-chain-wide innovation (emerging)
- Rise of bio-business method patents (or business bio-method patents)
- The IP rich are getting richer
- Some battles have begun (e.g., suits from GreenShift and its subsidiary, GS CleanTech).

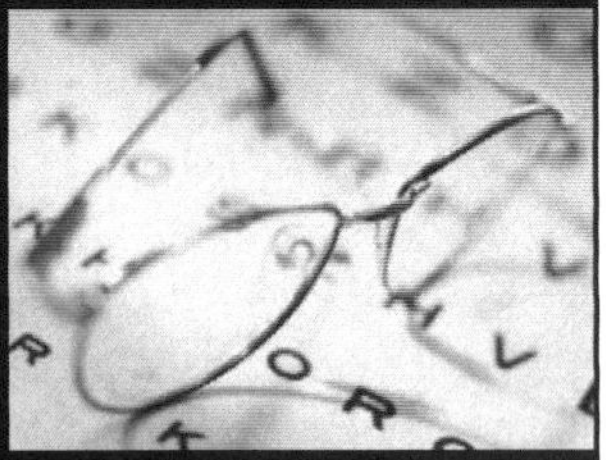

University Trends

- Universities file much less than industry in general. E.g., all universities combined fewer patents each year than IBM. Even Kimberly-Clark gets more than MIT and most other universities.
- University filings are about 10% of the biofuels space.
- US universities lead, but Chinese universities are moving rapidly.

IP Across the Supply Chain

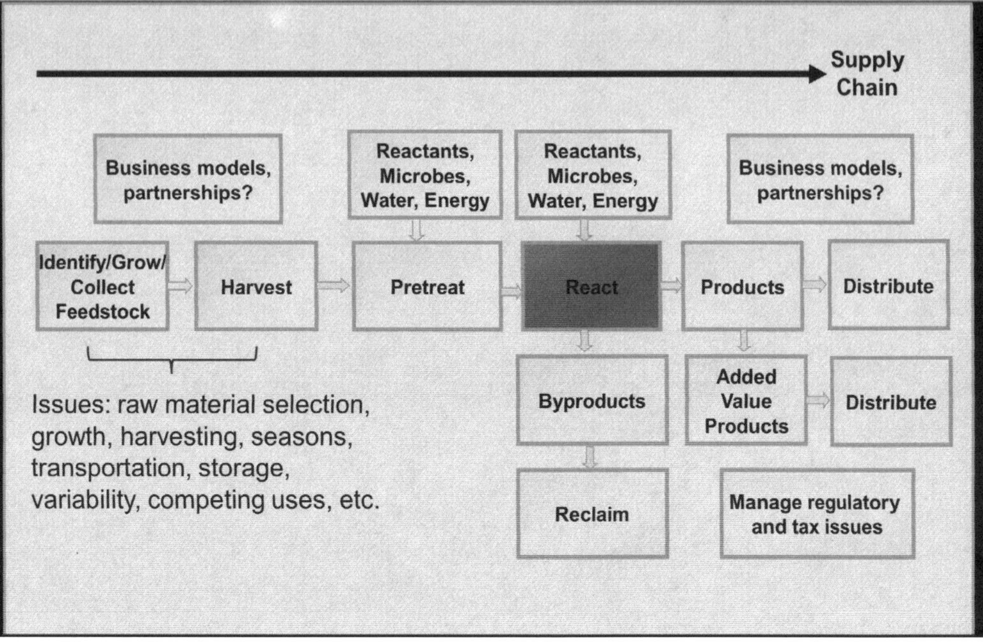

The Rise of Biofuels in IP

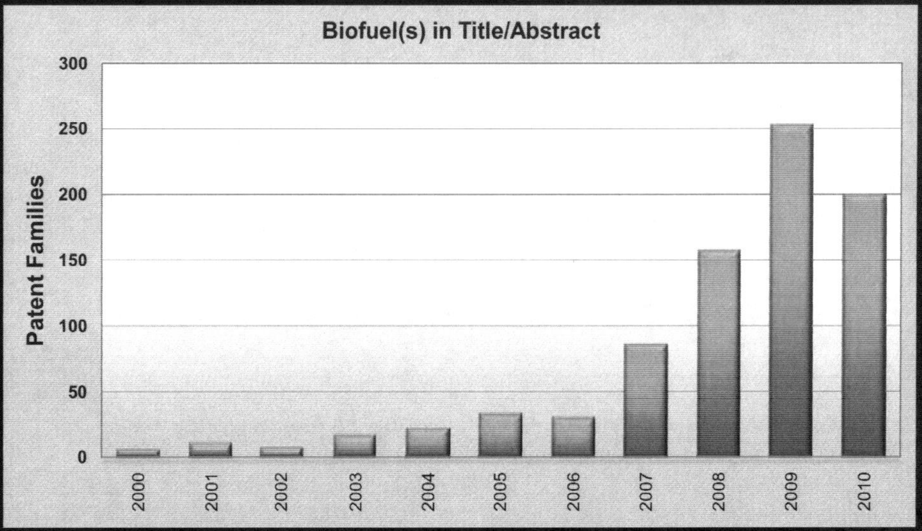

Reviving Old Technologies

- Gasification
  - Biomass, black liquor (Chemrec), municipal waste (Enerkem), etc.
  - Mixed Alcohol Synthesis (Range Fuels), packed bed, catalytic reactors (Fulcrum), or fermentation (INEOS and Coskata).
- Fischer Tropsch upgrades, catalytic processes to deliver many chemical products (Syntroleum, Greatpoint, etc.)
- Pryolysis (e.g., UOP Honeywell, Dynamotive) – updating uses of bio-oil (conversion to useful fuels)
- Acid and enzymatic hydrolysis – many players
- Fermentation – very hot
  - Discovery of new microbes (e.g., thermophilic strains)
  - Improved enzymes (Novozymes, Codexis, Novo Nordisk, Genencor/Danisco, Dyadic, etc.)
  - Reactor designs, control systems
  - Upgraded with genetic engineering /synthetic biology

Name the Leaders....

- Every year for the last 3 years, what company has had the most patent families that mention biofuel(s) in the abstract and title?
  - Hint: It's an automotive company.
  - Hint: It's not General Motors.
  - Background: This company has 3.5% of the 825 patent families, and almost twice as many as the second place company, Chevron USA.
  - Answer: TOYOTA

Search strategies

- General key words
- US and international classifications
  - Helpful, but inadequate for the biofuels area
  - A new European class can help
- By company
- Specific technologies

New EPO Technology Tagging

European Patent Classification Y02E50/17

Y - GENERAL TAGGING OF NEW TECHNOLOGICAL DEVELOPMENTS

Y02 - TECHNOLOGIES OR APPLICATIONS FOR MITIGATION OR ADAPTATION AGAINST CLIMATE CHANGE

Y02E - REDUCTION OF GREENHOUSE GASES [GHG] EMISSION, RELATED TO ENERGY GENERATION, TRANSMISSION OR DISTRIBUTION

Y02E50/00 - Technologies for the production of fuel of non-fossil origin

Y02E50/10	. Biofuels
Y02E50/11	. . CHP turbines for biofeed
Y02E50/12	. . Gas turbines for biofeed
Y02E50/13	. . Bio-diesel
Y02E50/14	. . Bio-pyrolysis
Y02E50/15	. . Torrefaction of biomass
Y02E50/16	. . Cellulosic bio-ethanol
Y02E50/17	. . **Grain bio-ethanol**
Y02E50/18	. . Bio-alcohols produced by other means than fermentation
Y02E50/30	. Fuel from waste
Y02E50/32	. . Synthesis of alcohols or diesel from waste including a pyrolysis and/or gasification step
Y02E50/34	. . Methane (not used, see subgroups)
Y02E50/34B	. . . production by fermentation of organic by-products, e.g. sludge
Y02E50/34D	. . . from landfill gas

Y02E50 Families Since 2000

Based on 1st 2500 of 3834 families

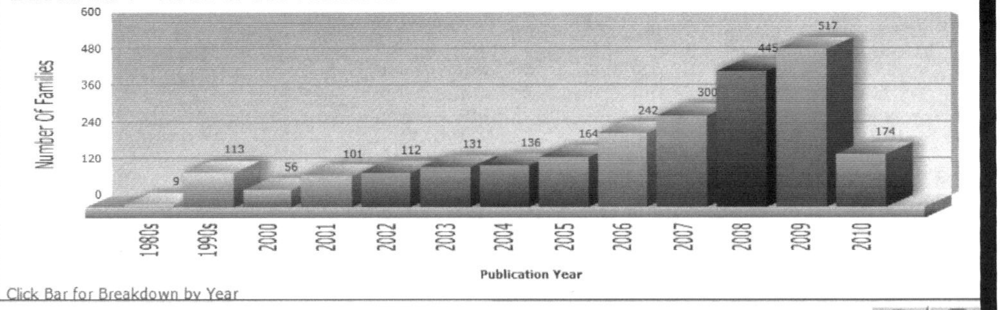

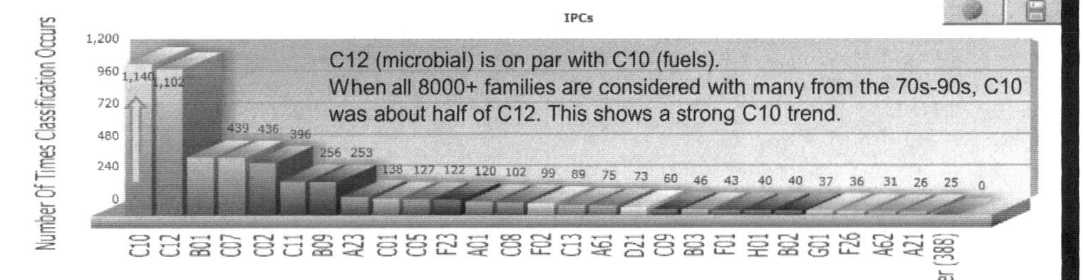

C10 deals with fuels; C12 with biochemistry, enzymes, microbes

Selected Results
Company-Oriented Searches

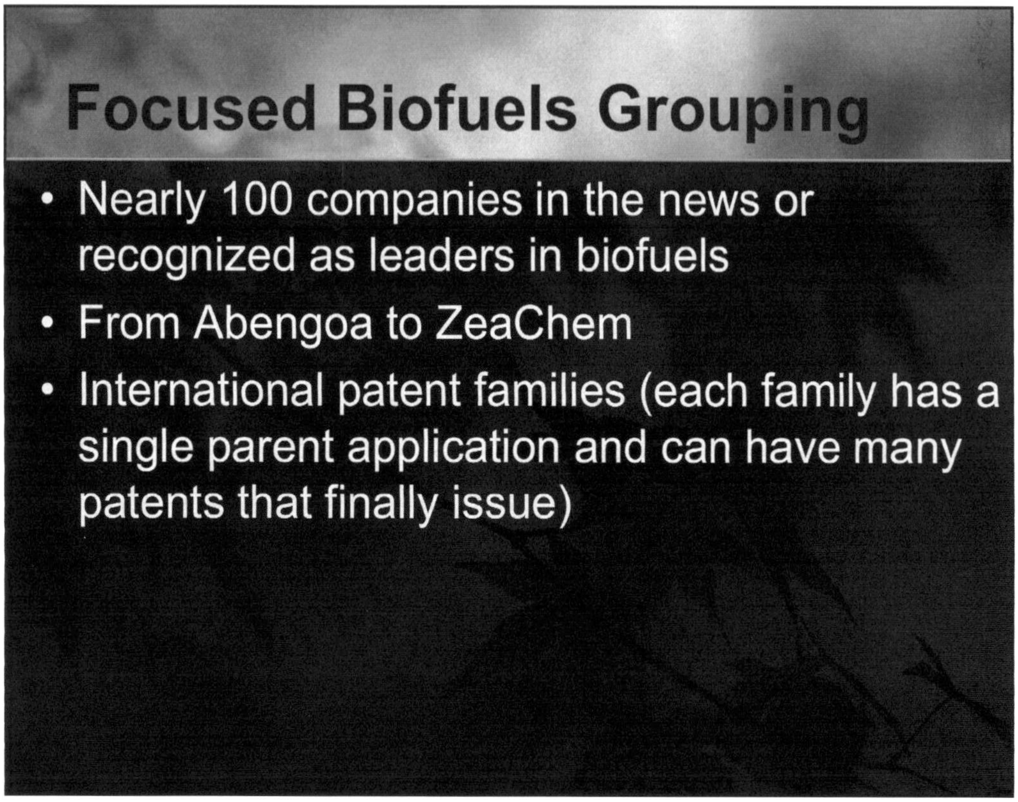
Focused Biofuels Grouping
• Nearly 100 companies in the news or recognized as leaders in biofuels
• From Abengoa to ZeaChem
• International patent families (each family has a single parent application and can have many patents that finally issue)

US Docs, Focused Corps.

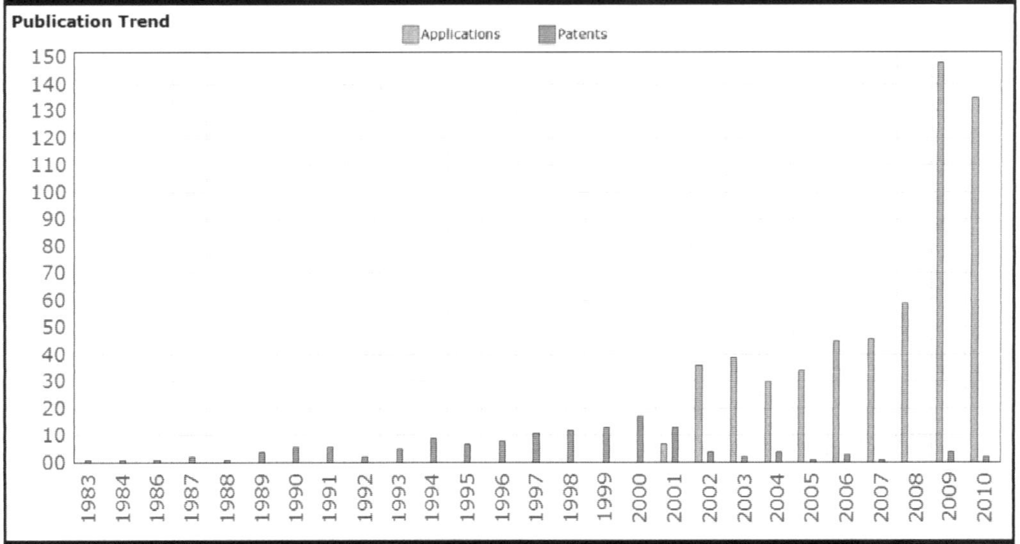

Observations

- Many of these biofuels patents are taking much longer than normal to issue as patents, if they succeed at all.
- Some applications may be part of a hasty "land rush"

- Pioneering IP in new fields often takes years longer to grant.
- The future, like most of the patent applications in this area, is still pending.

Forward Citations

- A metric to help see which patents are relevant to newer work
- Later patents cite older ones either because either the applicant or examiner felt they are "material" to patentability
- Lots of citations is generally believed to correlate to "value" of the patent
- Most biofuels patents aren't cited–a common issue in young fields. This will change.

The Forward Citing Universe

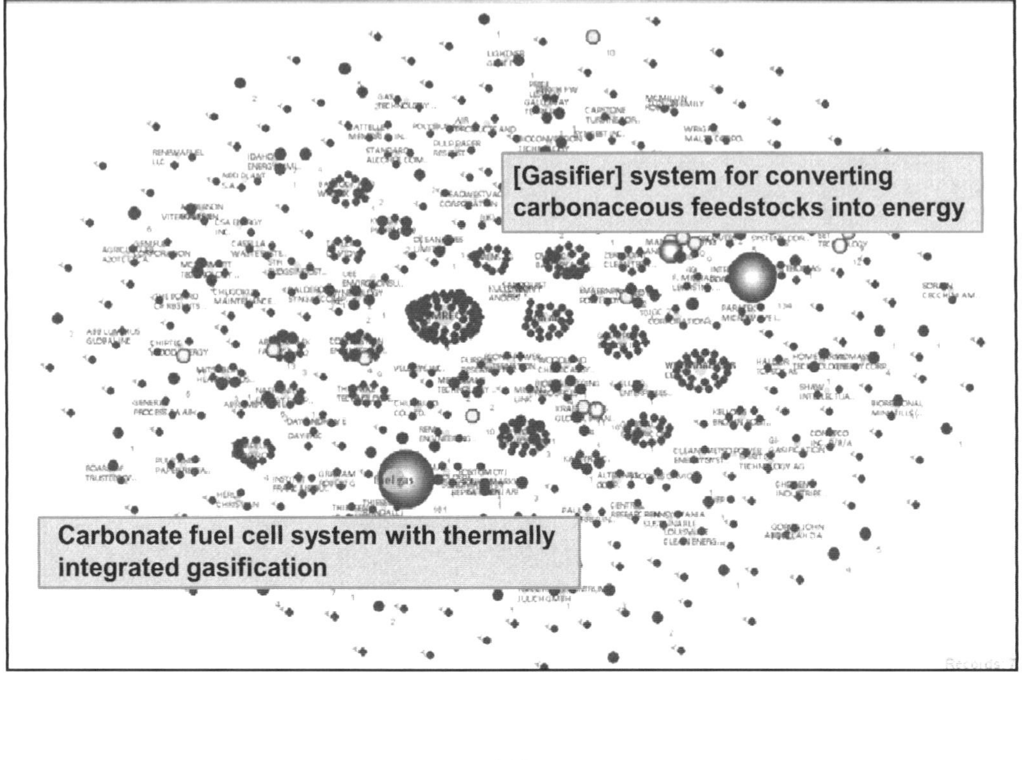

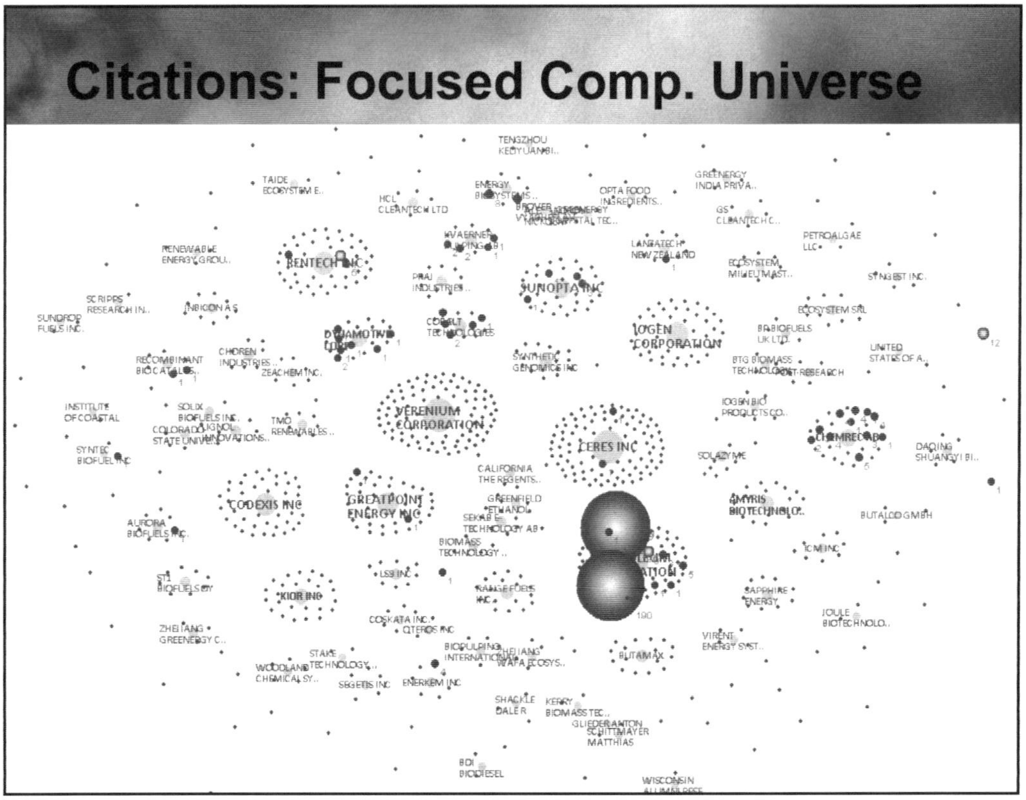
Citations: Focused Comp. Universe
VERENIUM CORPORATION
CERES INC
CODEXIS INC
GREATPOINT ENERGY INC
KIOR INC
IOGEN CORPORATION
SUNOPTA INC
RENTECH INC
CHEMREC AB
AMYRIS BIOTECHNOLO..
SOLAZYME
BUTAMAX
SAPPHIRE ENERGY
JOULE BIOTECHNOLO..
BUTALCO GMBH
WISCONSIN

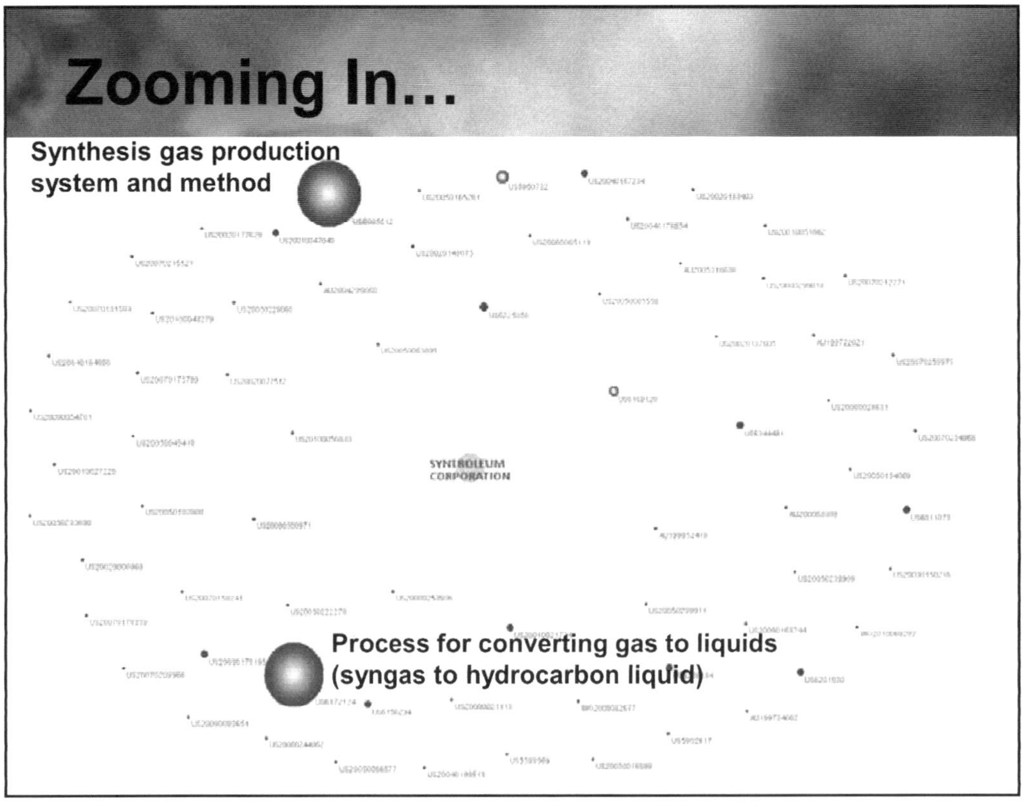
Zooming In...
Synthesis gas production system and method
SYNTROLEUM CORPORATION
Process for converting gas to liquids (syngas to hydrocarbon liquid)

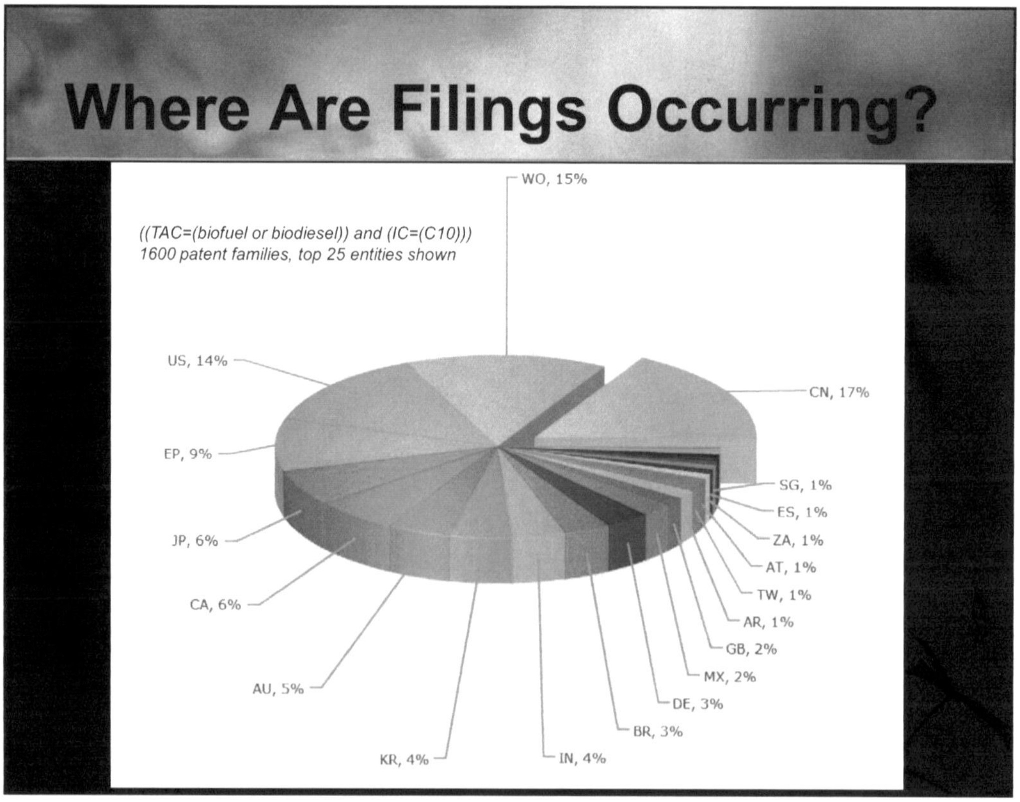
Where Are Filings Occurring?
((TAC=(biofuel or biodiesel)) and (IC=(C10)))
1600 patent families, top 25 entities shown
WO, 15%
US, 14%
CN, 17%
EP, 9%
SG, 1%
ES, 1%
ZA, 1%
JP, 6%
AT, 1%
TW, 1%
CA, 6%
AR, 1%
GB, 2%
MX, 2%
AU, 5%
DE, 3%
BR, 3%
KR, 4%
IN, 4%

Gasification of Biomass
Selected Results

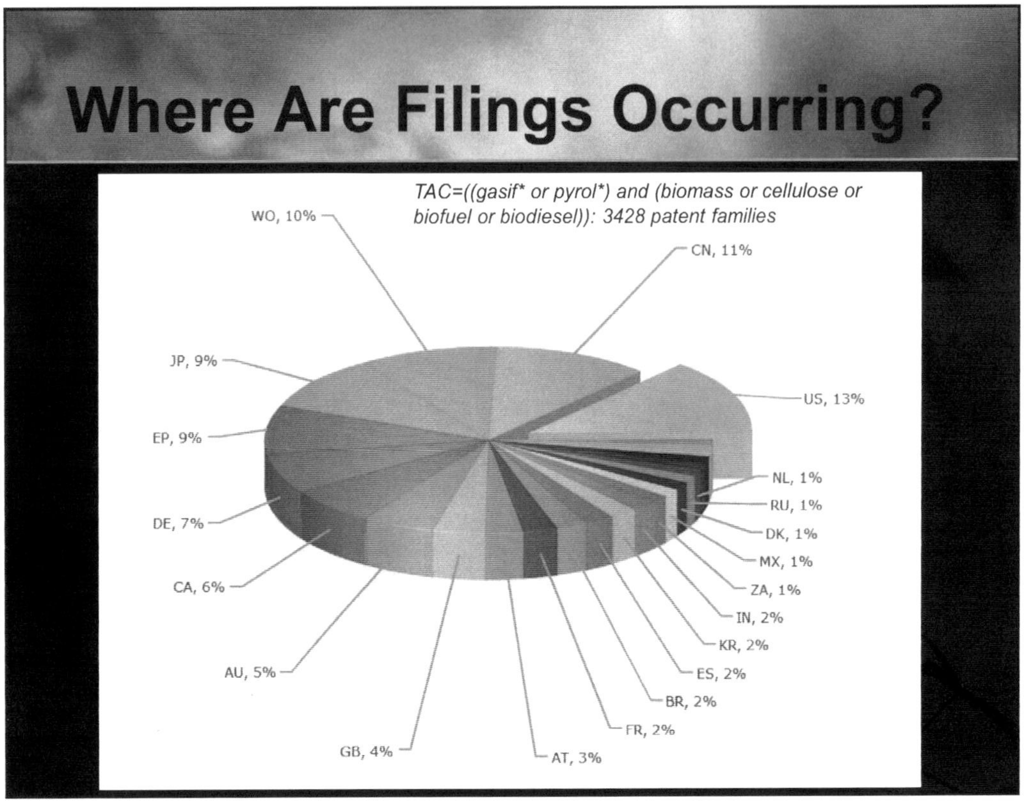
Where Are Filings Occurring?
TAC=((gasif* or pyrol*) and (biomass or cellulose or biofuel or biodiesel)): 3428 patent families
WO, 10%
CN, 11%
JP, 9%
US, 13%
EP, 9%
NL, 1%
RU, 1%
DK, 1%
MX, 1%
ZA, 1%
IN, 2%
KR, 2%
ES, 2%
BR, 2%
FR, 2%
DE, 7%
CA, 6%
AU, 5%
GB, 4%
AT, 3%

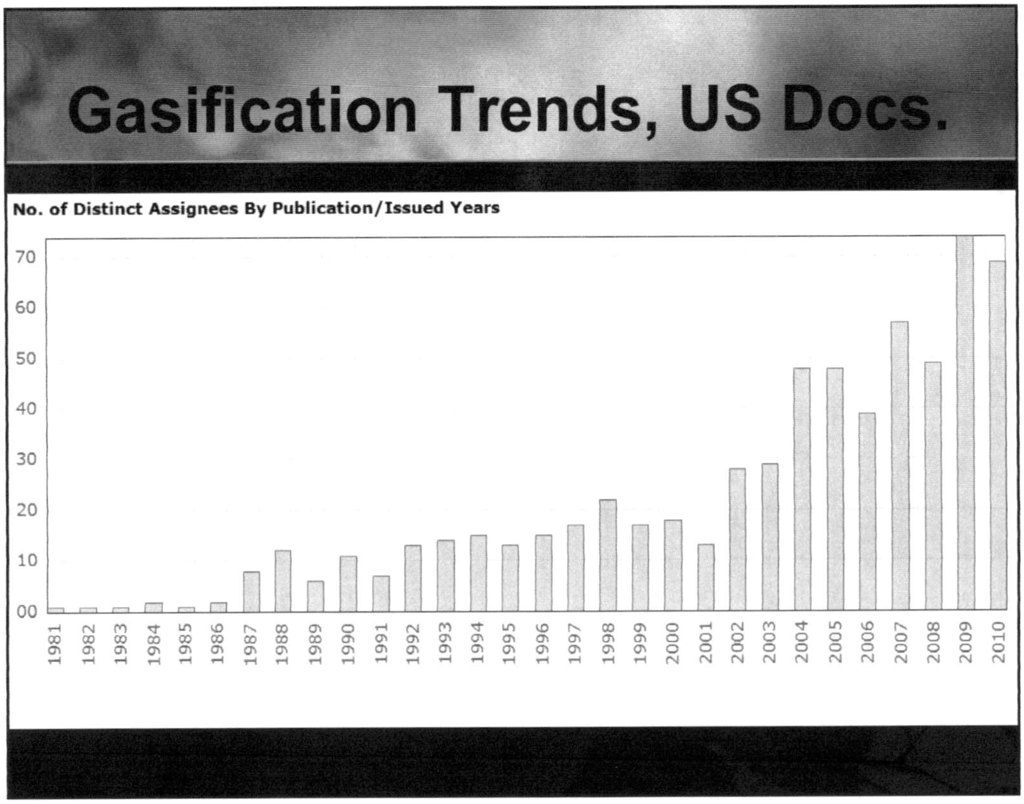
Gasification Trends, US Docs.
No. of Distinct Assignees By Publication/Issued Years
70
60
50
40
30
20
10
00
1981
1982
1983
1984
1985
1986
1987
1988
1989
1990
1991
1992
1993
1994
1995
1996
1997
1998
1999
2000
2001
2002
2003
2004
2005
2006
2007
2008
2009
2010

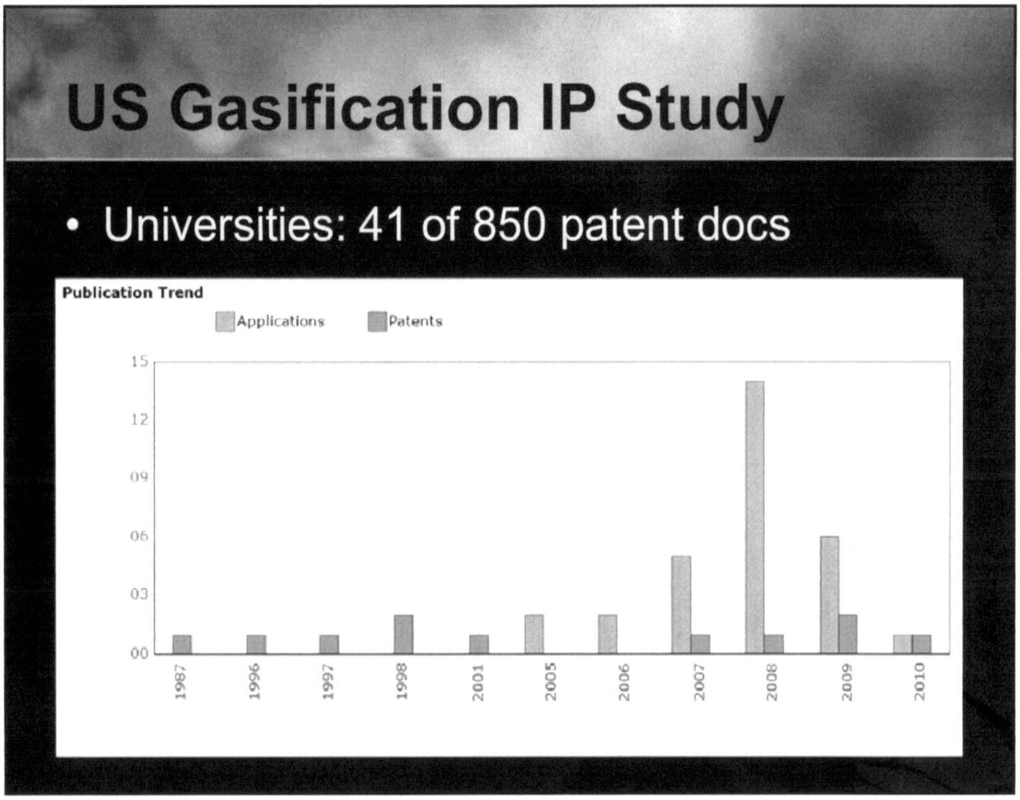
US Gasification IP Study
• Universities: 41 of 850 patent docs
Publication Trend
Applications
Patents
15
12
09
06
03
00
1987
1996
1997
1998
2001
2005
2006
2007
2008
2009
2010

Forest Products Industry
Paper & lumber producers, large suppliers

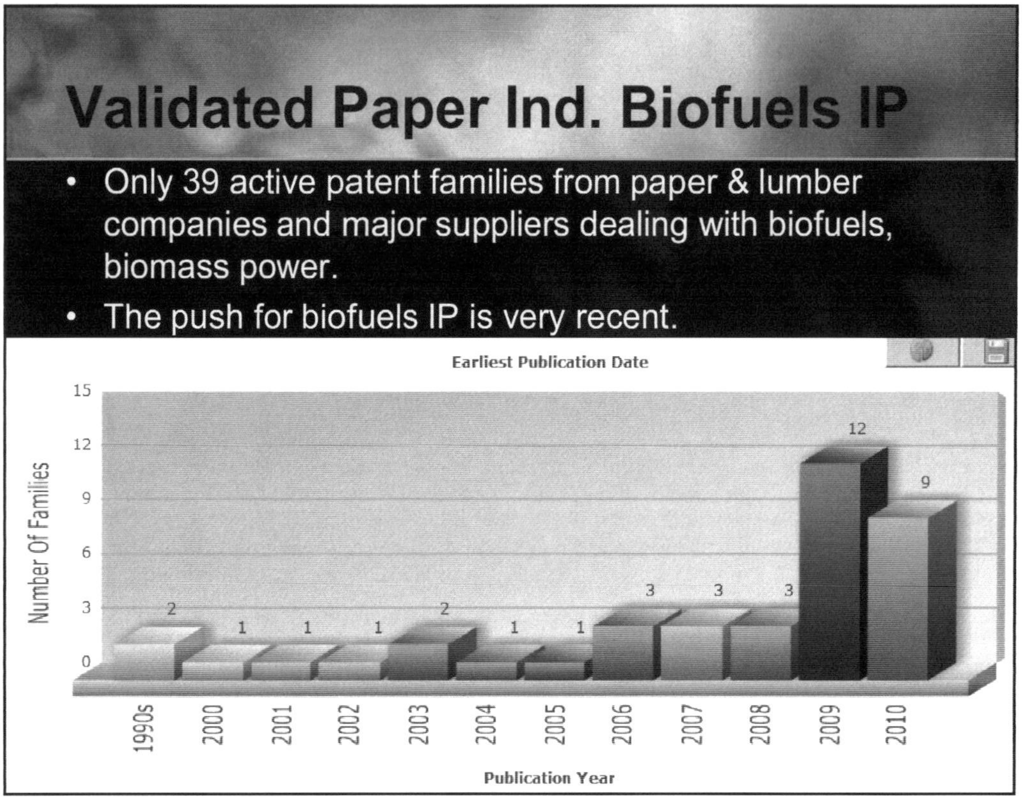
Validated Paper Ind. Biofuels IP
• Only 39 active patent families from paper & lumber companies and major suppliers dealing with biofuels, biomass power.
• The push for biofuels IP is very recent.
Earliest Publication Date
Number Of Families
15
12
9
6
3
0
2 1 1 1 2 1 1 3 3 3 12 9
1990s 2000 2001 2002 2003 2004 2005 2006 2007 2008 2009 2010
Publication Year

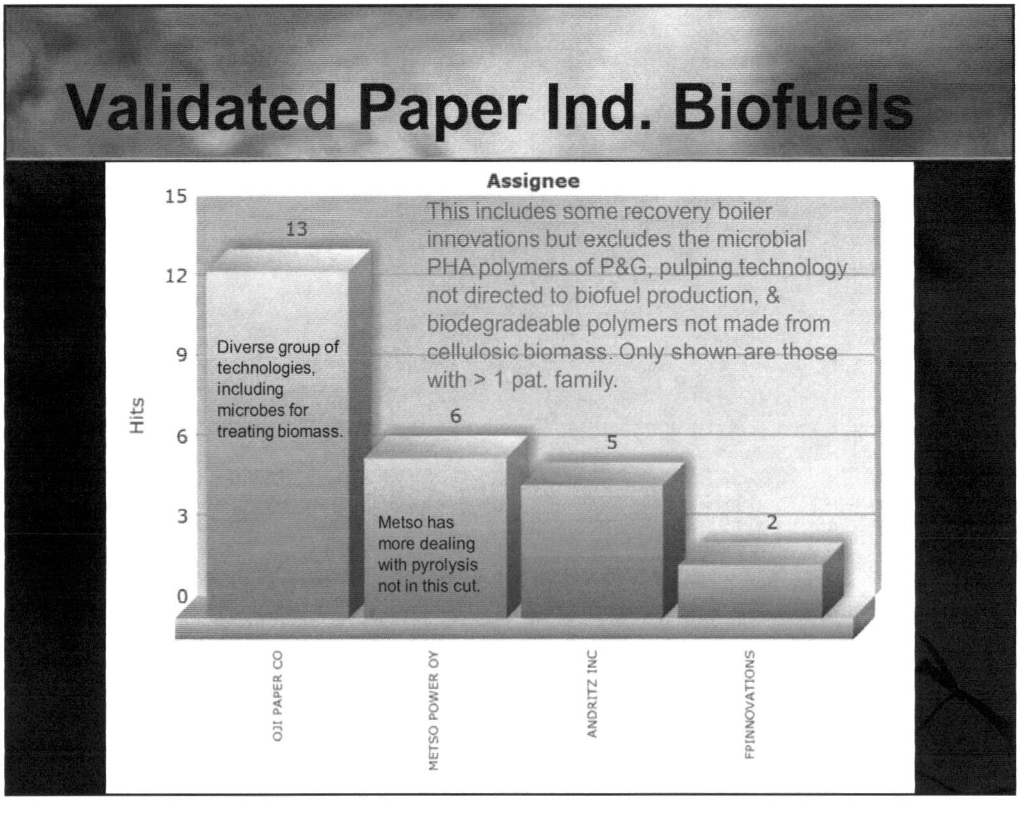
Validated Paper Ind. Biofuels
Assignee
This includes some recovery boiler innovations but excludes the microbial PHA polymers of P&G, pulping technology not directed to biofuel production, & biodegradeable polymers not made from cellulosic biomass. Only shown are those with > 1 pat. family.
Hits
15
12
9
6
3
0
13
6
5
2
Diverse group of technologies, including microbes for treating biomass.
Metso has more dealing with pyrolysis not in this cut.
OJI PAPER CO
METSO POWER OY
ANDRITZ INC
FPINNOVATIONS

Big Energy & Auto
Patent Families (Easily) Tied to Biofuels

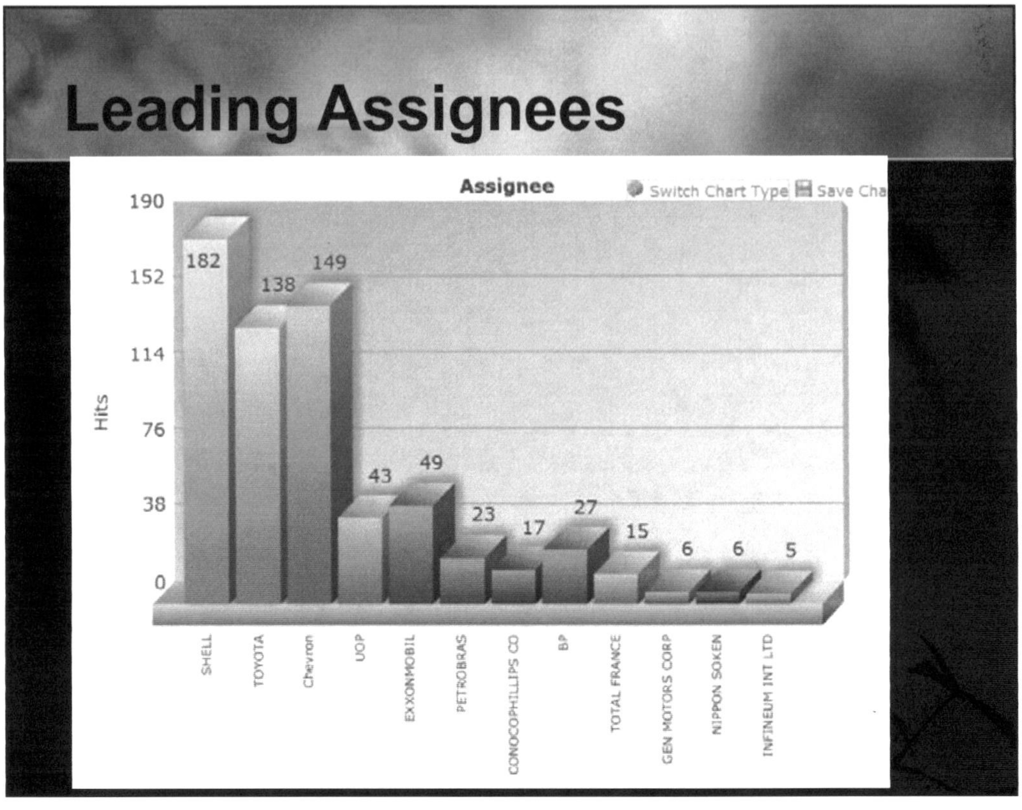
Leading Assignees
Assignee
Switch Chart Type
Hits
190
152
114
76
38
0
182
138
149
43
49
23
17
27
15
6
6
5
SHELL
TOYOTA
Chevron
UOP
EXXONMOBIL
PETROBRAS
CONOCOPHILLIPS CO
BP
TOTAL FRANCE
GEN MOTORS CORP
NIPPON SOKEN
INFINEUM INT LTD

Moving Past Low-Value Fuels

Biomass to Value-Added Chemicals

Fuels or Chemicals?

- $2/gallon fuel verrsus $5/lb chemicals?
- From ethanol/diesel to butanol, jet fuels, etc. or non-fuels
- ZeaChem: 2-carbon products (ethyl acetate, acetic acid and ethanol)
- LS9: diesel fuels and surfactants
- Amyris: fanesene for cosmetics and other industries
- Solazyme: renewable oils for the food and fragrances industries.
- Braskem is making products such as ethylene from bio-ethanol. WO09070858
- Aviation & specialty fuels (Syntroleum, COP, UOP, etc.)

Resource: http://biofuelsdigest.com/bdigest/2010/06/30/when-life-gives-you-lemons-make-citric-acid-and-ethyl-acetate/

Top 12 Chemicals from Biomass

- 2004 Doe Report: ***Top Value Added Chemicals from Biomass***

Four Carbon 1,4-Diacids (Succinic, Fumaric, and Malic) ....
2,5-Furan dicarboxylic acid (FDCA)...
3-Hydroxy propionic acid (3-HPA)..
Aspartic acid ....
Glucaric acid ....
Glutamic acid ....
Itaconic acid..
Levulinic acid ....
3-Hydroxybutyrolactone..
Glycerol...
Sorbitol (Alcohol Sugar of Glucose) ....
Xylitol/arabinitol (Sugar alcohols from xylose and arabinose) ..

Example: Levulinic Acid

Source: DOE, 2004 (http://www1.eere.energy.gov/biomass/pdfs/35523.pdf)

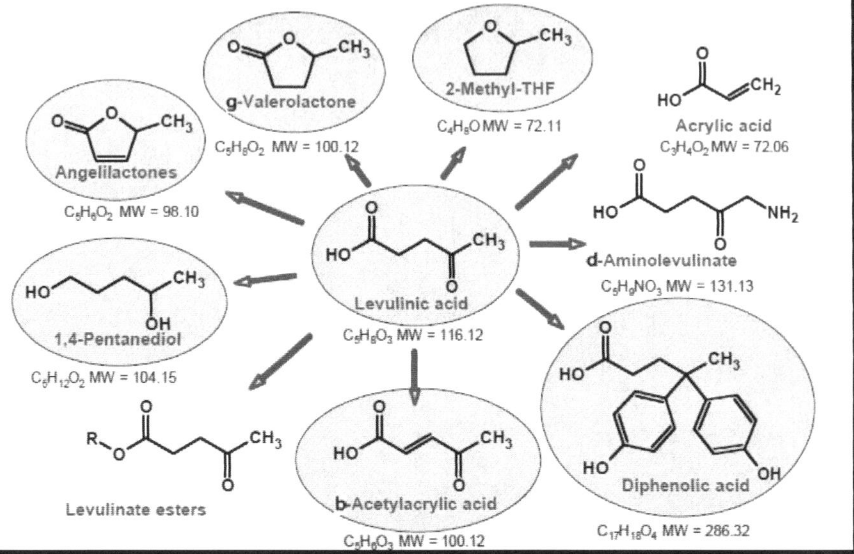

Prior art at least to 1931: Thomas, et al., "Studies on Levulinic Acid. I. Its Preparation from Carbohydrates by Digestion with Hydrochloric Acid Under Pressure". *J. Am. Chem. Soc.* Jun. 8. **1931**. 2324-2328. vol. 53.

Analysis of the L.A. Estate

- Information for live presentation only. Contact Jeff Lindsay for details.

What To Do with Glycerol?

- Solazyme: use it as food for algae to produce oils (US 2009004715)
- Segetis: Create other useful chemicals in a thermochemical process using ketal chemistry (US2009004715A)
- Cognis: Use it as fuel
- PrimaFuel: Convert it in gas phase to monohydric alcohols like propanol (US2009005614)

Expanding Patent Strategy
Business Method Patents

Oppty: Business Method Patents
Mostly for the US, these are unconventional patents on ways of doing business, often involving software or computers in the process stops. They survived recent legal challenges (the Bilski case).
Often associated with US class 705.

Blimps and Biofuels?

- Xyleco, US2010064746: Mobile Ethanol Prod.
  - "instead of transporting biomass to the production facility, the facility is transported to the biomass"
- Claim ... mass, comp ... first site w ... nter to a secon ... s at the secon

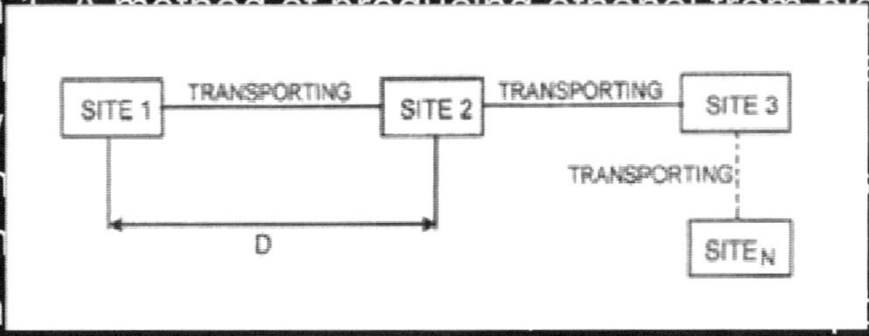

- Claim ... rting is performed with an airborne vehicle.
- Claim 8. The method of claim 7, wherein the airborne vehicle is a plane, helicopter or **blimp**.

A Classic Business Method

- US 20090076913 , Endicott Biofuels, "Method for Obtaining Biodiesel, Alternative Fuels and Renewable Fuels Tax Credits and Treatment"
- 1. **A method for obtaining U.S. Federal tax credits** under Title 26 Sections 40A and/or 6426 for ester based fuels, and/or a method for obtaining Renewable Identification Numbers under the EPA Clean Air Act as amended by the Energy Independence and Security Act of 2007, comprising:(A) passing one or more carboxylic acid components and an alcohol component countercurrently through an esterification zone . . . having a tax payer use the product of step (B) as a basis for a claim for U.S. Federal tax credits under Title 26 Sections 40A and/or 6426, and/or for U.S. Federal Renewable Identification Numbers under Environmental Protection Agency Clean Air Act as amended by the Energy Independence and Security Act of 2007.

Custom Blends at the Gas Station?

- GS Cleantech: US2009005903
  - A method and related system blend fuels and dispense a blended fuel to a vehicle at a fueling station. In one aspect, the method comprises selecting the desired proportion of at least one of the fuels available at the service station, blending that fuel with another one available, and delivering the blended fuel to the vehicle. Preferably, the first fuel is a non-renewable, petroleum based one, while the second fuel is a non-petroleum based, renewable
  - US Class 705, others

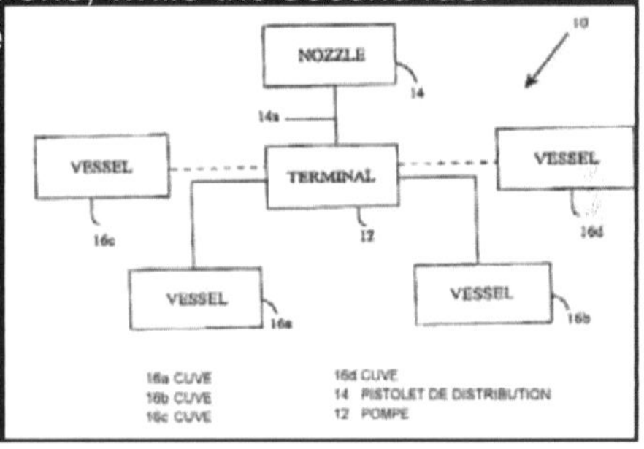

Collection of Biomass

- Iogen: US2008038815, "Method and System for the Large Scale Collection, Preparation, Handling and Refining of Ligno-cellulosic Biomass"
- Claim 1: A biomass collection and refining system, comprising: a biomass refining plant; a plurality of collection points; and at least one conduit loop connecting the collection points and the refining plant in a continuous circulating arrangement for transporting a slurry of water and biomass from the collection points to the refining plant.

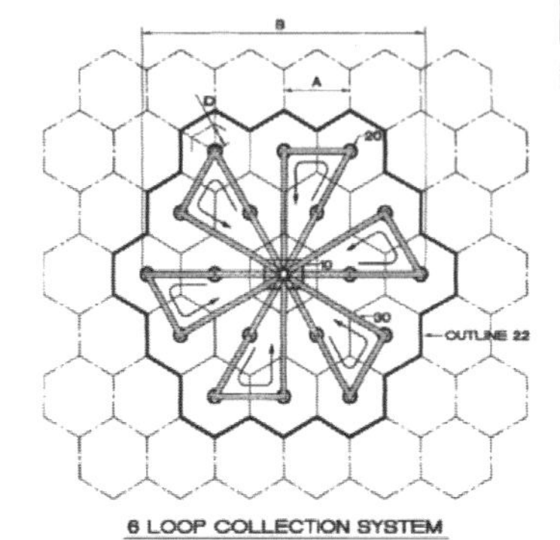

6 LOOP COLLECTION SYSTEM

Weyerhaeuser's Intercropping Patent

- **US 7,412,932,** "Methods of Managing Timberland"
- Deals with ways of planting and harvesting trees and a second energy crop on the same land.
- May relate to Weyer.'s joint venture with Chevron, Catchlight Energy

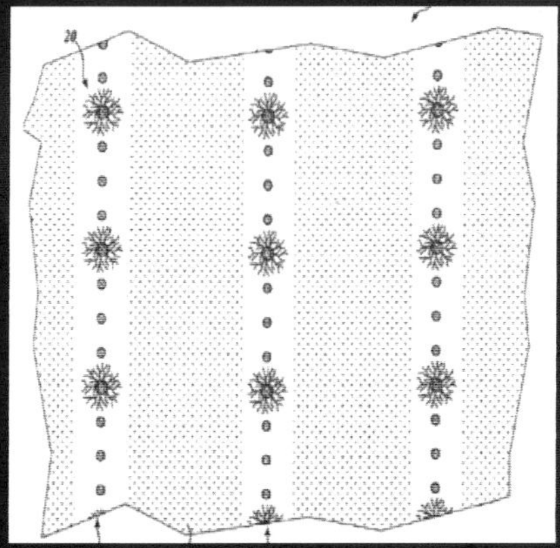

Selected Companies

Trends & Observations on Estates

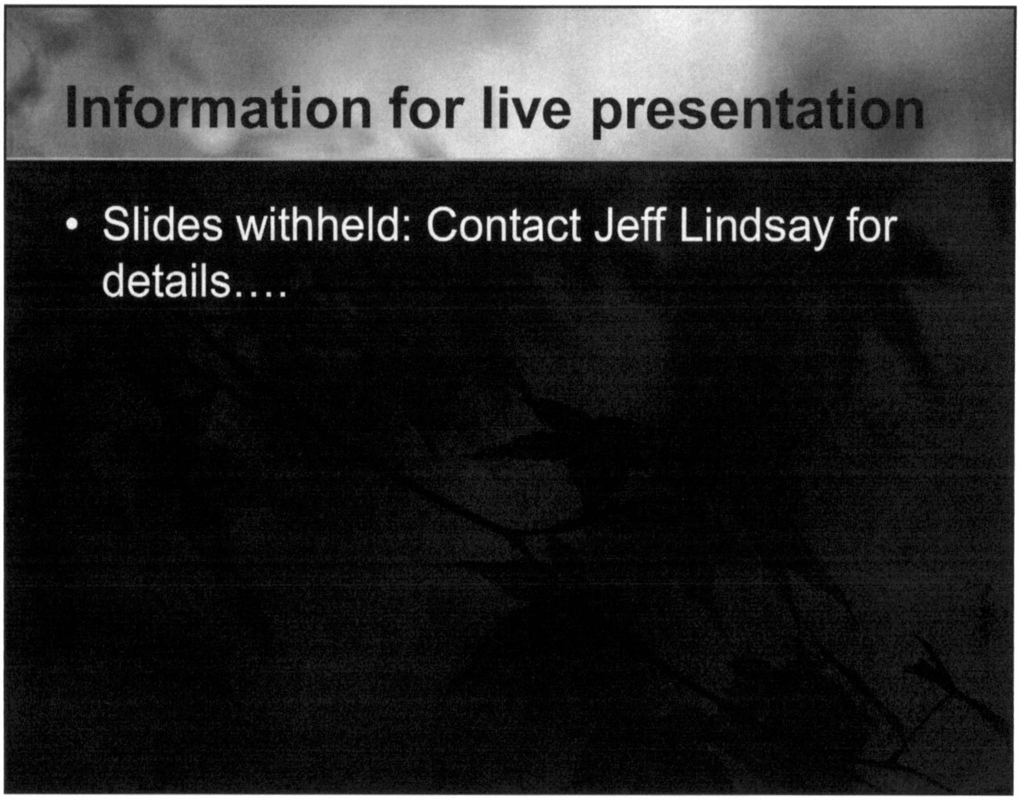
Information for live presentation
• Slides withheld: Contact Jeff Lindsay for details….

Final Observations

Green IP at Risk

- The G77 Green Tech Transfer Proposal calls for compulsory licensing of green technologies in exchange for climate change commitments
- Ouch.

Proactive IP Strategy Needed

- Companies & universities cannot wait for invention disclosures to trickle up.
- Define IP targets, whitespaces, threats, and create broad intellectual assets (not just patents) for offense and defense.
- IP wars will occur: you cannot ignore IP.
- Good IP is no guarantee of success. Some of the most promising biofuel efforts may rely on old, proven technology.

Standard Disclaimers Apply...

- Nothing herein constitutes legal advice. Any suggestion otherwise will get you sued.
- Do your own diligence. This is an incomplete snapshot and a work in progress.
- We are not accountable for anything presented in old, outdated, preliminary versions of this report.
- Today's presentation is an old, outdated, preliminary version.

Thank you! Questions?

- For further details, contact:

Jeff Lindsay, Ph.D. (US patent agent, #41,155)
Director of Solution Development
Innovationedge
1526 South Commercial St.
Neenah, WI 54956
jlindsay@innovationedge.com
Twitter: @jefflindsay, @bioproducts
Blogs: InnovationFatigue.com, SharpIP.com, and Innovationedge.com/blog

For the AIChE Forest Bioproducts Division, see http://forest-products.org.

(557a) Peering Into Supercritical-Water Biomass Gasification with Neutron Radiography

Andrew A. Peterson, Department of Physics, Technical University of Denmark, Lyngby, Denmark, Frederic Vogel, Catalytic Process Engineering Group, Paul Scherrer Institut, Villigen, Switzerland and Jefferson W. Tester, School of Chemical and Biomolecular Engineering and Cornell Center for a Sustainable Future, Cornell University, Ithaca, NY

Hydrothermal technologies have shown promise as high-efficiency means of converting lignocellulosic feedstocks into conventional fuels. In particular, supercritical-water gasification (SCWG) processes have been demonstrated to achieve 99% gasification of woody biomass to methane and hydrogen gases, although many challenges remain to develop commercializable processes. An obstacle to process development and understanding is due to the robust materials requirements necessary to contain the high-temperature, high-pressure reactive mixture involved with SCWG, which hinders the ability to observe the dynamics within process vessels. In this paper, we will report on our use of neutron radiography to visualize the performance of equipment within SCWG, and how the use of this tool can lead to more effective process design for biomass gasification.

See more of this Session: Reactor Engineering for Biomass Feedstocks
See more of this Group/Topical: Sustainable Engineering Forum

(557b) Supercritical Water Gasification of Phenol

Chad Michael Huelsman and Phillip E. Savage, Chemical Engineering, University of Michigan, Ann Arbor, MI

Biomass gasification is a clean and renewable energy alternative that can help mitigate growing global energy needs and concerns over fossil fuel depletion and greenhouse gas emissions. Hydrothermal gasification in supercritical water has recently received attention as a promising technology, but little is known about the chemistry that leads to the formation of gas and undesirable byproducts such as char under these harsh conditions. In order to better understand biomass gasification chemistry in supercritical water, a systematic study of the effects of process variables on product composition and yield was conducted. The process variables tested were temperature, water density, reactant loading, and reaction time. Phenol was studied as a model compound because phenolic compounds are difficult to gasify and represent one of the main obstacles to complete gasification of biomass. Phenolic lignin is a major constituent of lignocellulosic biomass, and phenols are known to form from both lignin and cellulose under high temperature and pressure conditions. Reactions were carried out at 500–700°C and 22–67 MPa in quartz batch reactors to eliminate catalytic effects from metal walls. Gas- and liquid-phase reaction products were identified and quantified using gas chromatography with mass spectrometry, flame ionization, and thermal conductivity detection. Reaction of phenol in supercritical water led to the production of little gas, but instead a swath of heavier molecular weight compounds, primarily polycyclic aromatic hydrocarbons (PAHs), formed, and more than 20 of these were identified for the first time in this study. These compounds represent char precursor molecules, and it is expected that even larger molecular weight compounds are formed via ring-fusion, polymerization, and cross-linking reactions, which ultimately leads to the char that was observed on reactor walls. Studying the formation of these molecules provides insight into how to suppress char formation and improve gas yield.

See more of this Session: Reactor Engineering for Biomass Feedstocks
See more of this Group/Topical: Sustainable Engineering Forum

(557c) Reaction Rate Parameters for Supercritical Water Gasification of Various Biomass Species

Yukihiko Matsumura[1], Yasunao Yamashita[2], Shuhei Inoue[1], Yoshifumi Kawai[3], Tomoaki Minowa[4], Yoji Noda[5] and Yoshihisa Shimizu[6], (1)Division of Energy and Environmental Engineering, Hiroshima University, Higashi-Hiroshima, Japan, (2)Department of Mechanical Science and Engineering, Hiroshima University, Higashi-Hiroshima, Japan, (3)Chuden Plant Co. Ltd., Hiroshima, Japan, (4) Biomass Technology Research Center, National institute of Advanced Industrial Science and Technology (AIST), Higashi-Hiroshima, Japan, (5)Toyo Koatsu Co., Ltd, Hiroshima, Japan, (6)The Chugoku Electric Power Co., Inc., Hiroshima, Japan

Supercritical water gasification is a promising technology for the effective use of wet biomass including manures, food wastes, and sewage sludge. The authors have shown that the reaction kinetics in supercritical water gasification with and without activated carbon catalyst can be experessed using Arrhenius type first-order reaction rate. In this study, the reaction paremeters for various biomass species including chicken manure, swine manure, cattle maure, soy-bean hydrogen fermentation residue, food waste hydorgen fermentation residue, waste grass, and seaweed ethanol fermentation residue were determined, and compared. The result showed that the bioamss species can be categorized into some groups in terms of its reactivity in supercritical water and the effect of activated carbon catalyst.

See more of this Session: Reactor Engineering for Biomass Feedstocks
See more of this Group/Topical: Sustainable Engineering Forum

(557d) A Decoupling Methodology for Wood Gasifier Modeling: From Laboratory Scale Kinetic Measurements to Pilot Plant Simulation. Application to a Dual Fluidized Bed Gasifier

Olivier Authier[1], Guillain Mauviel[1], Monique Ferrer[1], Az-Eddine Khalfi[2] and Jacques Lédé[1], (1) LRGP, ENSIC, CNRS-Nancy Université, Nancy, France, (2)Group I87, EDF R&D Chatou, Chatou, France

Different types of modelings have been developed in literature for gasifiers optimization. They could be simple equilibrium approach - in order to predict an outlet gas composition - or detailed kinetic approach reliable for gasifier scale-up. Until now, there are few available models taking into account close couplings between the chemical reactions, heat and mass transfer and hydrodynamics phenomena [1]. Indeed, such an approach requires the good knowledge of a great number of elementary processes. The determination of reliable kinetic constants is one of the main key points. Unfortunately, many values reported in the literature are only valid in the conditions where they have been measured [2]. In addition, analysis of all the phenomena is made complicated by the fact that they all occur simultaneously inside the gasifier. That is why it seems interesting to study the main involved elementary phenomena such as chemical reactions independently of each other.

Among the possible wood gasification technologies, the Dual Fluidized Bed (DFB) gasifier operating with steam as gasification agent may be used to produce a high quality syngas [3]. Typically, wet wood pellets enter the gasifier where they undergo heating and several physicochemical transformations. After pellet drying, primary pyrolysis produces char (carbon-rich residue) and volatile products (condensable vapors and gases) resulting in up to 80% wood weight loss [4]. Primary products are then subjected to secondary reactions occurring outside the pellets:

*vapors thermal cracking by homogeneous reactions occurring in the gas-phase and also through catalytic reactions with the bed material (e.g. olivine),

*gas reactions (water-gas shift and steam methane reforming),

*char-steam gasification.

In the present study, the main involved chemical reactions are decoupled and studied independently at the laboratory scale in thermal conditions close to those encountered in the DFB gasifier. For that purpose, original facilities are developed to follow the wood, vapors and char conversions.

Solid (i.e. wood pyrolysis) and gas/solid (i.e. char-steam gasification and vapors catalytic thermal cracking) reactions are carried out with an image furnace, a source of controlled concentrated radiation [4]. The solid sample (i.e. wood, char or olivine pellet) is submitted to heat flux densities representative of those prevailing in a fluidized bed (i.e. from 0.2 to 0.8 MW/m2). The sample is placed inside a transparent quartz reactor swept by an inert (N2) or reactive (N2/H2O, N2/H2O/volatile products) gas. The gaseous species issued from the sample surface are rapidly quenched by mixing with the sweep gas. The gas-phase temperature is significantly lower than that of sample surface in order to study only the reactions occurring at the sample level (quenching). All the products formed by the reactions are recovered and analyzed. Kinetic data for wood pyrolysis and char-steam gasification are extracted from a model taking into account couplings between chemical reactions and both internal and external heat transfers. Kinetic data for vapors catalytic cracking are derived from the experimental values of gas yields determined as a function of the olivine surface temperature. Experiments related to vapors thermal cracking in the gas-phase are carried out inside a Continuous Self Stirred Tank Reactor (CSSTR) where the gas phase temperature and composition are uniform [5,6]. The CSSTR cracking conditions (temperature: 973-1173 K, gas residence time: 0.3 s) are close to those encountered inside the gasifier. Kinetic data for vapors thermal cracking are simply derived from the experimental values of products yields (i.e. gases, vapors and water) measured as a function of the CSSTR temperature.

Chemical controlling steps in the DFB gasifier are then discussed on the basis of a characteristic time analysis. The gasifier is modeled at both pellet and gasifier levels (i.e. coupling between pellet conversion, secondary reactions, solid and gas-phase hydrodynamics and interfaces transfers). The reactor is represented by three different zones in series (i.e. dense, splash and freeboard zones). Finally, the decoupling methodology is validated from the comparison of model results with experimental measurements performed at the 8 MW (fuel power) DFB plant in Güssing/Austria. The good agreement between predicted gas composition (at the gasifier outlet) and temperature (in the fluidized bed) with experimental data proves the reliability of the decoupling methodology.

[1] A. Gomez-Barea, B. Leckner, Progress in Energy and Combustion Science, (2010), doi:10.1016/j.pecs.2009.12.002

[2] O. Authier, M. Ferrer, G. Mauviel, A.E. Khalfi, J. Lédé, 18th European Biomass Conference and Exhibition, Lyon (France), (2010).

[3] C. Pfeifer, R. Rauch, H. Hofbauer, Industrial & Engineering Chemistry Research, (2004), 43, 1634.

[4] O. Authier, M. Ferrer, G. Mauviel, A.E. Khalfi, J. Lédé, Industrial & Engineering Chemistry Research, (2009), 48, 4796.

[5] O. Authier, M. Ferrer, G. Mauviel, A.E. Khalfi, J. Lédé, 17th European Biomass Conference and Exhibition, Hamburg (Germany), (2009).

[6] S. Baumlin, F. Broust, M. Ferrer, N. Meunier, E. Marty, J. Lédé, Chemical Engineering Science, 60,

(2005), 41.

See more of this Session: Reactor Engineering for Biomass Feedstocks
See more of this Group/Topical: Sustainable Engineering Forum

(557e) Can the Effluent of An Efficient Biomass Carbonizer Be Combusted?

Javier Ábrego[1], Toshiaki Hanaoka[2] and Michael J. Antal Jr.[1], (1)Hawaii Natural Energy Institute, University of Hawaii at Manoa, Honolulu, HI, (2)Biomass Technology Research Center, National Institute of Advanced Industrial Science and Technology, Hiroshima, Japan

Biocarbon (i.e. biomass charcoal) is the most widely used of all biofuels: it cooks our food, cleans our water, reduces ores (e.g. SiO2) to pure metals and semi-metals (e.g. Si for photovoltaic cells), and can be used as a substitute for fossil coal to generate electrical power. An efficient biomass carbonizer (e.g. the UH Flash Carbonization process) retains as much as 70% of the carbon in the biomass feedstock in the biocarbon product. Recognizing that the carbon in the biomass carries its energy; it is legitimate to wonder if the effluent of the Flash Carbonization process can be combusted. The impacts of the feedstock moisture content on the flame temperature are of special interest. In this paper we present experimental measurements of the flame temperature within a catalytic afterburner located at the outlet of a Flash Carbonization reactor, and compare these measurements with theroretical predictions.

See more of this Session: Reactor Engineering for Biomass Feedstocks
See more of this Group/Topical: Sustainable Engineering Forum

(557f) Design of Exergy Recuperative Fluidized Bed Drying System for Biomass

Muhammad Aziz, Chihiro Fushimi, Yasuki Kansha, Kazuhiro Mochidzuki, Shozo Kaneko and Atsushi Tsutsumi, Institute of Industrial Science, The University of Tokyo, Tokyo, Japan

An innovative drying process based on self-heat recuperation technology utilizing both latent and sensible heat has been developed aiming for drying-energy saving of biomass particles. Drying processes are divided into three main subsequent stages, which are pre-heating, evaporation and superheating. A state of the art of this technology is on the exergy elevation and its recuperation through heat exchange and heat pairing for each sensible and latent heat. Hence, intensification of heat and mass transfer inside the drier becomes one of the most crucial matters that must be paid attention to obtain the optimum drying performance and efficiency. Furthermore, fluidized bed drier with internal heat exchanger is selected as the main drier conducting the second stage of drying which is evaporation because of its characteristics on high heat and mass transfer and good solids contact between particles leading to shorter drying time and its uniformity. In this system, the exhausted vapor from the biomass is once superheated and subsequently it is compressed by the compressor to certain temperature rise required for heat exchange. Next, the compressed vapor is re-circulated and utilized as the heat source for superheating, evaporation and pre-heating subsequently. The current study is focused mainly to the experimental study on the main drying process which is evaporation. The objective of this study is observing the optimal drying condition for biomass particles inside the fluidized bed drier applying self-heat recuperation technology. On the other hand, the biomass samples, which are wood chip and paddy straw, generally have a characteristic of peculiar shape resulting in difficult fluidization. Hence, a fine and inert particle, which is silica sand, is inserted inside the fluidized bed to enhance the fluidization performance of the biomass particles and improve the heat transfer and its uniformity across the bed. The effects of inert particle diameter, biomass to inert particle mass fraction, fluidization velocity and bed temperature corresponding to vapor pressure elevation to the drying performance of biomass particles are observed. From the experimental results, it is shown that decrease of inert particle diameter leads to increase of heat transfer across the bed resulting in lower moisture content of the dried products. Furthermore, higher pressure elevation of the exhausted vapor also brings to better drying performance as the temperature gradient increases leading to higher drying rate of the biomass samples. In addition, there is no significant effect of the fluidization velocity to the drying performance and energy efficiency in dynamic fluidization regimes.

See more of this Session: Reactor Engineering for Biomass Feedstocks
See more of this Group/Topical: Sustainable Engineering Forum

(580a) Bioproducts From Biomass Hemicellulose and Lignin

Xiao Zhang, Chemical Engineering and Bioengineering, Washington State University, Richland, WA

Despite several decades of intensive research effort, producing biofuels and biochemicals from lignocellulosic biomass feedstock is still not an economically viable proposition at present. Lignocellulosic biomass is mainly composed of cellulose, hemicelluloses and lignin. The majority of the existing bioconversion processes, either through themochemical or biochemical route, treat biomass feedstock as a whole at severe conditions. It is recognized the thermo and chemical degradation rates are significantly different among cellulose, hemicelluloses and lignin. As a consequence, the value of hemicelluloses and lignin from biomass are compromised or diminished in these bioconversion processes. An economically feasible biomass conversion process requires maximizing the potential value of all biomass components. Fractionation of main biomass components prior to themo- or biochemical conversion provides a new process configuration to achieve this goal. This presentation will discuss some recent developments in biomass fractionation process to obtain hemicelluloses and lignin from biomass and potential for biofuels and bioproducts production.

See more of this Session: Biorefinery - Improved Utilization of Co-Products From Biorefinery
See more of this Group/Topical: Forest and Plant Bioproducts Division

(580b) The Model Biorefinery: A Case Study

Robert Mellon and **Brian L. Cooper**, Hazen Research, Inc., Golden, CO

Biomass based chemical refining represents many new opportunities in engineering and process design. The population of companies seeking to develop their technology is large and diverse, from pure liquid fuel producers to integrated multi-product refineries. A common flow sheet among many of these organizations shows a hybrid biochemical/thermochemical process, whereupon the carbohydrate values in a biomass are removed and the resultant lignin portion is utilized for co-products or process heat. As a company with capabilities in both areas, Hazen Research has conducted a number of experiments to generate data for a hypothetical hybrid process utilized for the biorefining of Lodgepole Pine (Pinus contorta). The presentation will focus on the biochemical operations, biomass pretreatment, hydrolysis and fermentation, as well as thermochemical experiments including pyrolysis and gasification. A cost-benefits analysis will be presented that demonstrates the potential for the hybrid process operations model.

See more of this Session: Biorefinery - Improved Utilization of Co-Products From Biorefinery
See more of this Group/Topical: Forest and Plant Bioproducts Division

(580c) Integrated Forest Biorefineries: Production Platforms and Products

Lew P. Christopher, Center for Bioprocessing Research & Development, South Dakota School of Mines & Technology, Rapid City, SD

Due to the increasing off-shore competition, global movement and incentives for green fuels and chemicals, the North American pulp and paper industry needs to create additional revenues and diversify their products and markets to remain competitive. To achieve this, the pulp and paper mills have to evolve into Integrated Forest Biorefineries (IFBR). This presentation will provide an overview of the IFBR production platforms and research needs in the area of: 1) adding value prior to pulping; and 2) new values from residuals and spent liquors. Products derived from the IFBR platforms employing both biochemical and thermochemical conversion processes as well as models of integrated technological schemes for the complex utilization of forest biomass for biofuels, biochemicals and market pulp will be discussed. The main technological challenges that need to be overcome to maximize the IFBR value through integration of processing steps and utilization of waste streams will be summarized.

See more of this Session: Biorefinery - Improved Utilization of Co-Products From Biorefinery
See more of this Group/Topical: Forest and Plant Bioproducts Division

(580d) Can Cellulose Crystallinity Be Accurately Measured?

Sunkyu Park[1], David K. Johnson[2], Seong H. Kim[3], Ashutosh Mittal[2], Rui Katahira[4], Anna Barnette[3] and Junyeong Park[1], (1)Forest Biomaterials, North Carolina State University, Raleigh, NC, (2) Biosciences Center, National Renewable Energy Laboratory, Golden, CO, (3)Chemical Engineering, Pennsylvania State University, University Park, PA, (4)National Bioenergy Center, National Renewable Energy Laboratory, Golden, CO

Although measurements of crystallinity index (CI) have a long history, it has been found that CI varies significantly depending on the choice of measurement methodology. In this study, the concept and measurement techniques of crystallinity index were investigated using five different methods from x-ray diffraction and solid-state 13C NMR. It was found that the most popular and simple method, using measurement of just two heights in the x-ray diffractogram, produced significantly higher crystallinity values than did the other methods. Data in the literature and our experimental results support this observation. The analysis of CI measurement becomes complicated when biomass samples are used due to the amorphous contribution from hemicellulose and lignin. New method with sum frequency generation vibration spectroscopy will be introduced, which can selectively analyze cellulose structure without any interference from amorphous components.

See more of this Session: Biorefinery - Improved Utilization of Co-Products From Biorefinery
See more of this Group/Topical: Forest and Plant Bioproducts Division

PREDICTING COATING COLOR FORMULATION FOR PAPER AND PAPERBOARD WITH STATISTICAL EXPERIMENTAL DESIGN

A.K.Ray, and Sanjay Tyagi #*

Department of Paper Technology, Indian Institute of Technology Roorkee, Saharanpur Campus, Saharanpur- 247001, India
 # Central Pulp and Paper Research Institute, Saharanpur – 247001 (U. P.), India

ABSTRACT

Coating of any substrate such as paper is a multivariate problem. It is extremely difficult to quantify the parametric effect on the response and inter dependency of parameters. Coated paper and paper board are considered as layered constructions. The surface of the sheet is coated with a thin pigment layer which significantly improves its printing and properties like gloss, brightness, smoothness, opacity, specific scattering coefficient and surface strength. In this present study attempt have been made to carry out experiments in laboratory air knife coater for coating duplex board obtained from an Indian mill using statistical experimental design based on Plackett- Burman design methods. The development of regression estimation, optimization techniques and algorithms for application have also been made. The pigment, the binders and other ingredients used are China clay, precipitated calcium carbonate (PCC), ground calcium carbonate (GCC), Latex, carboxy methyl cellulose (CMC), starch and NaHMP (sodium hexa meta phosphate) with variable solid contents. Equipment variables studied were air pressure, number of passes through calender nips, knife angle, calendering pressure, knife angle and calendering pressure at fixed temperature. The output variables like brightness, gloss, smoothness, and surface strength (IGT) were studied and the corresponding regression equations were developed. The predicted properties for each model derived from statistical design are compared with the experimental data evaluated in the laboratory using standard tests as per TAPPI specifications. The predicted values are found to be with in acceptable range of accuracy allowed for engineering estimates. The regression obtained between observed and predicted values are also within reasonable limits.

Keywords: coating color formulation, paper board, statistical design, china clay

INTRODUCTION

Pigment coating is an excellent approach for the improvement of paper quality in terms of surface smoothness, brightness, gloss and printability properties. In addition it generally increases surface strength and decreases porosity. Coated paper is more uniform and more receptive to printing ink for better graphic reproduction. An even, better homogeneous, optically acceptable, continuous smooth surface with less irregular contour is obtained after coating that must interact with ink to provide accessible, attractive and persuasive graphical information during printing [1- 4]. Coating of paper represents a multivariate system and quantitative estimations on the effect of each parameter on the process are very difficult [4]. The set of variables is generally large but yet incomplete because not all of them have been analyzed. The situation becomes more complex by the interaction of different variables. Interaction effect cannot be separated from the main effect of a variable. The lack of co-ordinate, planned and complete investigations (which is due to the complexity degree of the problem) gives rise to several misconceptions on the effect of some variables. For the production of certain grade of coated paper, the components of the coating formulation and the process variables are fixed by trial and error methods. A limited number of experiments are sometime carried out and a step-by- step procedure is followed by fixing the level of each component in the formulation one at a time. Experience based estimates made by the plant engineers play a major role in such an exercise. Step by step elimination of the variables suffers from the drawback of eliminating the interaction effect at the same time.

The best way is to develop a mathematical model on the basis of theoretical arguments and mathematical analysis and then to use the experimental data for finding the most probable value of the unknown coefficients in the proposed model. However, the method cannot be adopted in cases where there are large number of variables present making them difficult to keep constant while one is under study. The best compromise in such instances is the use of regenerated equations. From a theoretical basis, the variables were identified. The range of values or the region of experimentation was also fixed to lie within the feasible range. The number of experiments necessary to completely study the effect of the ten selected variables is very large. It is therefore essential to restore two methods of experimental design by means of which the points at which the experiments have to be conducted are selected so that to give maximum information in a limited number of experiments.

For the multivariate problems like coating, the experimental design model by Plackett- Burman can be of extreme importance[4- 6]. The data generated are normally subjected to regression analysis. General equations, comprising only linear terms can be developed for different properties. The whole exercise can results in the development of a methodology for a sequence of steps that can be used to study the coating processes and formulations in terms of components with respect to their effect on the properties.

An attempt has been made in this present study to evaluate the process for the input- output relationship in coating colour formulation in the form of multiple non linear regression equations with base paper made from indigenous raw materials using Plackett- Burman Statistical design. Percentage of clay, PCC, GCC, latex, starch, CMC and % solids, air pressure, knife angle and number of nips or passes of calendar rolls are the variable investigated. The effect on properties such as brightness, gloss, smoothness and surface strength (IGT) can be studied and regression equations can be developed for the same.

Experimental design and its application

Plackett- Burman design [5-7] has been found to be the best for the analysis and design of the above coating problem. This method is in fact a two level saturated orthogonal design. This design is based on the fractional replicates of a full factorial design. For e.g. for K=3 (N=4), K=7 (N=8), K=15 (N=16), K=31 (N=32) where K is the number of factor and N is the number of experiments. The peculiarity of this design is that we can analyze the effect of maximum possible variables affecting the system because the numbers of experiments to be conducted are less in comparison to the other models. For a design model to be orthogonal the following conditions are necessary

(1) the number of times each factor is fixed to each of its level must be same.

(2) The number of times every two factor, each at any one of its levels are encountered must be the same.

(3) Number of observations must be divisible by the square of the number of level defined as $N=nl^2$. (Where N= integer)

In the above design the ranges of the variables is fixed. +1 value is assigned to the value of the maximum order and -1 to the minimum value. These are the two levels on which the experiments are carried out.

To construct the saturated design for K=11, 19, 23 one may use the row listed below the respective row taken from the table to produce a first column. A second column is generated from the first by placing its last element in the first positions and moving the remaining elements down one position. A third column is produced from the second one by placing the last two elements in the first two positions and shifting the remaining elements down, and the process is continued until column K is generated. The elements of the last row are "-1".

Sign Combinations for Saturated Designed, K =11, 19, and 23 are tabulated below

K	N	Sign combination
11	12	+ + - + + + - - - + -
19	20	+ + - - + + + + - + - + - - - - + + -
23	24	+ + + + + - + -+ + - - + + - - + - + - - - -

According to the sign combinations given above the following design matrix was made. The ten variables for coating/ calendaring systems are choosen which are defined as follows: X_1=%Clay; X_2 = %GCC; X_3 = %PCC ; X_4 = % Solids; X_5 = %Latex, (on pigment); X_6 = %CMC, (on pigment); X_7 = % starch (on pigment); X_8 = Air pressure through Air-Knife Coater ; X_9= Knife Angle (to the normal); X_{10} =Number of Nips (or passes through the calender nip

Design Matrix

N	X_0	X_1	X_2	X_3	X_4	X_5	X_6	X_7	X_8	X_9	X_{10}
1	+1	+1	-1	+1	-1	-1	-1	+1	+1	+1	-1
2	+1	+1	-1	-1	+1	-1	-1	-1	+1	+1	+1
3	+1	-1	+1	+1	-1	+1	-1	-1	-1	+1	+1
4	+1	+1	+1	+1	+1	-1	+1	-1	-1	-1	+1
5	+1	+1	-1	-1	+1	+1	-1	+1	-1	-1	-1
6	+1	+1	-1	+1	-1	+1	+1	-1	+1	-1	-1
7	+1	-1	+1	+1	+1	-1	+1	+1	-1	+1	-1
8	+1	-1	-1	+1	+1	+1	-1	+1	+1	-1	+1
9	+1	-1	-1	-1	+1	+1	+1	-1	+1	+1	-1
10	+1	+1	-1	-1	-1	+1	+1	+1	-1	+1	+1
11	+1	-1	+1	-1	-1	-1	+1	+1	+1	-1	+1
12	+1	-1	-1	-1	-1	-1	-1	-1	-1	-1	-1

)

Range of Experiments

X_1 = 50 - 70 X_6 = 0.5 - 1.0
X_2 = 15 - 25 X_7 = 5.0 - 10
X_3 = 10 - 20 X_8 = 50 - 60
X_4 = 40 - 45 X_9 = 0^0- 2^0
X_5 = 15 - 20 X_{10} = 4 - 7

Experimental Procedure

The experimental procedure comprises of preparation of coating colour, applying of coating colour on paper surface by air knife coater followed by drying and supercalendering in laboratory calender. While the coating and calendaring are usual, the most important is the preparation which is as under

Preparation of Coating Color

On the basis of the experimental design 12 different formulations of coating color were prepared. The composition of each of these formulations is shown in table. In addition to these components several other additives in small quantities, were used. Sodium Hexametaphosphate (1.0% on pigment) is used as dispersant. Sodium hydroxide is used to maintain pH (8.5 -9.0) of the coating color.

NaHMP, used as dispersant, was dissolved in warm water to which the pigment was added. Starch was cooked at 92-95 °C. Carboxy methylcellulose was dissociated in warm water with continues agitation. The solids content of SBR Latex was found to be 40% the general procedure followed in preparing a coating formulation is illustrated by taking case from the formulation that were prepared. Details of coating color preparation is given in elsewhere [7]

Components	**Actual weight**
China clay	70 gms.
PCC	15 gms.
GCC	20 gms
Latex	39.375 ml (15% of pigments)
Starch	10.5 gms (10% of pigments)
CMC	0.525 gms (0.5% of pigments)
Water	175.6 ml (40% of basis)

Results and Discussion

Methods of development of Regression models [4,6]

The techniques of regression analyses are employed for converting raw data into a useful form to obtain a relationship between the variables being studied. The utility of regression equation lies in its ability to make prediction for the point's intermediate between those for which the experiment was carried out. The experimental procedures includes determination of sample correlation coefficient, selection of functions based on least square techniques, finding the maximum of the functions of several variables followed by statistical analysis of regression equations. This is than subjected to the test of significance using Student's t test and Fisher's F test for fitness of the data with analysis of variance. The details of method of investigation using transformation of raw data to a new scale by normalizing all random variables and the entire methodology for the matrix solution of the coefficient of the regression equation are discussed [4,6].

The developed statistical model

Using the experimental data and Placket Burman design the following matrix has been developed. This has subjected to experimental data and the regression equations have been developed for gloss, brightness smoothness and IGT surface strength.

Design matrix and experimental data

N	X_0	X_1	X_2	X_3	X_4	X_5	X_6	X_7	X_8	X_9	X_{10}	Y_B	Y_G	Y_S	Y_{SS}
1	+1	+1	-1	+1	-1	-1	-1	+1	+1	+1	-1	78.7	76.7	16.4	2.9
2	+1	+1	+1	-1	+1	-1	-1	-1	+1	+1	+1	79.5	78.2	12.9	2.1
3	+1	-1	+1	+1	-1	+1	-1	-1	-1	+1	+1	77.4	80.2	12.1	3.2
4	+1	+1	-1	+1	+1	-1	+1	-1	-1	-1	+1	83.6	72.3	15.2	2.6
5	+1	+1	+1	-1	+1	+1	-1	+1	-1	-1	-1	79.2	70.1	18.8	2.7
6	+1	+1	+1	+1	-1	+1	+1	-1	+1	-1	-1	76.7	71.7	16.6	1.6
7	+1	-1	+1	+1	+1	-1	+1	+1	-1	+1	-1	83.4	70.5	22.1	1.7
8	+1	-1	-1	+1	+1	+1	-1	+1	+1	-1	+1	79.9	75.6	11.8	3.1
9	+1	-1	-1	-1	+1	+1	+1	-1	+1	+1	-1	78.5	72.2	15.5	2.7

10	+1	+1	-1	-1	-1	+1	+1	+1	-1	+1	+1	76.2	76.7	15.2	2.2
11	+1	-1	+1	-1	-1	-1	+1	+1	+1	-1	+1	75.8	75.0	11.4	1.7
12	+1	-1	-1	-1	-1	-1	-1	-1	-1	-1	- 1	74.7	76.8	12.9	1.4

The regression equations developed from statistical experimental optimization method and same obtained from the observed Vs predicted values from different properties are shown in the following table. It is evident from the table that observed Vs predicted values correlates reasonably well with R2 values above 0.8 for brightness. Better correlations are obtained for surface strength and gloss on an around 0.9. But it predicts very well for smoothness. The predicted data in terms of error are also estimated for sixteen number of observations.

Properties	**Regression Equations as developed from statistical models**	**Regression Equations developed graphically from data**
		Observed Vs Predicted, R^2
Brightness	$Y_B = 78.6 + 0.364X_1 + 1.309X_3 + 2.044X_4 - 0.660X_5 + 0.394X_6 + 0.236X_7 - 0.432X_8 + 0.332X_9 + 0.975X_{10}$	$y = 0.9699x + 2.5844$ $R^2 = 0.8118$
Gloss	$Y_G = 74.67 - 0.378X_1 - 0.389X_2 - 1.523X_4 - 1.600X_6 - 0.560X_7 + 1.090X_9 + 1.663X_{10}$	$y = 0.9041x + 7.383$ $R^2 = 0.9051$
Smoothness	$Y_S = 15.07 + 0.783X_1 + 0.583X_2 + 0.636X_3 + 0.974X_4 + 0.929X_6 + 0.863X_7 - 0.975X_8 + 0.625X_9 - 2.000X_{10}$	$Y = 0.9746x + 0.3364$ $R^2 = 0.9702$
Surface Strength	$Y_{SS} = 2.34 - 0.1.1X_2 + 0.188X_3 + 0.173X_4 + 0.260X_5 - 0.237X_6 + 0.132X_9 + 0.163X_{10}$	$Y = 0.921x + 0.1717$ $R^2 = 0.8978$

where Y_B, Y_G, Y_S and Y_{SS} represent the model Brightness , Gloss, Smoothness, and Surface strength respectively.

For better clarity the data are also shown in various figures 1 to 8. The values estimated from the model equations are also compared with their respective experimental data. The % error has also been estimated.

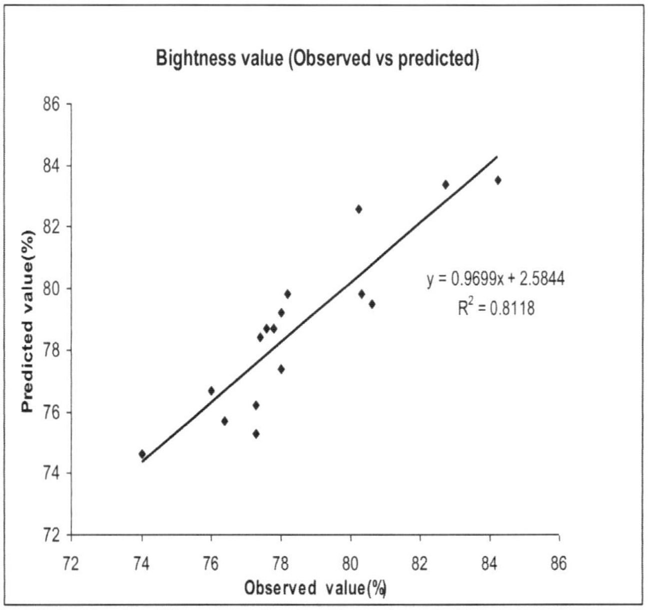

Fig. 1: Observed and Predicted value for brightness

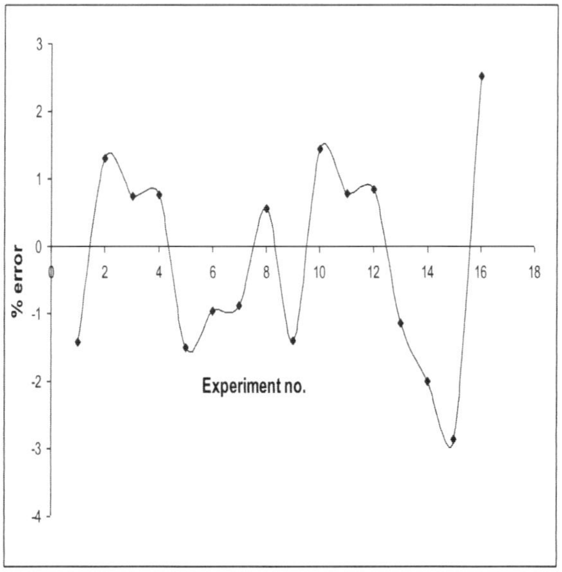

Fig. 2: Error (%) in observed and predicted values of brightness

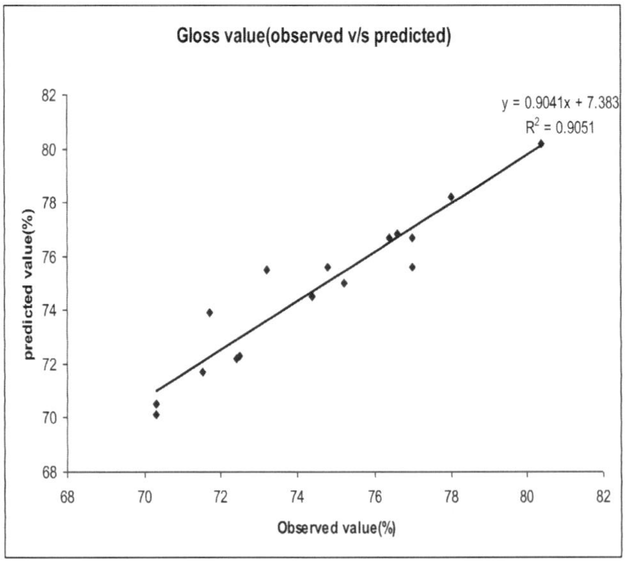

Fig. 3: Observed and predicted values of Gloss

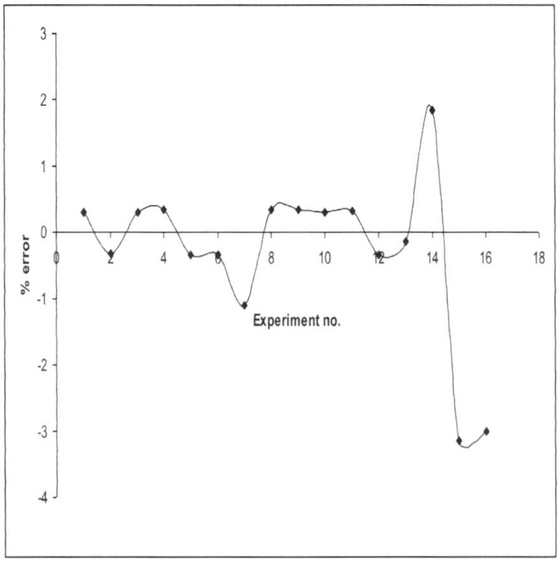

Fig. 4: Error (%) of observed and predicted values of gloss

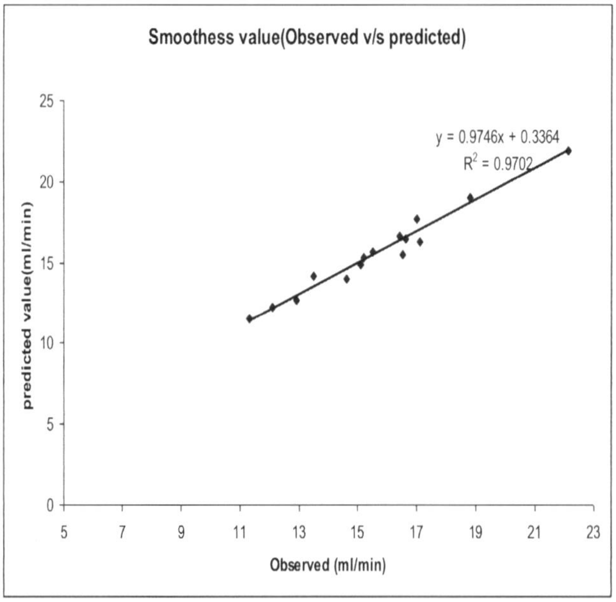

Fig.5: Observed and predicted values for smoothness

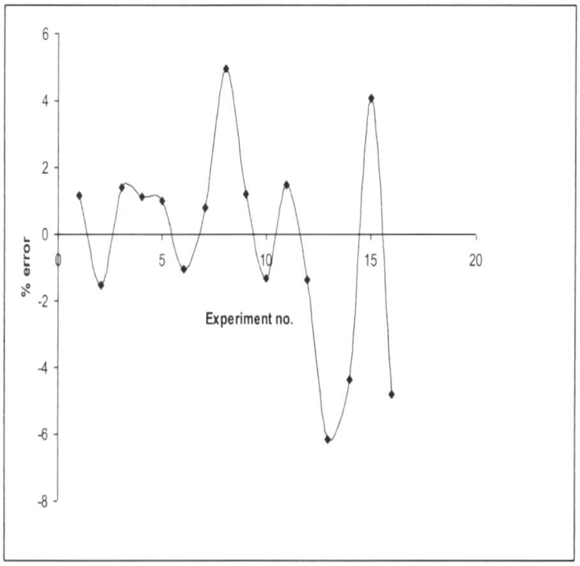

Fig.6: Error (%) of observed and predicted values of smoothness

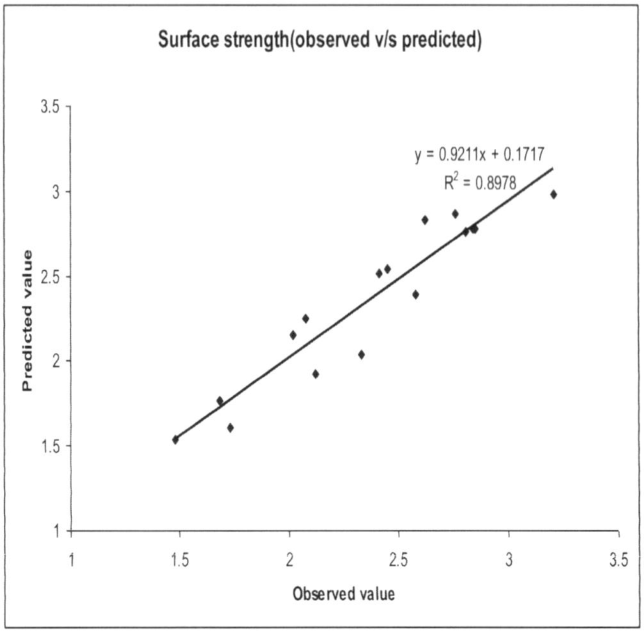

Fig. 7: Observed and predicted values for surface smoothness

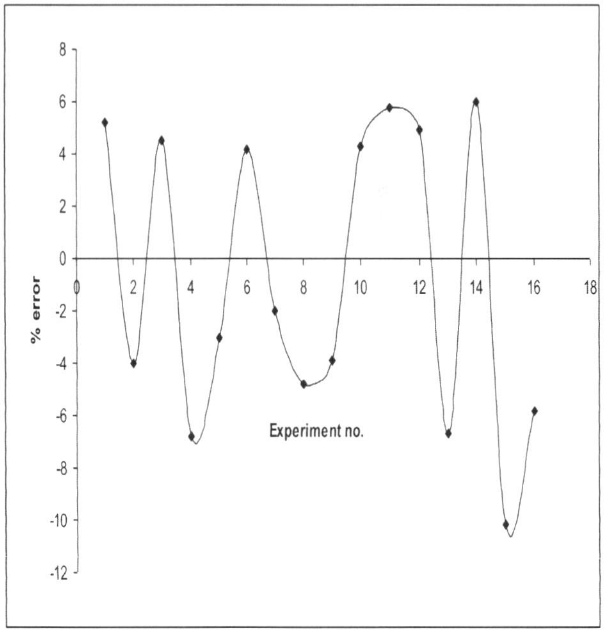

Fig. 8: Error (%) of observed and predicted values of surface smoothness

The multivariate regression models developed can predict the coated paper's properties fairly well and also indicating the relative importance of each factor on specific properties. The maximum and minimum percentage error for various properties obtained from the Plackett- Burman design are found as follows:

Brightness	:	-2.85 to + 2.53
Gloss	:	- 3.15 to + 1.85
Smoothness	:	- 6.17 to + 4.97
Surface strength	:	-10.15 to +6.0

The properties like brightness, smoothness, gloss and surface strength of a coated duplex board have been estimated on the basis of Plackett -Burman saturated orthogonal design for quantifying the effect of several variables such as pigments like Clay, PCC and GCC, total solids amount of starch used, impact of latex, effect of CMC, impact of air knife pressure, air knife angle, number of times calendering and their interactions.

The methodology adopted is a systematical procedure consisting of an experiment design experiment methods, regression techniques and the synthesis of these methods into a single methodology for application to coating studies.

The generated multivariate regression models from statistical experimental design can predict the coated paper's properties fairly well. This further indicates the relative importance of each factor on specific properties. The % error following the Plackett- Burman design is found as follows:

Brightness	:	-2.85 to + 2.53
Gloss	:	- 3.15 to + 1.85
Smoothness	:	- 6.17 to + 4.97
Surface strength	:	-10.15 to +6.0

The methods thus appear to be significantly important for accurate and efficient engineering estimates for the multivariate problem like coating of paper and paper board. This takes care for optimizations of coating color formulations.

References

1. Casey J. P. Pulp and Paper Chemistry and chemical Technology Vol. IV 3rd. Ed.
2. Kocurek M. J. Pulp and PAPER manufacture Vol. 8
3. Leonarard J. Calbro Handbook of Coating Additives.
4. R. P. Canard, TAPPI vol. 57, No. 11, Rheological Properties of Coating Colors at High Shear Rates, by Polymeric Organic Dispersants for Pigments, by Robert. D. Athley, TAPPI, vol. 58, No. 11
5. Matskovasky M. Vladimir. and Repyey P. Alexander. Experiment Optimization in Chemistry and Chemical Engg., MIR Publishers, Mascow (1982)

6. High yield Sulfide Pulping based on a Plakett-Burman design, by Broderick, J. L. Valade, and J. Paris, CPP 94:9 1993.
7. Ray A K, D. Dutt, Tyagi C H, Goel A., Batra A and Garg, G., Development of Coating color formulation for duplex board based on Plackett Burman statistical design, Cellulose Chemistry and Technology, 39, 3 – 4, 265-275 (2004).

(580f) Products From Hot-Water Extraction of Woody Biomass

Shijie Liu[1], Thomas E. Amidon[2], Christopher Wood[2] and Alan Shupe[2], (1)Paper and Bioprocess Engineering, SUNY College of Environmental Science and Forestry, Syracuse, NY, (2)SUNY College of Environmental Science and Forestry, Syracuse, NY

Lignocelluylosic biomass is mainly composed of cellulose, hemicelluloses and lignin. The majority of the existing bioconversion processes, either through themochemical or biochemical route, treat biomass feedstock as a whole in at severe conditions. It is recognized the thermo and chemical degradation rates are significantly different among with an order of hemicelluloses, lignin and cellulose. As a consequence, the value of hemicelluloses and lignin from biomass are diminished when going directly to cellulose. Replaceing the pretreatement with hot-water extraction, a host of products can be produced from the hemicellulose fraction. This presentation will provide an update to the hot-water extraction process and its product separation from woody biomass.

See more of this Session: Biorefinery - Improved Utilization of Co-Products From Biorefinery
See more of this Group/Topical: Forest and Plant Bioproducts Division

"Distilling" the issues of biorefinery deployment
Abstract/expanded outline

Eric R. Bober, Nexant, White Plains, NY

Introduction

The concept of the refinery has been around for decades. Refining has become a common practice and refineries are now a familiar site across North America, alongside our freeways and on our shores. This development has focused on petroleum refineries – but all that may be about to change, perhaps soon and perhaps significantly.

The concept of a biorefinery is quite more modern, and has gained popularity and become part of our vocabulary only more recently. Will biorefineries catch on and join their sisters, petroleum refineries, in familiarity and popularity? Will we see biorefineries along our highways and shores? Will biorefining become a major source of value-added products, like fuels and chemicals?

This paper will take a look at the development and deployment of biorefineries. There are many issues affecting this development, so it will try to "distill" them into the key factors. These key factors will be assessed to the extent practical in this forum so we can develop a feel for:

- what a biorefinery is,
- how it can be used,
- how likely widespread deployment is,
- by whom,
- where, and
- how it could all evolve.

I. The "old-fashioned" refinery
   A. Define "refinery" – Wikipedia defines refinery as "a production facility composed of a group of chemical engineering unit processes and unit operations refining certain materials or converting raw material into products of value." How can we "refine" this definition?
   B. Refineries in the petroleum sector
      1) The petroleum refinery takes crude oil and fractionates all of the prospective molecular feedstocks for additional processing and blending to generate many different fossil fuels based products. The facility is able to utilize and maximize the value of the different components in the feedstocks and intermediates.
      2) History of petroleum refining
      3) Statistics on petroleum refineries in North America
   C. Traditional refinery technologies and economics
      1) Processes – include reforming, cracking, coking, distillation, alkylation, etc. Diagrams and pictures, flow charts of petroleum refineries
      2) Products – include several intermediates, LPG, gasoline, a range of liquid fuels, asphalt base, petrochemicals, etc.
      3) Basic petroleum refinery economics

II. Adapting the refinery concept to biomass feedstocks
   A. Drivers of biorefining and biorefinery "enablers"
      1) Environment, supply, economics
      2) Role of government to enable development and deployment of a fledgling technology-based sector
      3) Illustrative drivers – several quotes to be shown, including:
         a) Chairman & CEO of Chevron recently observed: "the world needs every molecule of energy we can find, in every potential form"
         b) U.S. Energy Secretary Steven Chu, in a Department of Energy program brief: "Developing the next generation of biofuels is key to our effort to end our dependence on foreign oil and address the climate crisis – while creating millions of new jobs that can't be outsourced."
   B. Biorefinery concept
      1) The notion of a biorefinery is comparable to the current petroleum refineries, where various products are created from petroleum. The biorefinery starts instead with a biomass feedstock and uses various processes to generate "some combination" of fuels, power, chemicals, and "primary product" (e.g., forest products like pulp or lumber). The biorefinery is able to utilize and maximize the value of the different components in the biomass feedstock and intermediates. The "some combination" is determined by market prices of feedstocks and products, conversion costs, and other strategic factors. A U.S. Department of Energy program brief summarizes: "Integrated biorefineries are similar to conventional refineries in that they produce a range of products to optimize use of the feedstock and improve process economics." The concept is to fractionate the sources of value contained in the biomass feedstock. A biorefinery could end up with fractions of:
         a) lignin
         b) cellulose
         c) hemicellulose
         d) hydrocarbons
         e) residues (energy source)

         for further processing and conversion. In some ways, the current kraft pulp&paper mill is a basic, current-generation biorefinery. Now, we are looking at next generation biorefinery developments and deployment.
      2) Potential biorefinery feedstocks
      3) Potential biorefinery products
   C. Biorefinery feasibility
      1) Technical – promising
         a) Commercialized technologies
         b) Commercializing technologies
         c) Developing technologies
      2) Economics – needs to be proven
         a) "primary product" like pulp or paper
         b) low value/high volume products
         c) high value/low volume products
         d) captive, integrated products (e.g., heat sink, steam, etc.)

e) incentives, subsidies, etc. – to build critical mass, prove technology concepts, bring down production costs, improve comfort with concepts and technologies, and to overcome two primary obstacles: risk aversion and lack of capital

3) Is it feasible?

III. Practical applications of biorefineries – there are many real commercially-driven opportunities for deployment of the biorefinery concept:

- where current operations are not realizing full value of biomass feedstocks
- where primary products alone provide insufficient returns or bleak long term outlooks
- where by-products and waste streams are not utilized or monetized
- where clean, renewable energy is needed, or wanted
- where displacements for imports are sought.

The concept is to produce a portfolio of products from biomass feedstocks, ranging from "primary products (e.g., pulp)" to fuel to chemicals to many additional value added products.

A. Who may deploy?

1) Biomass handlers and processors, forest products companies, agricultural companies
2) Oil&gas companies, utilities
3) Chemical companies
4) Technology providers, venture capital
5) Investors

B. What may it look like? (examples)

1) sawmill that brings in logs, makes lumber, dries it in kilns, gasifies waste wood and perhaps additional biomass to get heat to run the kilns, electricity to power the site, and green power to sell to the grid
2) sawmill that brings in logs, makes lumber, dries it in kilns, gasifies waste wood and perhaps additional biomass to get heat to run the kilns plus renewable diesel fuel, which could be used for the sawmill's equipment or harvesting operations
3) pulp&paper mill that, in addition to pulp&paper, produces renewable diesel fuel, paraffin wax, green electricity, and steam for the pulp&paper mill, by adding gasification, primary and secondary syngas cleanup, and a cobalt catalyzed Gas-To-Liquid process
4) pulp&paper mill that also produces renewable diesel fuel, gas for its lime kiln, and steam and hot water for the mill by adding gasification of under- or un-utilized forest biomass to syngas, syngas cleanup, Fischer-Tropsch catalyzed process to convert to liquids
5) stand-alone, single-purpose biorefinery to convert unmerchantable timber and logging residues in a concentrated acid hydrolysis process to sugars, which are fermented and distilled to fuel grade anhydrous ethanol

C. Where may it be deployed?

1) near raw materials, including on current primary products sites (e.g., pulp mill)
2) near markets
3) where hosting conditions attract it
4) near current petroleum refineries, utility plants, chemical plants

D. When could it realistically happen? And how does this meet federal targets and mandates (e.g., 21 billion gallons of advanced biofuels by 2022)? Local mandates?
   1) "early deployers" – now to 2011
   2) "1st to be 2nd"ers – 2013 to 2015
   3) "mainstreamers" – 2020 to 2022

E. Why will deployment happen?
   1) natural evolution of sustainable forestry, farming, and biomass handling
   2) supply side drivers – fossil fuels are a depleting resource
   3) economic drivers – biomass value chain needs to improve its profitability to survive. Current players, partners, or new-comers could make it happen

F. How can the concept come to fruition - decision-making, financing, & deployment
   1) development by current players
   2) partnerships and joint ventures – upstream/downstream approach
   3) investment or acquisition by outsiders – oil&gas, chemical, financial

Conclusions and recommendations

September's Chemical Engineering Progress magazine, in an Update on Biorenewables – An Opportunity to Create Next-Generation Processes and Products, says "Despite the uncertainty of such predictions, the switch from fossil to biorenewable carbon feedstocks seems to be inevitable." To add value to these biorenewable feedstocks, industry will need to "establish integrated biorefineries capable of efficiently converting a broad range of biomass feedstocks into affordable biofuels, biopower, and other products," as proposed in a U.S. Department of Energy program brief. Drivers are in place, technologies are being commercialized, key players are recognizing the opportunity, and deployment of the biorefinery is nearing reality.

References

1. O'Reilly, David J., Chevron Corporation Chairman and CEO, October 25, 2007, "Securing California's Energy Future" at Town Hall Meeting, Los Angeles, California
2. Chu, Steven, U.S. Energy Secretary, March 2010, "Economy Through Product Diversity: INTEGRATED BIOREFINERIES" in U.S. Department of Energy, Energy Efficiency & Renewable Energy, Biomass Program
3. Update, AIChE JOURNAL Highlight, Biorenewables – An Opportunity to Create Next-Generation Processes and Products in September, 2010 *Chemical Engineering Progress*

(629b) Simulation and Uncertainty Optimization of Fuel Ethanol System

Zhiqiang Zhang, Department of Chemical Engineering, Tsinghua University, Beijing, China, Shanying Hu, Tsinghua University, Beijing, China and Dingjiang Chen, Dept.of Che.Eng., Tsinghua University, Beijing

Biomass energy system is a complex system that uses a variety of raw materials producing products in various processes. In the economic environment, sectoral policies, distribution and price of raw materials, market price of fossil energy, and manufacturing techniques are changing frequently. Therefore, how to choose the combination of the materials, products and techniques, how to determine the scale and layout of an enterprise, how to design an industry chain to realize the energy conservation and emission reduction, improve the overall efficiency is an urgent problem in the progress of current biomass energy industry. This paper has studied the following aspects:

(1)Developing a distributed simulation platform Bio-ESAS of fuel ethanol; (2)Making a life cycle MFA (Material flow analysis), EFA(energy flow analysis) and VFA(value stream analysis) of fuel ethanol system using different food, cellulose, sweet sorghum, sugar cane, cassava other raw materials, and simulating the main process with Bio-ESAS to identify the key factors affecting the system efficiency; (3)Constructing probability models and quantitative prediction model of uncertain factors impacting fuel ethanol system such as market prices, raw material supply, etc; (4)Constructing a multi-objective and uncertain-optimization model for fuel ethanol system, then solving; (5)Exploring the development roadmap of fuel ethanol in China.

See more of this Session: Biorefinery - Sustainability, Energy and Environmental Issues
See more of this Group/Topical: Forest and Plant Bioproducts Division

(629c) Polymer Induced Flocculation and Separation of Particulates From Extracts of Lignocellulosic Materials

Gustavo Duarte[1], **Bandaru Ramarao**[2] and Thomas Amidon[2], (1)Department of Paper and Bioprocess Engineering, State University of New York, Syracuse, NY, (2)State University of New York, Syracuse, NY

Biofuels from lignocellulosic materials like wood are renewable and sustainable alternatives to petroleum and other fossil fuels. Wood can be grown and harvested without adding to the carbon load of the atmosphere and thus can be part of the solution to the problem posed by global climate changes. Recently much interest has developed on the concept of the forest product biorefinery, where wood is hydrolyzed prior to conventional pulping and papermaking processes and the hydrolyzate consisting of hemicellulose sugars are to be used as a feedstock for biofuels or bioplastics. The purification of the hydrolyzate stream and the separation of fermentable sugars from it thus constitutes an important step in biorefinery processes. The separation of particulate material from wood hydrolyzates is considered in this paper. Sugar maple hardwood was extracted with hot water at 160°C. The extracts contain hemicelluloses (primarily xylooligomers, xylose and xylan), acetic acid and smaller amounts of lignin. The colloidal stability of the extracts plays a critical role in the separation and purification of the wood extracts. Here, we report the size and charge of the particles in the extract measured using standard instruments based on light scattering and microelectrophoresis. Particles were found to be in the size range from ~220 nm to 270 nm. Zeta potential measurements showed them to be negatively charged. By treating the extracts with a cationic flocculating agent poly-DADMAC, it was possible to preferentially precipitate out the colloidal fraction containing lignin and lignin derived compounds. Upon the addition of Poly-DADMAC the turbidity of the suspension reduced from 920 NTU to 4 NTU in a 24-hour period and particulates sedimented from the extract. The lignin concentration was reduced in the supernatant, while the sugar content remained unchanged. The addition of an indifferent electrolyte hindered the effectiveness of the polyelectrolyte. The optimum pH for the effectiveness of the polymer was found to be around 4.5. In order to accelerate the sedimentation of the particles, kaolin was added to attach to the flocs after adding the polyelectrolyte. Kaolin helped accelerate the separation of the particulates by almost an order of magnitude. This work shows that flocculation with a cationic polyelectrolyte followed by separation by sedimentation, centrifugation or microfiltration can be an effective technique to purify biorefinery wood extracts.

See more of this Session: Biorefinery - Sustainability, Energy and Environmental Issues
See more of this Group/Topical: Forest and Plant Bioproducts Division

MODELING, SIMULATION AND CONTROL OF ACTIVATED SLUDGE PROCESS IN EFFLUENT TREATMENT PLANT OF TROPICAL HARDWOOD AND BAMBOO BASED KRAFT PAPER MILL

A.K Ray*, Anirban Roy** and Neha Roy**
Indian Institute of Technology, IIT Roorkee,Roorkee,India
Birla Institute of Technology, Meshra Ranchi, India

ABSTRACT

Bamboo and tropical hardwoods are important raw materials in many Asian countries including Indian pulp and paper Industry. The effluent characteristics are quite different from those of softwood based industry. In this present investigation, detailed studies are made on the effluent characterization of these raw materials, models of the primary and secondary clarifiers, optimizing both process and design parameters and then simulating the effluent treatment plant (ETP) with the data stipulated by regulatory bodies such as state and central pollution control boards and ministry of environment and forests. The industry data are then compared with the data from simulations and the target values. The problem is further extended to control the entire ETP plant parameters to the desired values. In order to attain the above objective various multi-loop control strategies are attempted. As an example, application of ratio control for effluent flow and treatment chemicals in primary clarifier, discharge of effluent to streams and returned activated sludge flow and air to influent flow has been shown. Feed forward, feed back and cascade are also tried. Hybrid control strategy using PID feed back control for dissolved oxygen and feed forward control based on conductivity for nutrient addition has also been analyzed. Finally, system is designed and the responses are tested against those from the existing single loop (single input, single output) negative feedback or feed forward configuration. The results are found to be very promising.

Key Words: Effluent, modeling, optimization, control, single loops, multiple loops, simulation

Introduction

Ever increasing costs of chemicals, energy and other inputs and strict legislation of the pollution control authorities for treatment of the generated effluents due to many sources including the chemical reaction between the reactant and pulp in various stages such as pulping and bleaching has forced the paper Industry to optimize various processes and operations to be effective and work in a most economical way. Bleach plant effluent, in fact, presents the highest pollution load in paper industry compared to any other operation.

The data are however valid for softwood pulps. For hardwood and bamboo these values are lower while for agri-residues these are higher. For a mills a typical set of data are shown below [1]. The detailed account of various sources from an integrated pulp and paper mill are also shown in respective tables.

Indian Standards for paper Industry Effluents:

Parameters	Minas Value	CPCB*
pH	-	6.5-7.5
TS/SS	100-200	50
BOD,mg/l	30-50	30
COD,ppm	150-210	250
Color(Pt-Co)	150	-
Water,m	200-250	200/175
DO,mg/l	-	4
AOX,kg/t	-	2.0

* Oil and grease, sulphides, total residual chlorine, chloride and sulphide in mg/l are 10.0, 2.0, 1.0, 1000.0 and 1000.0 respectively.

Values of three categories of mills

Parameters	Large mills(>24000 T/year)	Small Mills(≤24,000 t/year)	
		Agro based	Waste paper based
Volume,m^3/t	200(100 new mills)	200(100 ,new mills)	75
pH	7.0-8.5	5.5-9.0	5.5-9.0
BOD_5 at 20°C,mg/l	30	50	50
COD,mg/l	350	-	-
SS,mg/l	50	100	50
TOCl,kg/t	2	-	-
DO,mg/l	4	4	4
Color,pt-Co unit	150	-	-

Characteristics of Combined waste waters from pulp and paper mills(Large)

Parameters	Range (Avg.)		
Flow	167-281(220)	COD,mg/l	840-1660(1118)
SS,mg/l	620-1120(764)	COD/BOD	2.95-4.37(3.8)
BOD_5,mg/l(at 20°C)	240-380(295)	Colour(pt-Co unit)	300-655
pH	6.6-10	SAR	2.0-5.3(3.5)
Pollution Load, kg/t of paper			
SS	168	BOD	65
COD	246	-	-

Normal Design Features of Activated Sludge Process

Parameter	Values	Parameter	Values
Depth, m	4-11	Toxicity reduction,%	>70
Retention time, day	0.1-0.4(≈2-10hrs.)	Phosphorous reduction%	Increases
BOD_5 load,g/m^3,d	1000-4000	Nitrogen reduction,%	Increases
BOD_5reduction,%	75-95	Sludge reduction,	0.6-0.8

		kg SS/kg BOD_5	
COD reduction,%	40-70	Sludge Conc.,g/m^3	4-6
Sec. Sedimen. Surface Load,m/h	0.8-1.0	Sludge time, day	3-4
MLSS,mg/l	3000-4000	-	-

It produces more sludge ,requires more depth, demands higher amount of nutrients ,BOD_5:N:P=100:5:1,(100:4.3:0.6) more sensitive to organic shocks but requires less retention time, and less sensitive to temperature effects with optimum temperature range 30-35 ^oC and handles much more BOD load, more BOD and COD reduction

All weak wastes generated in pulp and paper industry, that is, those less than 4,000 mg/l are susceptible to biological waste treatment for BOD_5 reduction. Among the various methods like oxidative and facultative ponds, and aerated lagoons, activated sludge process and its modifications, the activated sludge process is still the economical way for treating paper mill effluents. Eight criteria should be used to select the best bio-treatment option for a particular mill among the many variables. These are efficiency of removal of BOD_5 and suspended solids, amount of operator attention required, sensitivity to climatic conditions, resistance to shock loading conditions, power required, nutrient demand, sludge yield and characteristics, space requirements

Mathematical models, optimization and simulation:

Roy et al.[1] developed detailed models for optimization and simulation of an effluent treatment plant. Some of the works are published elsewhere[1] with various developed algorithms.
Steady state material balance relationships with kinetic rate and flocculation in aerator involves some 14 number of steps , specification of area of aerator, equipment specification through Lagrange Multiplier and numerical analysis using Newton Raphson method, finding out of area of clarifiers with two different ways. The recycle ratio, SVI, sedimentation velocity, residence time in the reactor , net yield of MLVSS ,are determined optimally through considerations of both effluent quality as well as optimum flocculation conditions. The constraints for various parameters are obtained from industry. The parameters considered are: BOD values for different flows keeping in mind the regulatory norms. Further, parameters like oxygen requirements, specifying aeration equipment based on total kW, number of aeration units, spacing between aerators, checking neutralization requirements, evaluating nutrient requirements ,estimating requirements for N and P. some control parameters such as the sludge age ,volumetric sludge load, S_{SVI} , circulation Number, CN are determined through algorithms. Practical design values of surface overflow rate (SOR) from different industries are also considered for preliminary design estimates .

For optimization of entire plant objective functions are developed and constraints are defined. The objective function is total annual operational cost for a Greenfield mill and an already existing mill. The cost models of each cost contributing factors are developed. Calculations are made through 47 steps. Costs models of aerator, primary and secondary clarifiers are taken from Mishra et al.[11] Smith[22].Methods of estimating cost indices, scale -up factors and optimum pipe diameters are taken from Perry[15] and Peter and Timmerhaus[17] and Coulson and Richardson .Both laminar as well as turbulent flows are considered during calculation. Programs are developed in C^+ language and simulation is carried out. The Annual cost model as objective function requires minimization through optimization techniques for various situations including uncertainty of cost data, varying recycle ratio of sludge by iterative procedure and simulation technique. Sensitivity analysis of some parameters in cost equation has also been carried out.

Simulation for plant of varying capacities

Different plant sizes, 2.5 MGD, 5 MGD ,7.5 MGD and 10 MGD are taken for simulation study. Recirculation, r, Residence time, t, Reactor Vol., V, F/M Ratio, Q'', Q', ΔX_{NV}, ΔX_t, Oxygen required, Power Required HP, No. of Aerators, Neutralization, Sludge Age, S_{SVI}, and circulation number are estimated. Capacity and dimension of all equipments, the capital investments and operational charges for the above three different EPT plant capacities as well as paper mill capacities ranging between 200-400 TPD are found out.

Mathematical model for Bio-Chemical kinetics

Study of kinetics of aerobic biological treatment yields the rate at which microorganisms degrade a specific waste, and therefore, provides the basic information required for sizing biological aerobic reactors. The reaction is known as The Henri- Michaelis- Menten Reaction for simple single-substrate-enzyme–catalyzed reactions or for multi-substrate-multi-enzyme or when applied for cellular systems, the reaction is called Monod Reaction[2, 23]. The equation can be described by single differential equation

$d\upsilon /dS = =K \upsilon^a (1-\upsilon)^b$

where $\upsilon=\mu_g/\mu_m$, S= the rate limiting concentration, K, a, b are constants. For Monod reaction a=0,b=2 and $K=1/K_s$ and the equation can be as under:

$\mu_g = \mu_m S/(K_s+S)$

where μ_m = the maximum growth rate when $S >> K_s$

If the recycle ratio of sludge is r and excess sludge ratio js w, then the steady state mass balance on active solids in the process can be expressed as $(1-w)Fx_e + wFx_e = Vx_a[\mu_{max}S_a/(S_a+K_s) - k_e]$ .On simplification one can get,

$1/\theta_s = \mu_{max} S_a/(S_a+K_s) - k_e$

where θ_s is the main residence time of the activated sludge (ratio of active-solids retention to active solids effluent rate), also called sludge age. This is required to achieve a specified BOD level s_a from given waste , provided the kinetic constants are known for that system.

$\theta_s = Vx_a/[F(1-w)x_e + wFx_e]$

In most sewage –treatment plant the sludge age is of the order of 6 to 15 days. Appropriate values of μ_{max} ,K_s ,k_e Y should ways be measured for a specific kinds of waste of interest.

If Y is assumed to be constant, the steady –state models for substrate in the aeration basin;assuming substrate is not separated in the clarifier($s_a=s_e=s_r$) and if the first term on the equation is the rate of sludge production, one can write

$Vx_a = YF\theta_s(s_o-s_a)/1+k_e\theta_s)$

The volume of the aerator for a given recycle ratio α can be calculated from

$V= F\theta_s(1+r-rx_r)$ which is obtained from active solid balance on the aerator and the equation for mean residence time.

In more simplified way Michaelis Menton enzymatic kinetics and Monod equationfor activated slude process can be written as:

$-ds/dt = k_S\ SX\ /\ (K_S + S)$ where ds/dt = rate of substrate use (mass/volume time), k_S = maximum rate of substrate use [mass/ (mass microbes) (time)] = rate constant, K_S = subsrate concentration when the rate of use is half the maximum rate (mass/volume), X = mass of microbes (mass), and S = substrate concentration (mass/volume).

This is a variable order kinetic expression that depends on the magnitude of the substrate concentration in comparison to K_S. if S is large in comparison to K_s, the reaction becomes a zero order, and the equation reduces to $DS/dt = Xk_s$

If K_S is large in comparison to S, the equation is reduces to

$-dS/dt = (k_S/K_0)\ XS = K_1\ S$

which is a first order rate expression in terms of substrate. The value of K_1 in terms of TOC/ (gMLVSS) (h) at 25^0C ranges from 0.375-0.528 for pulp and paper mill effluent. This is true if the continuous steady state reactor operating substrate concentration (S_e) is considerably below 500 mg/l (BOD_5). It is customary to express substrate removal rate per mg/l of MLVSS present in the reactor.

If $X_{v,\ a}$ be this MLVSS concentration, then the equation can be rewritten as

$(1/\ X_{v,\ a})\ (dS/dt) = -\ kS$ where $K_1 = kX_{v,\ a}$

or $dS/dt = -\ kX_{v,\ a}S$

k is the substrate removal rate constant. For time t equal to residence time in the continuous reactor, concentration S corresponds to S_e and one can get

$(dS/dt)_{\text{cont. reactor}} = -\ kX_{v,\ a}\ S_e$

A mass balance on the system begins with the process of relating process efficiency to kinetic coefficients, hydraulic time, and mean cell residence time. Rate of accumulation of bacterial mass= rate of flow of bacteria in – rate of flow of bacteria out + net growth within the system

More explicitly, Change of substrate in reactor = increase due to influent flow-decrease due to effluent flow- decrease due to reaction.

Under steady state conditions, considering Q, the volumetric flow rate,

- Change of substrate in reactor = 0
- Increase due to influent flow = QS_o
- Decrease due to effluent flow = QS_e
- Decrease due to reaction = $kX_{v,\ a}\ S_e\ V$

Substituting all the terms one can get

$(Q/V)\ [S_o- S_e)/X_{v,\ a}] = kS_e$

$t = V/Q$

Consequently, the following equation for rate kinetics can be obtained

$(S_o- S_e)/X_{v,\ a}\, t = kS_e$

$(S_o - S_e)/X_{v,a}\ t$ = substrate removal rate. It corresponds to rate of removal of substrate (eg BOD), mg/l in continuous reactor per mg MLVSS.
$(S_o - S_e)/X_{v,a}\ t = k\ (S_e - S_n)$ where S_n is the concentration of non-biodegradable matter.
The substrate can be expressed also in terms of TOC.

Importance of control:

As for any industrial process, the need is obvious for efficient monitoring and automatic control systems of biological waste water treatment processes in order to optimize their operation and maintain the wastewater degradation at a desired level at the output of the process unit. Today most of the effluent treatment plant runs use simple control. or no automatic control at all.
The efficiency of any control system depends highly on the design of the control and monitoring techniques and the care taken in its design. Indeed, a control algorithm will prove to be efficient if it is able to incorporate the major elements of common knowledge about the process while being able to deal with the missing information (lack of on-line measurements ,uncertainty on the dynamics and reaction kinetics, and so on) in a robust wayside. such that this missing information will not significantly deteriorate the control performance of the process. Control techniques in current use include simple PLC-techniques, time control, manual control, rules of thumb, or simple proportional control. Activated sludge process is a multivariable process. There are several inputs and several outputs.
There are many attempts to design controllers for this process such as classical PID, nonlinear control, adaptive control, predictive control etc. Some of the later controllers have been claimed to provide better performance than PID.

The dynamic behaviour of activated sludge processes is governed by numerous and intricate biological and physico-chemical mechanisms which are not yet completely understood. Different mechanistic models of these processes have been reported in the literature [1,2, 3]. These models are always very complex and often require a lot of a priority experimental information and knowledge that are difficult to gather on a real wastewater treatment plant (WWTP). That is why the control of such processes using classical control tools is difficult and sometimes impossible.

Control strategy

The parameters which the state of the sludge is monitored are volumetric sludge load, or sludge volume index, S_{SVI} ($\approx 120 \leq 200$), circulation number, F/M ratio or sludge load(0.1-0.4),sludge content(2.5-4kg/m^3),floc load(gCOD$_{in}$/g of activated sludge), SRT(5-15 days) or the amount of nutrient in relation to the amount of biomass, type of nutrients, temperature, pH(7.0-7.5),oxygen content(1.5-2.0 mg/l), and concentration of toxic substances. These are again functions of the return sludge ratio. Sludge return ratio is again function of consistency of return sludge, target for sludge content in the aerator, and sludge settling properties. S_{SVI} is the volume in ml that a gram of the sludge occupies after settling for 30 min in a 1000 ml graduated cylinder
Volumetric sludge load = $S_{SVI} = S_H * C * SVI * 10^3$
F/M=kg BOD_5/d/kg MLVSS
Where MLVSS is the amount of active or organic solids in the aeration tank.

.
C = concentration of activated sludge
S_H = Hydraulic surface load
SVI = Sludge volume index
K = Safety factor

- Circulation Number, CN = $(Q_r/Q_w) + 1$
- $(Q = rQ_F)$ = quantity of return sludge
- Q_w = quantity of excess sludge wasted from the system
- F/M = gm COD in /g of activated sludge = kg BOD_7/day/kg MLVSS

Loops design:

Single loop control strategy

Generally in pulp and paper industry single control systems are available. These are temperature, pH, and oxygen concentration. Most bacteria for waste water treatment plants are of mesophiles type having temperature range: 25-45^{0}C [2, 23]. Optimal temperature range is 30-35 ^{0}C. pH range of pH from about 5 to 9 is tolerable but optimum around neutral 6.5-7.5. Toxicants (destroy bacteria population, toxic compounds like phenols, chloro-organics, cyanide, sulphide and heavy metals like Pb, Cu, Ni, Zn, Zinc)

Control of dissolved oxygen:

Control of dissolved oxygen has been in use since several years in order to optimize plants in terms of energy consumption through blower with better effluent quality

Temperature Control:

The rate of biochemical reaction depends on temperature as indicated by $k_1 = k_{20}\, Q^{t-20}$ where Q is the temperature coefficient(1.0-1.03). The effluents from pulp and paper industry can perform also at up to 35°C-37°C. Temperature must be maintained below 40°C. Adjustment of desired temperature in the aerator at the controller(TIC) as a set point and temperature in the aeration tank is the control variable with solenoid valve as actuators.

Control of pH:

pH is a difficult variable to control because of its inherent abrupt nonlinear characteristics in response to small changes, particularly so in the region of pH 7.0.This is true even when operating in the continuous mode at a fixed set point(Leslie).The reason is of course is the nonlinear pH titration curve. The problem is well known and has been discussed by Harriot, Shinskey , Mahuli et al.[9],Sung et al.[25] and Wang[27]. It is evident from their work that most pH control systems are run as a monitoring unit and control of pH is largely performed by manual operation though many techniques for pH control has been tried. The most frequently used conventional control systems are single loop feedback control in which adjustment of pH at its desired value (set point) is required at the feed inlet of aeration tank at the controller (FIC) and measurement of pH at the feed inlet of the aeration tank as controlled variables with metering pumps for acid and caustic as actuators.

Nagrath and Gopal [13] discussed the tuning of a pH PID controller when dynamic model of the process is not known., the open-loop response for a step input is determined experimentally or by simulation. If the response is S-shaped (characterized by two constants, dead time L and time constant T) , Ziegler- Nichols tuning method is applicable. These two constants are determined by drawing a tangent to the S-shaped curve at the inflexion points and finding its intersection with the time axis and the line corresponding to the steady- state value of the output. The S- shaped curve gives a response identical with the dynamics of first order with delay time.The values of the controller parameters , K_p,K_I,K_D can be calculated as

$K_c=1.2\ (\tau/\tau_d\ ),\ Kc\ /\tau_I\ =\ 2\tau_d\ ,K_d= Kc\ \tau_D=\ 0.5\tau_d$

While discussing about a waste treatment plant with acidic influent designed a single loop pH control strategy where it is desirable to control pH of the effluent to a set point in the range 4 to 7 .Taking into consideration of both flow rates(inflow and effluent) constant, with gain of the control valve 5.67 ml/s, the system has been tuned with the following values the closed loop step response is obtained

$\tau_d.=19.9$ s, τ =1.8 s, $K_P=0.11$, τ_I = 39.8 s ,τ_D=9.95s, $K_P=5.67$

The process transfer function,can be written as $G_P(s)= 5.6\ e^{-19.9s}/(1.8s +1)$

% overshoot=26%, Rise time=0.3 s and settling time =9.2 s

Ray et al.[21] developed PID and neural network model for controlling pH the wet end in paper making based on the guidelines of Harriott considering both equilibrium constant model and process identification technique and developed first order dynamics of the process as 7.0/(1.5 s +1 as approximate one through Matlap Simulation for continuous as well as discrete signals. The ANN controller is designed with 2 neurons in the first layer, 12 neurons in hidden layer and 1 neuron in output layer. In this network, logsig activation functions are used. The number of layers and number of neurons in different layers are decided by trial and error procedure. The gradient descent technique is used to train the network.
Mahuli et al.[9] discussed pH control using a statistical technique for the continuous online model adaptation considering both the nonlinearity and non –stationary nature of titration curve, but the operating range was zero. Sung et al.[25] presented a summary of previous works and suggested a pH control using an identification reactor and through an equivalent titration curve is firstly obtained from the identification reactor, and then the pH value is controlled in the main reactor using the obtained equivalent titration curve.

For wide range of pH control and to also to minimize its variation ,effective modeling and control strategy for pH of stock solution in approach flow system has been attempted by Wang [27] and Wright et. al. [28, 29] in various ways such as linear control system with fixed PID implementation scheme, an adaptive PID controller , discrete –time model for computer implementations and nonlinear controller with parameter identification. In the modeling the total charge of the thin stock in the mixing box has been expressed in terms of the ion concentration of each species. Further a single algebraic equation relating the hydrogen concentration to the total concentrations of each species in the thin stock has been developed. The dynamics of the system has been modeled as $dx/dt + Fx = (1-x)u$ where pH has been expressed as a function of x(x being dimensionless concentration ratio). pH is also a function of H^+ which in turn is also a function of x. As a result, the control of pH can be realized by manipulating x by the equation pH=h(x) where h (.) is the monotonic function representing the equivalent titration curve. In this model, pH is monotonically dependent on x. the relation between pH and x is given in eqn. pH=h(x) where h(.) is the monotonic function representing the equivalent titration curve. The structure of the open loop dynamic system is nonlinear and possibly unknown owing to difficulties in obtaining an accurate estimate for h(.).The closed loop system with a fixed PID controller for the pH similarly can be drawn. The PID controller parameters are based on a trial and error approach owing to difficulties in establishing an analytical solution for the nonlinear control. In order to simulate the overall process and to establish some realizable control strategies, it is assumed that the relationship between pH and x can be expressed as $x = h^{-1}(pH_s) = \sigma \tan(pH/4.1781 \; -1.8) + 0.5$ .The slope of the curve can be changed by adjusting the coefficient σ

Control of Oxygen Concentration: adjustment of desired oxygen concentration (set point) in the aerator at the controller (FIC) and measurement of oxygen concentration in the aeration tank as controlled variables with compressor as actuator.

Multiloop control system design:

Feedforward (Ratio)-Feedback control system

Most of the conventional control systems use combination of feedforward control to compensate for sudden disturbances, variable gain feedback control to offset the non-linearity of pH., large process capacities to damp out variations in pH.
Because of the tremendous sensitivity of the curve in the region of neutrality, it is always necessary to trim the ratio with a feedback loop. In addition, the nonlinearity of the measurement should be compensated by using a nonlinear controller. Shinskey designed such a system (feedforward- feedback) based on feed flow in which the flow signals are linearized to maintain loop gain constant over the full range of flow.
In most effluent treatment plants, flow is reasonably constant, yet pH can vary rapidly over a wide range. The feedforward system attempts to supply the reagent needed to balance the acid or base content of the waste stream. Shinskey further developed feedforward controller with adaptive nonlinear feedback which accommodates the logarithmic nature of the pH function by means of an equal percentage valve. The feedback controller biases the relationship as necessary to reach neutrality, leaving feedforward gain K and intercept r as manual adjustments.

Love [8] designed a simple ,reliable , microprocessor based system (using a DDC system) for pH control using dedicated software to eliminate non-linearities at source, conventional feedforward and feedback control loops, standard control loop hardware and firmware, and a process capacity with a small residence time.

A ratio control loop is used to adjust the opening of controllable stream (primary acid flow). A pH control loop adjusts the opening of control valve the secondary acid flow of controllable stream, being the remaining 10%. The wild stream (caustic) is set by a needle valve. The time constants of the mixing vessel constituting the process and the load, each is 1 min, and the distance velocity lag between and the electrode cell is of the order 1 s, the dynamics of which is negligible. The control loop consists of standard hardware elements using 4-10mA and 0.2-1.0 bar signals,. The caustic flow rate is measured with an orifice plate and a dp cell assembly, a P/I converter being used to generate a current output signal.

The H^+ ion concentration is measured using a standard glass electrode and calomel reference electrode assembly, whose mV output is given by Nernst equation $E=E_o+kT \log C_H$. A temperature compensated EIL amplifier/ transmitter generates the pH value from the electrode output according to $pH=(E_O - E/ kT$ and transmits it as a current signal.

Feedforward/ split range –Feedback control system

The more usual neutralization problem is when both the pH and the flow rate of a process stream vary simultaneously. In this example the principal differences are in the feedforward loop where both the pH and the flow rate of the process stream are inputs to the ration controller. Its output is applied to two control valves, in a duplex arrangement. The ratio controller makes a crude adjustment of pH , the primary caustic or acid flow rates being such as to make the process stream just alkaline, say. The Ph controller then makes a fine adjustment of pH by varying the secondary acid flow.

Control of substrate (bod) and dissolved oxygen:

Vanrolleghem [16] discussed various control strategies for DO control and respiratory rate control after showing relationships between substrate concentration S_s, DO concentration S_o and respiratory rate r_o. Traditional feedback, with potential feedforward from substrate concentration or respiratory rate, cascaded control of DO, respiratory rate control, by manipulating return sludge flow with possible feedforward from an influent flow measurement, and cascaded respiratory rate control were discussed. Various controlled variables, manipulated variable were also identified.

80 respiratory based control strategies were designed according to manipulated variable and control scheme. One of the manipulated variables was mass transfer coefficient, K_La is used for feedback, feedforward, feedforward-feedback control system designs. Similarly returned activated sludge flow rate, sludge waste flow rate, influent flow rate , influent flow distribution, cycle time, and sludge flow rate were also considered as manipulated variables.

Multivariate control strategy:

Norhaliza [14] Wahab et al.[12], Talat Mahmood [26] investigated the application of multivariable PID (MPID) controllers to wastewater treatment plants (WWTPs) with pre-denitrifying process. WWTPs have usually a number of interacting control loops. The controller's parameters are often left unchanged since the plant commissioning and hence poor performance is often observed due to inevitable changes in the plant and its influent characteristics. The model-based MPID tuning methods are complicated, cumbersome and expensive, and require a high level of control expertise to design, and implement. In this paper, a simple and novel tuning method is proposed along with three existing methods, which require only step or frequency response tests. The performance of these methods is compared and the parameters are tuned using a proposed optimization technique to meet standard quality effluent at a minimum cost. This is achieved using a nonlinear ASM1 model to determine the optimal values of the controller parameters. The results show that by effective and proper tuning of the MPID controllers, the closed-loop performance of WWTP improves and it could potentially lead to energy saving and improvement in the removal of nitrogen. 4 methods of tuning are compared. These are Davison method (M1), Penttinen–Koivo method (M2) ,Maciejowski method (M3) and a new method proposed (M4)

Stenstrom and Andrews [24] developed a technique for real-time control of the activated sludge process using specific oxygen uptake rate (SCOUR). Mathematical modeling and computer simulation techniques were used to develop multivariant control strategies using SCOUR and mean cell retention time as the controlled variables, and manipulating sludge recycle rate, sludge wasting rate, and wastewater feed point location. It is shown that SCOUR is a valid indicator of process efficiency and sludge activity for nonsteady state operation, and that multivariant control techniques using SCOUR are superior to techniques that use only food-to-mass ratio, or mean cell retention time. It is also shown that SCOUR can be used to detect shock loads of toxic and biodegradable materials. The value of forecasting flow rates for improving control is illustrated through the use of time-series models.

Adaptive Control systems:

The activated sludge process is a biological process used for secondary treatment of sewage and is probably the most widely used method in the waste water industry. However there are still a number of problems that are required to be solved and these may be broadly categorized in terms of external disturbances and internal complexities, the effects of which are both exacerbated due to the nonlinear behaviour of the process. To overcome the problem in the presence of inherent plant nonlinearities requires adjustment/adaptation of controller gains. The problem which has been identified in this work is that of minimising the effects of disturbances on the plant, which include changes in external loading conditions affecting the influent flow and concentration

Rangla et al. [20] and Dochain and Perrier [4] have carried out extensive investigation on control of BOD and DO in activated sludge process

For control of BOD and dissolved oxygen concentrations, design of adaptive linearizing controllers for activated sludge process has been attempted.

The controller design is based on the (nonlinear) dynamic balance equations and the knowledge of the process reaction scheme while not requiring any knowledge about the reaction kinetics nor about the on-line measurement of the biomass concentrations in the settler and the aerator. These are estimated on-line by using appropriate estimation algorithms (recursive least squares for the reaction kinetics, asymptotic observer for the biomass concentrations). Simulation results have shown good performance of the proposed control algorithm.
Adaptive control of pH is another example used by Leslie [7]. If process gain decreases loop become sluggish, with large overshoots and wide variation in pH. By using adaptive control process gain, controller gain could be adjusted to match process gain and maintain loop stability.

Model Adaptive Control

Moreno et a.[18] has developed non-linear predictive control of dissolved oxygen in the activated sludge process. Francisco et al.[10] has developed also general method for tuning MPC considering asymmetric bounded signals for disturbance and tested in MPC applied to a simulated activated sludge process. The closed loop responses show that obtained controllers is properly tuned, taking into account the large magnitude of influent disturbances. The use of asymmetric bounded signals allows for a more realistic selection of the tuning controllability criteria, providing better control performance. The methodology proposed here is general, so any other chemical processes or performance criteria could be considered. Finally it is important to show that the developed method is particularly suitable for its inclusion in the resolution of the Integrated Design optimization problem, which determines the optimum controller and the optimum plant
at the same time.

Dynamic Modeling

Jon Bolmstedt [3] developed dynamic modeling of an activated sludge process at a pulp and paper mill. The activated sludge basin operates at neutral pH conditions and at temperatures around 35°C. Dynamic modeling has also been attempted by many investigators. [1]

Design of control systems of activated sludge system industrial approach:

The designs of activated sludge are available for industrial practice in Kokurek [5] and Levigne.[6]. These designs are modified and proposed as follows:
For control of activated sludge process the important control variables are considered by various authors pH, DO, recycled sludge(r)/biomass concentration, F/M(food, BOD/microorganism) and speed control of agitator. Optimum level for disposal of sludge and sometimes sludge blanket level in secondary clarifier are also considered as the underflow solids from primary clarifier should be at the maximum density. This is often controlled using a sludge blanket level control system. Two types of simple designs of control systems are proposed.

Design I: For the above control the manipulated variables should be air header pressure or air flow rate or agitator speed which can control oxygen concentration(DO), oxygen transfer coefficient ,K_La . The air header pressure

set point is to be supervised by the dissolved oxygen (DO) controller. The average concentration of DO and the difference between two DO sensors should be controlled. The difference between the DO sensors is the rough estimate of the uptake rate of oxygen by the microorganisms.
At the inlet of the plug flow reactor, the oxygen uptake rate is high; consequently the DO there would be zero. The two sensors should be located at the half way and the ¾ down the reactor. The average value is used to control the header pressure set point, and the difference is used to control the return sludge ratio.
If the slope is increasing, then insufficient microorganisms are being returned and hence the ratio should be increased. If the average value of the two signals decreases, the microorganisms donot have sufficient oxygen to promote growth and more air is needed. The DO set point is important because its value determines the biological diversity of the mixed microorganisms culture.
The current ratio of Zoogleal to filamentous microorganisms is required is required for good settling in the secondary clarifier.
The optimum level of DO is determined by the clarity of the secondary effluent. If the suspended solids increase, then the DO set point decreases to promote additional growth of filamentous microorganisms.

Design -II:

Control of returned activated sludge and control of oxygen content in the aerator demands control of ratio of returned activated sludge and air to influent flow All the flows(influent, returned and air) are measured by respective transmitters but controllers are only used to control air flow and returned sludge through flow controllers. The pH can be measured and controlled independently.
To control the biological reaction the ratio of TOC to microorganism (TOC=F/M) should be regulated. This is accomplished by computing the mass ratio of TOC entering the reactor, the adding returned activated sludge to maintain the correct ratio. Flow ratio controller set point is to be adjusted based on the product of tOc and the flow rate of the primary effluent.

Sensor selection:

The On-line sensors are required for sensing , and measuring the control parameters in aerator are generally include Oxygen, MLSS, pH, Redox, Temperature, Nutrients, N &P,sludge blacket turbidity, conductivity, suspended solids etc.
For conductivity measurements probe reading as function of conductivity of liquid indicates a linear relationship. Liquid resistivity has similar type of relationship but having different slope. pH and conductivity of different standards can be calibrated or modeled prior to install in the plant. For example the relationship between ,pH ,measurement through ISO 6588 and IEC 60641-2 and conductivity measurement through ISO6587 and IEC 60641-2 can be modeled with least square regression equations.
pH though widely defined as $pH=-\log_{10}[H^+]$ is not however , amenable for measurement and control. Hence a pH meter is expected to show a nonlinear response to a step or sinusoidal changes in solution concentration since pH

depends on the logarithm of the concentration as already indicated:$E=E_o+kT \log C_H$, $pH=(E_O - E/ kT$

The relationship between specific conductivity and ionic concentration was expressed by Kohlrauch using the following empirical equation.

$\Lambda= \Lambda_o-bC^{0.5}$

Where $\Lambda= 1/R=l/V$, Λ_o and b are constants for a specific ion, C is the equivalent concentration of the ion in solution. For a solution of solution containing n different electrolytes, the conductivity $K_l = \sum C_i \Lambda_i$ where C_i is the equivalent concentration of the electrolyte i (g electrolyte/l) and $\Lambda_i$is its equivalent conductivity [(mS/cm)/(g/l)] calculated with the above equation at the total concentration of the mixture. This is however valid only for a mixture of pure solutions of strong electrolytes. In the presence of nonconductive solutes, the conductivity of the suspension, K_s, is lower than the conductivity Kl of the clear solution and can be calculated using Maxwell's equation

$K_s = K_l/F$ where F is the formation factor which takes into account the shape and concentration and concentration of the particles in the suspension.

For oxygen measurements photocolorimetric methods, electrochemical and electro-conductivity methods are generally employed.

For primary and secondary clarifier the parameters to be measured and controlled are level of suspended solids and density, rake speed. Back scatter near infrared level of suspended solids less than 2000ppm. Settling aids is manipulated in proportion to dry solids rate. Rake speed employs feedforward function of the dry solids loading rate. For sludge bed level measurements ultrasonic level sensor and /or rake torque or rake motor amps are used. Underflow density sensor is normally a nuclear density transmitter.

Conclusion

In this present investigation, detailed studies are made on the effluent characterization of a bamboo and hardwood based paper mill, modeling, simulation and optimization of aerator, primary and secondary clarifiers, their process and design parameters and the entire plant of varying capacities. The problem is further extended to control the entire ETP plant parameters to the desired values by considering various options for single loop and multiloop control strategies. Such as ratio control of effluent flow, returned activated sludge flow and air to influent flow has been shown. Feed forward, feed back and cascade are also tried. Hybrid control strategy using PID feed back control for dissolved oxygen and feed forward control based on conductivity for nutrient addition has also been analyzed.

References

1. Anirban Roy, Neha Bhaskar, Piyush Tiwari, and Swodha Jha – Modeling of Effluent Treatment Plant For Paper Industry,BCA Project conducted at BIT , Meshra2001.
2. Bailey, J. E. and Ollis, D. F. Biochemical Engineering Fundamentals, Second Edition, Tata McGra Hill Edition, 2010
3. Bolmstedt J.-Dynamic modelling of an activated sludge process at a pulp and paper mill,Master thesis,Lund, November 2000

4. Dochain, D and M. Perrier. Pulp & Paper Canada, 96: 12 (1995), pp. 128 – 132.

5. Kokurek, M.J, Pulp and Paper Manufacture, Vol. 10, TAPPI- CPPA joint publication, 1989.

6. Lavigne, J. R. Instrumentation application for the pulp and paper industry, Miller Freeman Publication, 1979

7. Leslie M. Zoss,Applied instrumentation in the process industries, Volume –IV, Gulf Publishing Company,1979

8. Love Jon,The Chemical Engineer,July 1983,pp.24-27

9. Mahuli , S.K.,R.Rhinehart, and R.Riggs,ion, Comput. Chem. Eng., Vol.17,pp.309-317,1993

10. Mario Francisco, Hernán Álvarez, Pastora Vega, Silvana Revollar , 20th European Symposium on Computer Aided Process Engineering – ESCAPE20, 2010.

11. Mishra N.P., Liang – Tseng Fan, L.E. Erickson and Ming-Ching T.Kuo,,II Ch, E, 1973Vol.XV(3), pp.18-29

12. N.A. Wahab, M.R. Katebi and J. Balderud, Proceeding of the 15th Mediterranean Control Conference on Automation and Control Athens (2007

13. Nagrath I.J. and M.Gopal, Control system Engineering, fifth edition, New Age International (P) Limited, Publishers,New Delhi,2007

14. Norhaliza A. Wahaba, Reza Katebia and Jonas Balderudb,Biochemical Engineering Journal,Volume 45, Issue 3, 15 August 2009, Pages 239-248

15. Perry R.H., and Cecil H. Chilton-Chemical Engineers' Handbook-5thand7th Ed., McGraw Hill. 1973.

16. Peter A. Vanrolleghem, Control of Activated Sludge Waste water Treatment by using Respiratory, Universiteit Gent, October 16,2002

17. Peters, M.S and Klaus D. Timmerhaus "Plant Design and Economics for Chemical Engineers- International Students Ed., McGraw- Hill Kogakusha Ltd., London, 1980.

18. R. Moreno, C. De Prada, J. Lafuente, M. Poch, G. Montague. Non-linear predictive control of dissolved oxygen in the activated sludge process. IFAC BIO 2, Ed. Pergamon Press(1992) pp. 289-298.

19. Ramalho, R.H.- Introduction to water treatment processes-Academic press, 1976.

20. Rangla, M.K. James, D.J.G, English, S.E, Burnham, K.J. Reeve, P.J., Stephens, R.I, Ware, A.J.

21. Ray Amiya K.Ray,Anirban Roy, Neha Bhaskar, Piyush Tiwari, and Swodha Jha, Design Philosophy Of Activated Sludge Process In Effluent Treatment Plant Of A Kraft Mill Based on Indian Mixed Hardwoods and Bamboo, 07 AIChE annual meeting held at Salt Lake City,Nov. 2007,paper 533a, pp.257
22. Smith, R.- JWPCF, 40(9), 154(1968).
23. Spring M. Springer, Industrial Environmental Control Pulp and Paper Industry, 3rd Edition, Tappi Press, 2000
24. Stenstrom M. K.,John F. Andrews, *Journal of the Environmental Engineering Division*, Vol. 105, No. 2, March/April 1979, pp. 245-260
25. Sung ,S.W.,I..B.Lee, D.R..Yang ' Ind.Eng.Chem Res..Vol.34,pp.654-664,1995

26. Talat Mahmood and Allan Elliott,J. Envir. Engrg. Volume 132, Issue 9, pp. 1019-1027 (September 2006

27. Wang Hong, Wang Ai Ping and Stephen R. Duncan- Advanced Process Control in paper and Board making.- Pira International,1997

28. .Wright, R.A. and C. Kravaris, Ind. Eng. Chem. Res. Vol.30,pp.1561-1572,199126.

29. Wright , R.A., M. Soroush and C. Kravaris, Ind. Eng. Chem. Res. Vol.30,pp. 2437-2444,1991

BioSUCCEED: Integrated Platform for Renewable Energy & Products

Lucian A. Lucia, Stephen S. Kelley, Hasan Jameel and Paul Michael Jett

Department of Forest Biomaterials, North Carolina State University, Raleigh, NC

ABSTRACT

BioSUCCEED (**Bio**products **S**ustainability, a **U**niversity **C**ooperative **C**enter for **E**xcellence in **Ed**ucation) is an integrated and cooperative effort among three Universities to address the needs of the quickly evolving biomass & bioenergy climate. This work will provide content on one of the most important disciplines that the BioSUCCEED program is addressing: thermochemical transformations of wood and associated residuals to fuels. We will provide an overview of the state-of-the-art in modeling and evaluation of thermochemical platforms, the gaps and challenges associated with its continued value, and several of the efforts that are being conducted in our laboratories that serve the basic mission of BioSUCCEED.

INTRODUCTION

The overall goal of BioSUCCEED is to education the next generation on the engineers and scientist who will continue the development and commercialization of sustainable biomass conversion technologies. To this end a series of educational units have been developed. In this case the thermochemical conversion and life cycle analysis of biomass will be highlighted.

It is also important to highlight the context of any technology. Thus, thermochemical conversion of biomass must be put in the context of other carbon based resources.

THERMOCHEMICAL PROCESSES – A HISTORICAL CONTEXT

Mankind has utilized fire and carbonaceous fuels for thousands of years for heating, whether for food, warmth, or altering other materials. In many parts of the world, wood or coal is still be used daily as a heat source in domestic and industrial applications. Historically, these solid fuels would be used in a direct combustion process, where the fuel material would be completely oxidized in the presence of air to liberate heat, carbon dioxide, and water vapor.

This fundamental process changed in 1665 when it was realized that a gas could be produced from coal by heating it in an oxygen deficient atmosphere. (Ahrenfeldt and Knoef 2005) By the late 18th century commercial coal gasification had been developed for practical applications. Processing facilities called gasworks would indirectly heat (pyrolyze) coal in sealed coke ovens, yielding a medium-high energy density "producer gas" (20-23 MJ/m^3) suitable for lighting applications, and coke that could be used as solid fuel. (Reed 2002; Ahrenfeldt and Knoef 2005; Basu 2006; Higman, van der Burgt et al. 2008) Improvements in materials of construction led to higher temperature coke oven designs, and coupled with the introduction of steam to the coke furnace chamber, allowed for the wide spread production of a medium energy density "blue water-gas" or "town gas". Where the original producer gas contained hydrogen, carbon monoxide, and coal volatiles; the composition of town gas was very similar to the modern concept of syngas: primarily hydrogen and carbon monoxide, with an approximate energy content of $12 MJ/m^3$. Town gas derived from coal gasification was initially used for lighting and domestic cooking, and finally for space heating as technology improved and gas prices fell. (Higman, van der Burgt et al. 2008)

Experiments to operate spark ignition engines with producer gas were attempted around 1880, and small scale coal and biomass gasification units were used to fuel many types of engines through World War II. At the height of World War II, there were approximately one million biomass, charcoal, or coal gasifier powered motor vehicles in use in Europe. Germany, Italy, and France, accounted for the majority of small scale gasifier installations used in both motor vehicles and tractors. (Egloff and Van Arsdell 1943)

Starting with dyestuff experiments in the gasworks of Badische Anilin und Soda Fabrik (now BASF) (Ludewig 1966), and with fuel synthesis experiments of Franz Fischer, Max Tropsch, Ruhrchemie A.G. and others through the 1920's and 30's, coke gasification became a significant feedstock process for synthetic motor fuels, lubricating oils, and an array of industrial chemicals in Germany. (Golumbic 1946)

In 1950, the Suid Afrikaanse Sinteriese Olie Limited or SASOL company was formed to realize the South African government's goal of fuel independence. SASOL employed coal gasification technology from Lurgi Gesellschaft, the Fischer-Tropsch synthesis process from Ruhrchemie AG, and plant engineering from the US company M.W. Kellogg. (Meyers 1984) SASOL's production of synthetic fuels, lubricants, and industrial chemicals from gasified coal began in 1955, expanding capacity over the years to a 2009 annual fuel production rate of 7.1 million tonnes. (Sasol 2009)

With the exception of SASOL, commercial interest in coal or biomass gasification for fuel synthesis was absent until the mid-1970's following the OPEC oil embargo. Texaco began research into coal gasification in 1948 and built a demonstration plant in 1957. It was not until 1978 that a pilot scale facility was brought online.(Meyers 1984) In a manner similar to that around each World War, concerns of crude oil scarcity and political unrest prompted researchers and engineers to actively pursue new projects. Indeed, many demonstration, pilot, or small industrial scale coal gasifiers were brought online between the energy crisis in 1973 and the early 1980's. In addition to Texaco's pilot gasification plant, designs by Shell, Combustion Engineering, British Gas/Lurgi, Koppers/Babcock & Wilcox, Davy AG, Foster-Wheeler, and others came online as a response to global concerns about energy availability and cost. (Meyers 1984)

Oil prices began to rise again in the late 1990's; coupled with the uncertainty around Middle East stability following 9/11 and rapid industrialization of China and India, led to a renewed interest in gasification in the last decade. (Higman, van der Burgt et al. 2008) The spate of gasification technologies developed in the 70's and 80's that were halted at the demonstration phase are being reexamined following the sharp rise in oil and energy costs in advance of the 2008 economic downturn. Clean coal technologies have received significant political and public attention in recent years as a means to reduce the United States' energy independence. (DOE 2008) Integrated Gasification Combined Cycle (IGCC) plants are seen as a core technology for clean coal initiatives, as well as a more energy efficient means of power generation. (Basu 2006; Higman, van der Burgt et al. 2008) General Electric Energy acquired Texaco's gasification process in 2004, bringing together technologies that could foster IGCC large scale commercial viability. (Basu 2006; Higman, van der Burgt et al. 2008)

MODELLING BIOMASS GASIFICATION – STATE OF TECHNOLOGY

The National Renewable Energy Laboratory (NREL) has been researching the production of alcohols from biomass since its beginnings as the Solar Energy Research Institute in the late 1970's. Over the course of the same time period, continuous work has been carried out in the area of biomass gasification. Bain brought the two research threads of work together in 1992, designing ASPEN models to simulate gasification systems from the Institute of Gasification Technology and the Battelle-Columbus Laboratory (Figure 1).

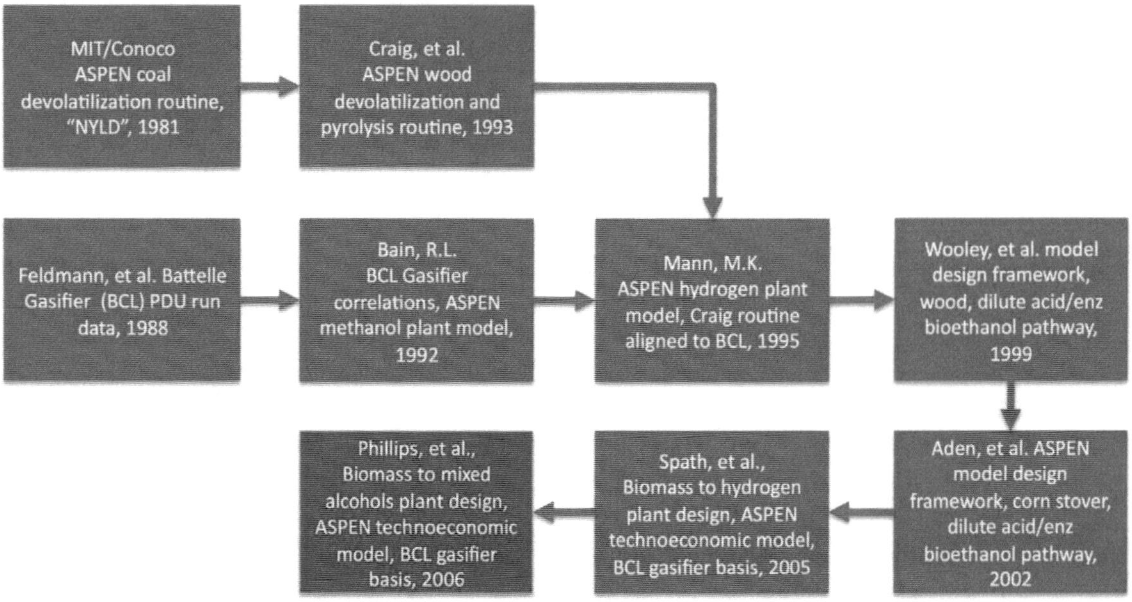

Figure 1. NREL mixed alcohols model lineage block diagram. Green block indicates model that was examined and exercised.

The ASPEN models were then used to evaluate case studies utilizing the two gasification technologies to provide syngas feedstock to methanol synthesis unit operations (Bain 1992). The ASPEN models developed by Bain simulated all of the major unit operations of a biomass to methanol gasification plant. Data from a total of 37 sets of gasifier runs (from a total of 142) were analyzed (Bain 1992). These gasification runs were made with a variety of woody feedstocks including cherry (*Prunus* sp.), mixed oak (*Quercus* sp.), birch (*Betula* sp.), maple (*Acer* sp.), pine (*Pinus* sp.), and mixed hardwoods. (Feldmann, Paisley et al. 1988).

The first biomass to fuels model from Bain was the predecessor to a gasification-derived hydrogen plant model developed by M.K. Mann in 1995. Mann's gasifier model was also based upon the BCL gasifier correlations from Bain, but incorporated a FORTRAN user model to more rigorously account for temperature dependent material balances across the gasifier. This was in contrast to the original Bain model, which relied on user input for derived yields at a particular gasification temperature.

In 2005, Spath, et al. presented a model of a biomass gasification for hydrogen production (Spath, Aden et al. 2005). Spath's process design started with the generation of full process flow diagrams (PFDs), informed by literature review and professional input from project engineering firms. Finished PFDs were translated to ASPEN models, the subsequent material and energy balances used to size equipment and construct economic analyses.

PROCESS MODELING AS AN ANALYTICAL TOOL

One of the great values of process modeling is as an instructional tool, albeit is complex one, that allows the student to ask questions about the impact of different feedstocks or operating conditions on the process or products.

For example, using different biomass feedstocks can have a significant impact on the Heat balance and materials flow through a process. The impact of changing the biomass feedstock on the operating conditions can be predicted (Table 1).

Table 1. General heat and material flows around the R202 char combustor.

	Char Flow	Char HHV	Olivine Enthalpy
Feedstock	Kg/hr	MJ/kg	GJ/hr
Sugarcane bagasse	25057	16.90	2.258
White oak	19019	23.55	2.276
Loblolly pine	16728	29.19	2.320
Torrified loblolly pine (light)	15235	33.16	2.351
NREL base poplar	**15342**	**32.52**	**2.353**
Torrified loblolly pine (medium)	15050	34.93	2.398
Spruce	17943	28.68	2.412
Torrified loblolly pine (heavy)	16000	35.11	2.469
Peat	18375	29.53	2.475
Lignite	24508	30.28	2.792

The effects of feedstocks on the downstream gas yields and composition can also be predicted (Figure 2).

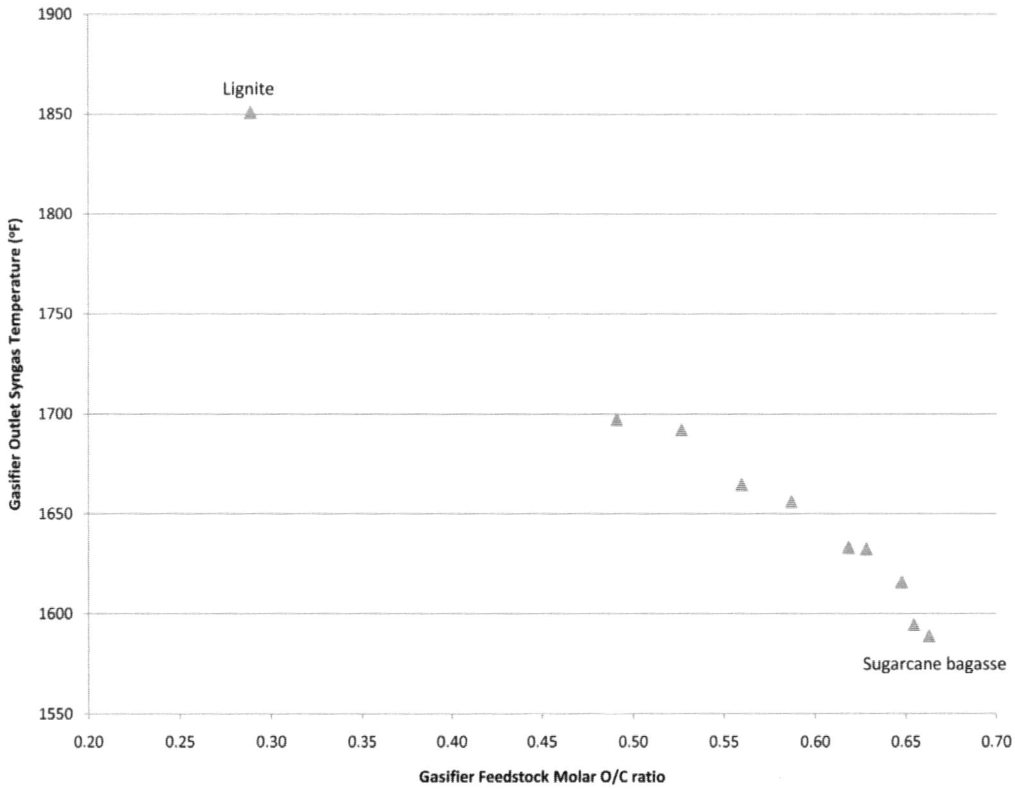

Figure 2. Gasifier outlet temperature as a function of feedstock O/C molar ratio.

CONCLUSIONS

Thermochemical conversion technologies can be used to produce fuels and power from biomass resources. Process modeling tools, such as ASPEN, can be used as an effective means to develop an improved understanding of processes. Process modeling can also be an effective tool for studying, teaching and communicating economics and life cycle concepts.

REFERENCES

Ahrenfeldt, J. and H. Knoef (2005). Handbook biomass gasification. Netherlands, BTG Biomass Technology Group.

Aspen Technology, I. (2009). EAP101 AspenPlus: Process Modeling. Burlington, MA, Aspen Technology, Inc.

Bain, R. L. (1992). Material and Energy Balances for Methanol from Biomass Using Biomass Gasifiers. Golden, CO, National Renewable Energy Lab.

Basu, P. (2006). Combustion and gasification in fluidized beds. Boca Raton, CRC/Taylor & Francis.

DOE. (2008). "Clean Coal Technology: From Research to Reality." 2010, from http://www.fossil.energy.gov/aboutus/fe_cleancoal_brochure_web2.pdf.

Egloff, G. and P. Van Arsdell (1943). Motor Vehicles Propelled by Producer Gas. The Petroleum Engineer. Dallas, Petroleum Engineer Pub. Co. **15:** 65-144.

EIA. (2010). "OPEC Revenues Fact Sheet Energy Data, Statistics and Analysis - Oil, Gas, Electricity, Coal." Retrieved May 2010, from http://www.eia.doe.gov/emeu/cabs/OPEC_Revenues/Factsheet.html.

Feldmann, H. F., M. A. Paisley, et al. (1988). Conversion of Forest Residues to a Methane-Rich Gas in a High-Throughput Gasifier. Columbus, OH, Columbus Battelle Division, Pacific Northwest Laboratory (PNL). **PNL-6570**.

Golumbic, N. R. (1946). Review of Fischer-Tropsch and related processes for synthetic liquid fuel production. [Washington], U.S. Bureau of Mines.

Higman, C., M. van der Burgt, et al. (2008). Gasification. Amsterdam ; Boston, Gulf Professional Pub./Elsevier Science**:** 435.

Ludewig, W. (1966). "Highlights in History of BASF." Transactions of the Institution of Chemical Engineers and the Chemical Engineer **44**(6): T237.

Mann, M. K. (1995). Technical and Economic Assessment of Producing Hydrogen by Reforming Syngas from the Battelle Indirectly Heated Biomass Gasifier. Golden, CO, National Renewable Energy Laboratory (NREL). **NREL/TP-431-8143**.

Meyers, R. A. (1984). Handbook of synfuels technology. New York, McGraw-Hill.

Reed, T. B. (2002). Encyclopedia of biomass thermal conversion : the principles and technology of pyrolysis, gasification & combustion. Golden, Colo., Biomass Energy Foundation Press.

Sasol (2009). annual review and summarised financial information 2009. Johannesburg, South Africa, Sasol Limited: 37.

Spath, P., A. Aden, et al. (2005). Biomass to Hydrogen Production Detailed Design and Economics Utilizing the Battelle Columbus Laboratory Indirectly-Heated Gasifier. Golden, CO, NREL: Medium: ED; Size: 161 pages.

AUTHOR INDEX

AUTHOR INDEX

AUTHOR INDEX

American Institute of Chemical Engineers
3 Park Avenue
New York, New York 10016-5991